ELEMENTS OF AIRCRAFT POLLUTION

G.J.J. Ruijgrok

Professor of aerospace engineering
Faculty of Aerospace Engineering
Delft University of Technology
Professor Emeritus, Brussels Free University

D.M. van Paassen

Lecturer
Faculty of Aerospace Engineering
Delft University of Technology

Delft University Press / 2005

DUP Science is an imprint of

Delft University Press
P.O. Box 98
2600 MG Delft
The Netherlands
Telephone: +31 15 2785678
Telefax: + 31 15 2785706
E-mail: Info@Library.TUDelft.NL

ISBN 90-407-2602-7

Printed in The Netherlands

ELEMENTS OF AIRCRAFT POLLUTION

CONTENTS

PREFACE

Within the last few decades, concern about the environmental conditions on Earth has grown steadily as it has been recognized that the exhaust of pollutant emissions from burning fossil fuels into the atmosphere cannot be allowed to continue unlimitedly. The two environmental problems currently causing most concern are the global changes in the climate due to the enhanced greenhouse effect and the depletion of the stratospheric ozone layer.
This has also raised the question on the effects of emissions from aircraft upon the atmosphere. In addition to carbon dioxide, CO_2, and water vapor, H_2O, the emissions of carbon monoxide, CO, unburned hydrocarbons (UHC), nitrogen oxides, NO_x, and smoke or soot are the essence of the problem of air pollution caused by aviation.
Although the total quantity of emissions from air traffic is very small compared to other man-made emissions, their impact on the environment may not be negligible. Important is the fact that aircraft emissions are the only significant anthropogenic source of pollutants in the upper troposphere and lower stratosphere. Also of importance is the fact that the demand for air travel is still growing.
Carbon dioxide and water vapor are important greenhouse gases. In addition, there is evidence that due to the exhaust of water vapor by aero-engines, the cloud cover of the Earth is increased to a certain extent by contrail-induced cirrus. Also the exhaust of nitrogen oxides is a matter of major concern because this compound may promote the enhanced greenhouse effect and as well as depletion of ozone in the stratosphere. An enhanced greenhouse effect means that the world is becoming warmer, while a depletion of the stratospheric ozone layer will cause a harmful level of ultraviolet (UV) radiation at the surface of the Earth.

This book, especially, is concerned with the numerous sciences and technologies which make up the air pollution problem from flight operations, giving special attention to the performance and emissions of airplanes at their cruise altitudes. Obviously, a good knowledge of the different subjects involved is a prerequisite for obtaining an understanding of human influence on the environment.
The contents of the book are a reflection of an annual course on propulsion, noise, and aircraft pollution, presented by the author to MSc Aerospace Engineering students at Delft University of Technology.

The author is grateful and happy to share the responsibility for all the constituent parts of the book with his coauthor Mr. D.M. van Paassen, lecturer/pilot at Delft University of Technology.

The book is an attempt to place the question of *aircraft pollution* in the context of aerospace engineering by bringing together the most relevant topics covering the problem. To this end, many sources have been consulted because much of the material could only be obtained from textbooks, reports, and articles. The references to the literature are indicated in the text and listed at the end of the book. It should be mentioned, however, that the current state of knowledge of the effects of aviation upon the global atmosphere is described comprehensively in the Intergovernmental Panel on Climate Change Special Report (Ref.1). This IPCC report also makes clear that the effect of aviation on the environment is a complex problem, drawing upon several widely-different scientific disciplines.

In the book the International System of units (SI units) has been used throughout. A point to note is that in international civil aviation the use of certain non-SI units is prescribed, such as foot instead of meter for flight altitude, and so these units also have been cited in the text.
No claim to be comprehensive is made for the book, but it is hoped that no major part of the problem has been omitted from consideration. Above all, it is hoped that the contents of the book will somewhat clarify the various subjects and may urge the reader to further study in this field.

Delft, February 2005

G.J.J. Ruijgrok
D.M. van Paassen

1 INTRODUCTION

1.1 AIR POLLUTION

When the natural composition of the atmosphere is changed, the air is considered to be polluted. Air pollution is recognized as one of the world's most serious environmental problems, and includes both natural sources and anthropogenic sources (pollutants from human activities).
Sometimes it only manifests itself by an oppressive odor, however, often it concerns environmentally harmful substances for human beings, animals and vegetation.
In brief, air pollution may appear in the following forms:

- Unpleasant stench or odor nuisance.
- Small droplets, gases, and fumes.
- Particulate matter, PM (any airborne, finely divided, liquid or solid material).

As depicted in Figure 1.1-1, the order of magnitude of particulates varies from 10^{-9} to 10^{-2} m. The smallest particulates are called *aerosols* (10^{-9} to 10^{-6} m), and are able to penetrate into our respiratory system. Aerosols involve extremely fine dust or droplets of microscopic size such as tobacco smoke and soot. By soot are meant the carbon-containing particles resulting from incomplete combustion of hydrocarbon fuels. Dust, having particulate sizes ranging from 10^{-6} to 10^{-2} m, refer to, for example, fly ash, coal dust, sand, and rain drops (Ref. 2).
This stuff, whether natural or man-made, stimulates the condensation of moist air a great deal. In the absence of condensation nuclei the air may remain supersaturated (see Chapter 4).

Natural sources of air pollution are:

- Volcanic eruptions, emitting a variety of gases and particles into the atmosphere. These include water vapor H_2O, carbon monoxide CO, carbon

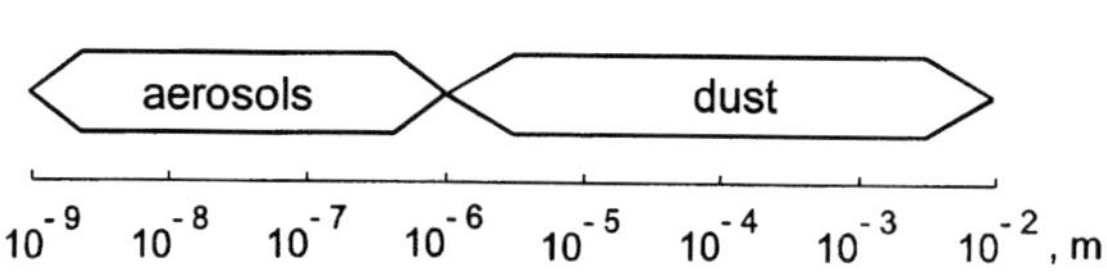

Fine particle concentrations are called $PM_{2.5}$, comprising particulate matter 2.5 micrometers in diameter and smaller.

Figure 1.1-1. Particulate sizes.

dioxide CO_2, sulfur dioxide SO_2, nitrogen oxides NO_x (NO and NO_2), mercury Hg, hydrogen sulfide H_2S, hydrochloric acid HCl, and dust particles.

- Saline droplets arising from the oceans.
- Organic byproducts of biological activities and vegetable substances, such as pollen, ammonia NH_3, nitrous oxide N_2O, methane CH_4, and hydrogen sulfide H_2S.
- Dust storms.

Examples of anthropogenic forms of pollutants are combustion products of fossil fuels for power and heat generation. Other forms are residues from refineries and industrial processes. Note that petroleum and natural gas contain no fossils, because of the extreme conditions under which they have been formed. Identifiable fossils may only be found in coal and peat.

In residential areas the exhaust gas emissions near ground level from road traffic will considerably contribute to air pollution through the combustion of gasoline, diesel, and liquefied petroleum gas (LPG).
Figure 1.1-2a shows the world population increases since the year one (Ref. 9). The explosive growth of human population of the twentieth century is obvious. Figure 1.1-2b portrays the growth in world population and global number of motor vehicles since 1970 and projections to 2030, showing that the current number of vehicles may double in the next 30 years.
The number of cars per 1000 inhabitants varies substantially per country. In 1992 the number of vehicles per 1000 inhabitants was approximately 550 in the

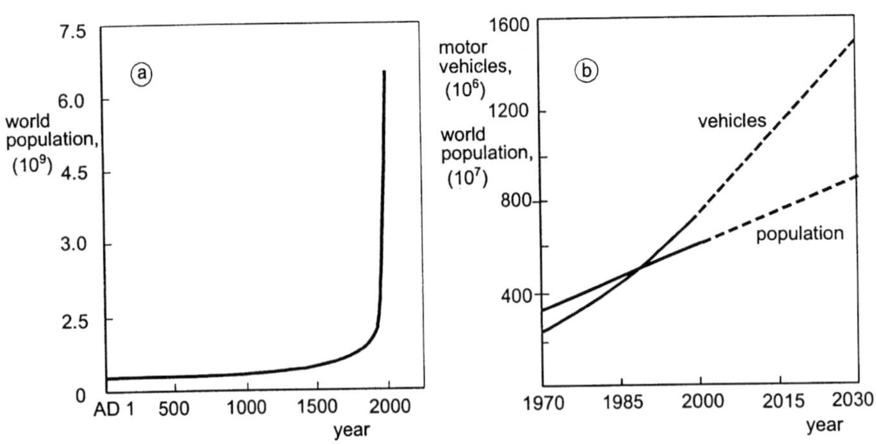

Figure 1.1-2. World population and global number of motor vehicles.

United States, in Western Europe 200 - 400, while in Africa it was only 9, in India 2, and in China 0.4 (Ref. 3). As noticed earlier, the exhaust gases of motor vehicles are particularly harmful because the emissions take place in the environment in which we live. In urban environments, where dispersion barely takes place, high concentrations of pollutants may occur.
The air polluting substances from automobiles and trucks are, besides water vapor H_2O and carbon dioxide CO_2, carbon monoxide CO, nitrogen oxides NO_x (NO and NO_2), aerosols, sulfur oxides SO_x (SO_2 and SO_3), lead P_b, unburned hydrocarbons (UHC) and soot particles. Substances that evaporate at temperatures less than about 100°C, are called *volatiles*, e.g., benzene C_6H_6.
Also, the emission of sulfur, nitrogen oxides, and dust by shipping delivers a noticeable contribution to air pollution.
It should be noted that when considering the environmental effects of electrical systems of transportation, the environmental impact of power generation for their propulsion should also be taken into account, as it contributes likewise to air pollution.

Besides primary pollutants, those emitted directly from the source, secondary pollutant species may be formed. The latter arise from mutual chemical reactions involving originally discharged pollutants. This is called *secondary air pollution*.

1.2 CHEMICAL COMPOSITION OF CLEAN AIR

Air is a mixture of gases. The total mass of the atmosphere is approximately 5.30×10^{18} kg. For comparison, the mass of the Earth is 5.98×10^{24} kg and the total mass of the water on the Earth is 1.39×10^{21} kg. More than 97% of the water is stored in seas and oceans. The atmosphere contains only about 0.035 percent of the water.
The bulk of the air mass is concentrated in the lower layers of the atmosphere as the air density becomes gradually lower with increasing height. For example, 50% of the atmosphere's mass is contained within the first 6 km and 95% is concentrated below a height of 22 km. If the total air mass would be converted to mean sea-level pressure and temperature, a layer of just about 8 km would remain.

In Table 1.2-1 the chemical composition of clean air near the Earth's surface is listed (Ref. 4). Evidently, the air we breathe at sea level consists for the greater part of the two diatomic gases nitrogen N_2 and oxygen O_2 (together about

constituent gas		content, percent by volume	molecular mass kg/kmol
Nitrogen	N_2	78.084	28.0134
Oxygen	O_2	20.9476	31.9988
Argon	Ar	0.934	39.948
Carbon dioxide	CO_2	0.0314	44.00995
Neon	Ne	0.001818	20.183
Helium	He	0.000524	4.0026
Methane	CH_4	0.0002	16.04303
Krypton	Kr	0.000114	83.80
Sulfur dioxide	SO_2	0-0.0001	64.0628
Hydrogen	H_2	0.00005	2.01594
Nitrous oxide	N_2O	0.00005	44.0128
Xenon	Xe	0.0000087	131.30
Ozone	O_3	0-0.000007	47.9982
Nitrogen dioxide	NO_2	0-0.000002	46.0055
Iodine	I_2	0-0.000001	253.8088
Water vapor	H_2O	0-3.0	18.0

Table 1.2-1. Normal composition of clean atmospheric air near sea level.

99.03% by volume). Oxygen, O_2, the second largest constituent, is one of the most reactive elements.

The remaining constituents, although present in relatively small quantities, may be nevertheless of great importance. For instance, the ample quantities of carbon dioxide CO_2 and water vapor H_2O form a kind of shield, blocking the heat radiation from the Earth into outer space and heating up the Earth's surface. This phenomenon is called the *natural greenhouse effect,* the condition that keeps an enclosure warm in the wintertime.

Ozone, O_3, is present everywhere in the atmosphere. However, the amount of ozone varies significantly, geographically, with seasons, and, typically, depends strongly on the altitude.

Ozone is an unstable gas, formed by reaction of molecular oxygen O_2 and atomic oxygen O. Atomic oxygen arises from dissociation of molecular oxygen, caused by ultraviolet solar radiation.

Simultaneously, ozone depletion takes place, forming molecular and atomic oxygen, caused by infrared radiation. This continuous process of formation and depletion results in a constant overall equilibrium concentration of ozone in the

atmosphere. Depending on the intensity of the solar radiation and the number of particles per unit volume, a maximum in the ozone concentration occurs at an altitude between 25 and 35 km.
Important is that the ozone at these high altitudes acts as a filter against solar ultraviolet radiation, which otherwise has many harmful effects on Earth's animal and plant life.

Water in the atmosphere can be present in the form of water vapor, an invisible gas, or as minute droplets and ice particles (clouds, precipitation).
The amount of water vapor varies with geographical latitude, and decreases substantially with increasing altitude. Generally, the highest degree of humidity occurs near the Equator. At an altitude of 10 km virtually no water vapor is found.

The composition of the air remains more or less constant up to an altitude of about 90 km. Above this altitude the composition gradually changes, mainly because of molecular dissociation, i.e., the break up of the molecules of the various gases into their atoms, called *radicals*, due to the increased intensity of ultraviolet sunlight. In consequence, the molecular mass of the air decreases with increasing altitude.

1.3 ARRANGEMENT AND NOMENCLATURE OF THE ATMOSPHERE

With respect to the chemical composition, the atmosphere may be classified into the *homosphere* and the *heterosphere*. The homosphere extends up to an altitude of 90 km, where the air is characterized by a nearly constant chemical composition (with the exception of water vapor and ozone). The region above 90 km is indicated as the heterosphere. Here, the molecular mass decreases from about 29 kg/kmol at a height of 90 km, to 18 kg/kmol at a height of 500 km. The dividing plane between the homosphere and heterosphere is known as the *homopause*.

In Figure 1.3-1 the typical variation of the average temperature with altitude is sketched. Based on this temperature distribution, the atmosphere can be divided into the following four layers:

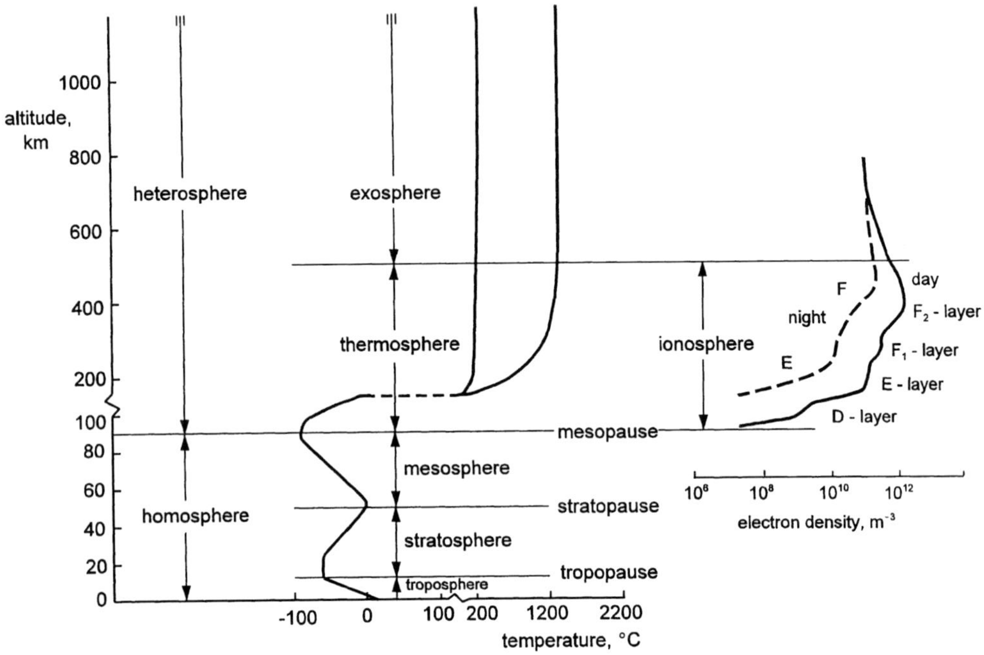

Figure 1.3-1. The nomenclature of the atmosphere.

- Troposphere.
- Stratosphere.
- Mesosphere.
- Thermosphere.

The *troposphere* is characterized by a decrease in temperature with increasing altitude. In this layer occur the phenomena which we call *weather*, i.e., the local state of temperature, pressure, humidity, cloudiness, wind, and precipitation.

The troposphere can be subdivided into two parts (Figure 1.3-2). The first part, up to about 2 km above sea level, is called the *planetary* or *atmospheric boundary layer* and the remaining part is named the *free troposphere.* This distinction is of significance when considering the effects of air pollution, since the atmospheric boundary layer is most directly influenced by the underlying land or sea surface.

The dividing plane between the troposphere and the *stratosphere* is called the *tropopause.* For average latitudes, this plane is located at a height of about 11

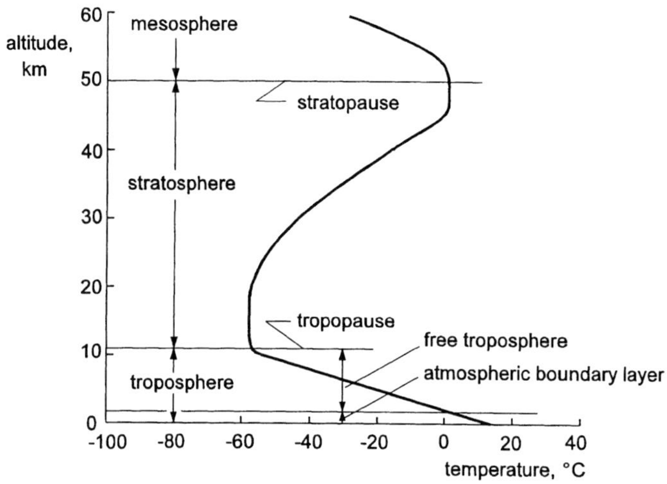

Figure 1.3-2. Temperature rates in the atmosphere.

km. Near the polar regions it is located at about 8 km, while at the Equator it reaches up to approximately 17 km (Figure 1.3-3).

In the stratosphere, at first, there is a nearly constant temperature of about -50 to -60°C up to an altitude of 20 to 25 km. From thereon the temperature increases to a maximum value of 0°C at an altitude of about 50 km. The dividing plane between the stratosphere and the *mesosphere* is called the *stratopause*. This altitude sometimes is referred to as the *ozonepause*. In fact, the temperature increase above 20 km is related to the absorption of ultraviolet radiation, involving ozone formation.

Typical for the mesosphere is a decreasing temperature, reaching a minimum value of about -90°C, at an altitude of 90 km. The latter altitude is called the *mesopause*.

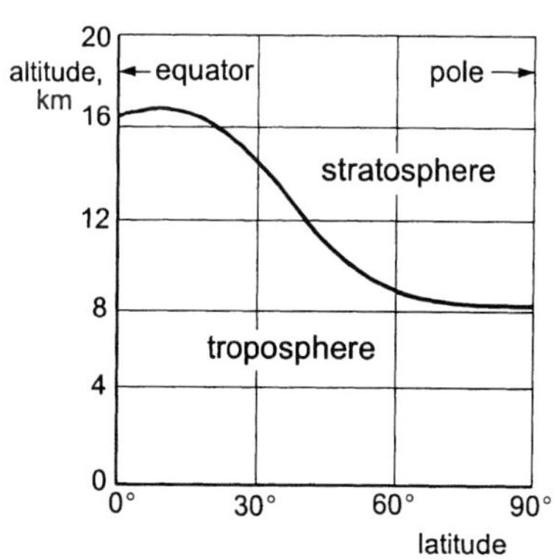

Figure 1.3-3. Impression of tropopause level variation.

altitude [km]	number of particles, m^{-3}
0	2.5×10^{25}
10	10^{25}
500	10^{14}
1000	10^{12}
1500	10^{11}
2000	10^{10}

Table 1.3-1. Number of particles per cubic meter for several altitudes.

The region between 90 and about 500 km is called the *thermosphere*. In the thermosphere, the temperature increases quickly with increasing altitude until at about 500 km the so-called *exospheric temperature* is reached. Depending on solar activity, the value of this temperature may vary between 200 and 1000°C. These temperatures are only an indication of the kinetic energy of the air particles. The temperature of an object at these very high altitudes is entirely determined by the heat balance between its absorbed heat radiation, its produced internal heat and its emitted heat radiation. Convective heat transfer is hardly possible due to the very low value of the air density.

From 90 km upwards, ionization processes occur, i.e., the generation of ions and the accompanying free electrons takes place. Therefore, in Figure 1.3-1 also is depicted the *ionosphere*, in which the highest concentrations of ions are found. Also the ionosphere extends roughly from a height of 90 to 500 km, Depending on electron density, the ionosphere is subdivided into four layers, designated D, E, F_1 and F_2. The curves in Figure 1.3-1 show that the extent of ionization increases with altitude up to approximately 300 km. Local maxima of the amount of ionized particles occur, corresponding to the four layers. At night, when no radiation is perceived, some of these sub-layers may be absent. The curves also vary with solar activity and time of the year.
As depicted in Figure 1.3-4, due to its ability to reflect electromagnetic radiation, the ionosphere is of paramount importance to long-distance radio communication using wavelengths greater than 10 m.
At altitudes above 500 km, where the air density decreases further, the number of particles per unit volume becomes extremely low. By way of illustration, Table 1.3-1 gives the number of particles per unit of volume for several altitudes.
Consequently, the particles will show large *mean free path lengths* of several hundreds of kilometers or even more. Hence, particle collisions are unlikely to take place.

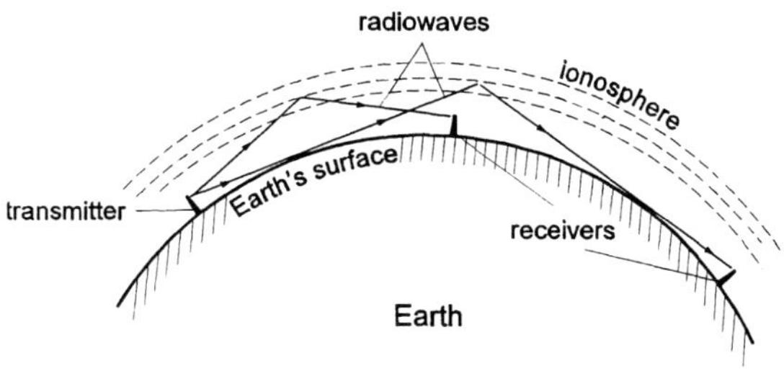

Figure 1.3-4. Long-range radio communication.

At a height of 500 km begins the region that is called the *exosphere*, where collisions between molecules are so rare that neutral particles move in trajectories which are subject to gravity only. Almost all the atmospheric gases are ionized, and the charged particle motions are strongly directed by the magnetic field surrounding them.

In the exosphere the distribution of the gases is controlled by *diffusion*, which process implies that a substance moves from a region where its concentration is high to one where its concentration is low. As a result, a transition from mainly atomic oxygen ions into primarily helium ions takes place at about 1000 km, and from helium ions to the lighter hydrogen ions at heights of the order of 3000 km.

Clearly, a knowledge of the Earth's magnetic field is a prerequisite to an appreciation of exospheric motions. To this end, we have to consider the *magnetosphere*, which region refers to the magnetic properties of the Earth. Near the Earth's surface, the magnetic field of the Earth may be represented by

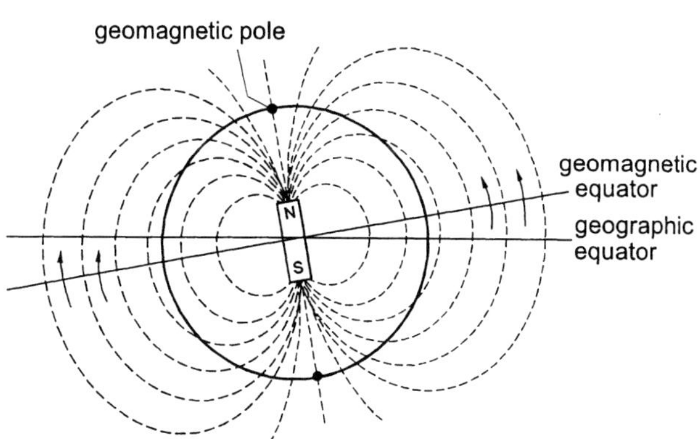

Figure 1.3-5. Magnetic field near the Earth's surface.

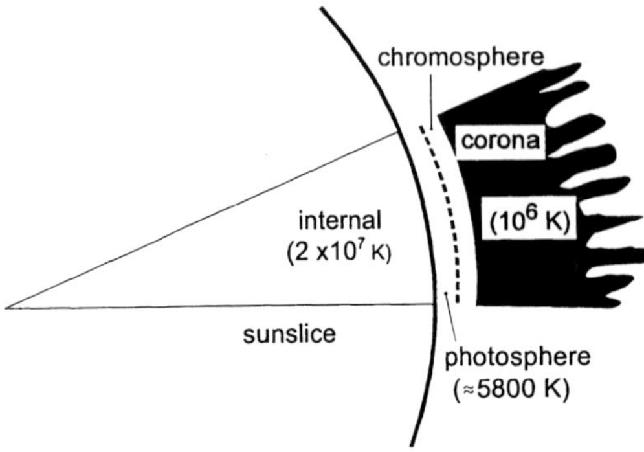

The apparent visible surface of the Sun, known as the *photosphere*, consists of a network of bright markings on a darker background. The effective temperature in the photosphere amounts to about 5800 K. The portion of the Sun's atmosphere nearest the photosphere, called the *chromosphere*, is a layer of ionized gases extending to several thousand miles. The energy radiated by the Sun is the result of nuclear processes in the interior, where helium is formed by the fusion of hydrogen nuclei. In the Sun's core, the temperature rises to about 20 million kelvin.

Figure 1.3-6. Structure of the Sun.

the dipole field of a bar magnet (Figure 1.3-5). The axis of the dipole is not aligned with the polar axis of the Earth, but is inclined to it at an angle of about 11.3°. Drawing a straight line through approximately the center of the Earth along the dipole axis yields the two geomagnetic poles at the points where this line breaks through the Earth's surface. A compass needle will follow the magnetic field lines, which run from one pole to the other.

Of course, the magnetic field of the Earth is three-dimensional and extends more or less spherically from the magnet, the lines of force converging at the two magnetic poles.

From rocket and satellite observations it is known that at very large heights the magnetic field is flattened on the side of the Earth turned toward the Sun and stretched out on the night side.

This transformation of the Earth's magnetic field arises from its interaction with the *solar wind*, which is a stream of charged particles. The solar wind is emitted from the outermost layer of the Sun's atmosphere, the solar *corona*, which consists of an extensive, very tenuous envelope of gases at high temperature (Figure 1.3-6).

Although the Sun has a strong gravitational field, its enormous high temperature contains enough kinetic energy that the particles can escape from the Sun into the interplanetary space.

The solar wind is composed almost of protons and electrons, which particles are electrically charged. Protons carry a positive charge and electrons a negative charge. Their travel speeds vary between 300 and 1000 km/s, depending on solar activity.

When the solar wind strikes the magnetic field of the Earth, a bow shock wave is generated, changing the shape of the Earth's magnetic field. The resultant

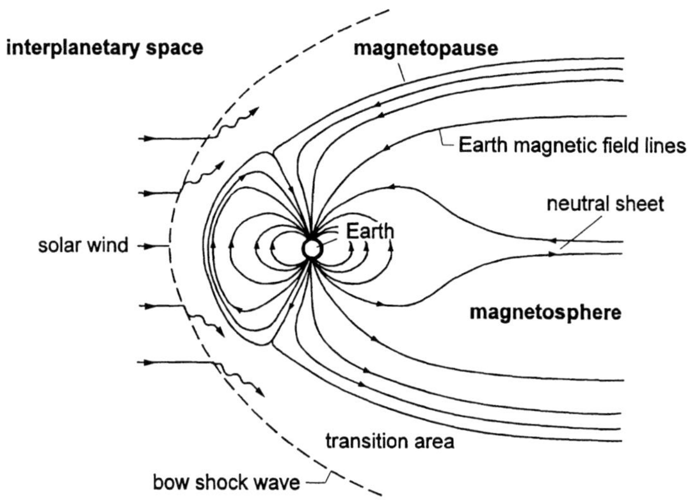

Figure 1.3-7. The magnetosphere.

distortion is sketched in Figure 1.3-7, giving a two-dimensional representation of the *magnetosphere*, the three-dimensional magnetic field that surrounds the Earth. Figure 1.3-7 also shows that the Earth's magnetic field is enclosed by the solar wind with the *magnetopause* as a quite definite boundary.

The electrically charged particles of the solar wind are trapped in the magnetosphere. The forces acting, cause them to follow spiral paths along lines of force. Charged particles are most abundant in a concentric series of belts, shaped like tires which surround the Earth, located over the Equator. These belts are known as the *Van Allen radiation belts*, named after the American physicist James A. Van Allen (b. 1914), who discovered their occurrence from information gathered by *Explorer I*, the first American satellite (1958).
The particles of the solar wind that carry no charges (neutrons) are also trapped by the magnetosphere, but at a lower level. They also penetrate the magnetosphere but, as they move toward the Earth, they collide with atoms of atmospheric gases, producing elementary particles.
The charged particles making up the Earth's exosphere are not able to escape from the magnetosphere, so that the magnetopause can be regarded to be the outermost part of the Earth's atmosphere.

The electromagnetic radiation coming from the Sun is not affected by the presence of the magnetosphere. As will be discussed in Chapter 5, their fate is especially governed by their frequencies.

1.4 HARMFUL EFFECTS OF ENGINE EMISSIONS

The various pollutants from burning fossil fuels may contribute to the following five principal effects of air pollution:

1. Effects on the climate.
2. Acidification.
3. Ozone layer breakdown.
4. Photochemical air pollution.
5. Local effects at ground level.

Each of these effects is briefly discussed below.

1. Effects on the climate are commonly indicated as the *enhanced greenhouse effect* or *global warming*. Anticipating our discussion in Chapter 6, global warming is the term used to describe the phenomenon of the increasing global average temperature at the Earth's surface. Analysis shows that this temperature has increased by about 0.6°C in the last 100 years, taking full account of diurnal and seasonal temperature changes as well as for being averaged across the world in northern and southern hemispheres and over land and sea.
   The atmosphere and the Earth receive energy through short-wave radiation by the Sun. At the same time, the Earth emits longer-wave, infrared radiation, of which the energy is partly absorbed by the molecules of atmospheric gases, such as water vapor H_2O, carbon dioxide CO_2, and ozone O_3. All the greenhouse gases are transparent to short-wave radiation, but partially opaque to long-wave radiation. Three or more atoms per molecule constitute the common property of greenhouse gases, with up to five atoms in the case of methane CH_4.
   Like the action of the glass walls of a greenhouse (Figure 1.4-1), the heat

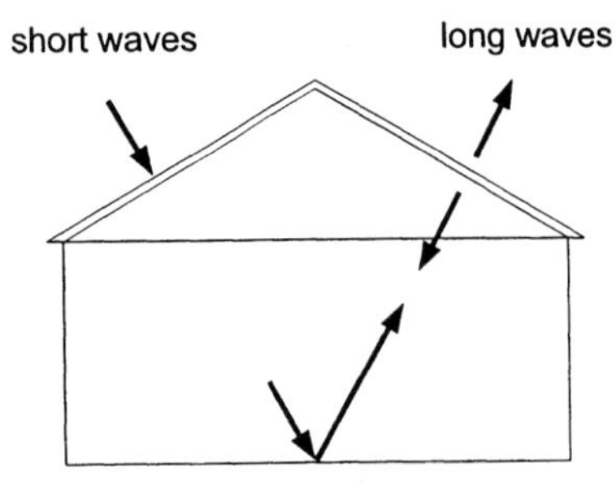

Figure 1.4-1. The greenhouse effect.

absorbed by a shield of greenhouse gases is re-emitted both upwards to space and downwards to the Earth's surface. Consequently, some of the heat that would otherwise be radiated into space is 'trapped' by these gases. As mentioned earlier, this phenomenon is known as the natural greenhouse effect. As long as the annual global energy radiated into space equals the solar energy absorbed by the Earth, there is a global equilibrium temperature on the Earth.

Especially, the increasing man-made production of carbon dioxide by burning fossil fuels and cutting down tropical rain forests cause a continuous increasing CO_2 content in the atmosphere. Also enlarged concentrations of H_2O, NO_x, CO, UHC, and soot from burning fossil fuels disturb the thermal equilibrium condition of the Earth, and may lead to an enhanced greenhouse effect and so to an increase of the Earth's mean temperature. By NO_x, CO, and UHC emissions the greenhouse gas ozone O_3 is introduced in the atmosphere via secondary air pollution. This is the so-called tropospheric or 'bad' ozone, formed by photochemical reaction of oxygen O_2 with nitrogen oxides, carbon monoxide, and unburned hydrocarbons.

The trapping of heat in the atmosphere may affect not only the Earth's surface temperature, but also weather conditions such as heavier rainfall, more serious floods, and greater numbers of thunderstorms and tornadoes.

For many coastal regions throughout the world the danger of storm surges may increase, especially in areas in which a rise in sea level and a greater risk of storms coincide.

2. A second issue of concern pertains to the occurrence of acid precipitation. Normal pure rainwater always has a slightly acid character because it takes up some of the atmospheric carbon dioxide and convert it into carbonic acid. Precipitation becomes an unnatural acidity by the presence of nitrogen oxides NO_x and sulfur dioxide SO_2 as released into the air when burning fossil fuels. The oxides NO_x and SO_2 readily combine with water vapor to form nitric and sulfuric acids. These acids then become condensation nuclei, thus producing acid precipitation. As soon as acid precipitation was detected, it was ascribed to anthropogenic pollution. It was found that, after evaporation, acid raindrops on plants and trees may leave behind high concentrations of acids, which may burn holes into the structure of the leafs. Evidently, acid precipitation may be harmful to forests and vegetation by accumulating acids in the soil. With high sulfur concentrations in the atmosphere, precipitation with sulfuric acid also can cause severe corrosion of mineral building materials. In recent years there has been an acceleration in damage of this type to old buildings and structures in city centers. Obviously, the only way to solve the problem is to stop or reduce the exhaust of harmful man-made products into the air.

3. Over the past twenty-five years, concern has grown over the effects of man-made chemicals on the ozone layer at very high altitudes in the stratosphere. Especially the introduction of chlorine and bromide compounds, and nitrous oxide N_2O into the atmosphere at ground level participates in the depletion of the ozone layer.
   Also due to a transport of nitrogen oxides to larger heights when emitted by aircraft in cruise, air traffic may participate in the catalytic destruction of ozone through which the risk is increased that harmful solar ultraviolet radiation penetrates to the Earth's surface.

4. In addition, nitrogen oxides not only may cause the phenomenon of acidification, but also the formation of tropospheric ozone by a photochemical reaction with oxygen. In this way, NO_x may promote an enhancement of the greenhouse effect. Also, ozone is the photochemical precursor of the OH radical, one of the most important oxidants in the global atmosphere. In chemistry, a radical is a group of atoms that is replaceable in a molecule by a single atom and is capable of remaining unchanged during a series of reactions. When in densely populated areas with high NO_x concentrations photochemical smog is formed, irritation of the eyes and mucous membranes may occur. Increased concentrations of ozone near ground level not only may have harmful effects on human beings but also can damage vegetation and forests, for instance by faster aging and poorer assimilation of nutrients and water.

5. Unburned hydrocarbons (UHC) as a result of incomplete combustion or insufficient mixing of fuel and air, carbon monoxide CO, and smoke (vapor made visible by the presence of tiny particles) have direct local effects in addition to those of ground level ozone, especially in urban environments. Acids from nitrogen oxides and sulfur dioxide can precipitate locally and damage vegetation or cause corrosion to buildings. Unburned hydrocarbons and soot particles are also detrimental to the living conditions of both humans and plants. Partial oxidation of the fuel may lead, via the separation of hydrogen atoms, to the formation of aromatic (e.g., benzene C_6H_6) and polycyclic aromatic hydrocarbons (PAH). These substances can affect the body's DNA (deoxyribonucleic acid) and can cause cancer.
   Local air pollution occurs mainly in a stable atmosphere, with little or no convective motions. Dispersion of pollutants hardly takes place, possibly leading to photochemical smog formation. This type of smog is formed when high concentrations of pollutants (smoke) are combined with fog. Automobile emissions of UHC, NO_x and VOC (volatile organic compounds), such as benzene and toluene may lead, under the influence of ultraviolet solar

radiation, to the formation of ozone O_3 at ground level. Ozone is a strongly oxidative and poisonous gas.

A summary of the various harmful effects of combustion emissions is given in Table 1.4-1.

Without doubt, the troposphere is of tremendous significance for the generation and dispersion of pollutant substances. In particular, the lowest 2000 meters of the atmosphere are of significance to accomplish the desired dispersion and dilution of air pollutants. Within this layer the horizontal and vertical motions of the air greatly affect the way in which pollutant substances disperse, and to what degree the atmosphere is self-cleaning.

A measure to reduce anthropogenic pollution near the ground from heavy industry is the use of high smoke stacks. In residential areas, the improvement of smoke disposal from domestic chimneys is an effective measure.
The replacement of coal by natural gas for heating and power generation purposes, and equipping motor vehicles with catalytic converters have also proven to be effective. Flue gases from installations using natural gas as a fuel contain relatively low amounts of sulfur dioxide SO_2 compared to the use of sulfurous coal.

Pollutant	Effects on environment	Effect on health
CO	Tropospheric ozone production	Toxic
UHC	Global warming Tropospheric ozone production Smog	Toxic
Smoke	Global warming	Respiratory problems Cancer
SO_X	Global warming Corrosive Acid rain	Toxic
NO_X	Global warming Tropospheric ozone production Tropospheric ozone depletion Smog Acid rain	Toxic
CO_2	Global warming	
H_2O	Global warming	

Table 1.4-1. Harmful effects caused by combustion emission.

1.5 MOTOR VEHICLE EMISSIONS

Worldwide, the transport sector accounts for about 30 percent of the global energy consumption, almost all of which comes from oil-derived products. Ground transport (road and rail) is responsible for about 85 percent of the total energy use in the transport sector. Therefore, the transport sector is a major contributor to greenhouse gases as it produces about 18 percent of all carbon dioxide released from fossil fuels (Ref. 3).
In addition, ground transport generates a large percentage of anthropogenic carbon monoxide (CO) emissions, unburned hydrocarbons (UHC), soot (particulate matter), nitrogen oxides (NO_x), and sulfur oxides (SO_x). Accordingly, road traffic contributes also to the acidification of the environment, and the formation of photochemical smog. Therefore, in confined places and congested streets, emission pollutants from ground transport may rise to levels that are hazardous to health.

At stoichiometric (ideal) combustion conditions, the combustion process consumes approximately 15 kg of air per kg of fuel, to form the combustion products carbon dioxide CO_2 and water H_2O. If the sulfur contained in the fuel is burned completely, sulfur dioxide is produced.
The propulsion of motor vehicles generally involves combustion processes with less favorable fuel to air ratios, and insufficient mixing of fuel and air. Furthermore, the process often takes place too fast at a relative (too) low temperature. Consequently, pollutants will be formed.
Carbon monoxide CO is the intermediate product in the combustion process of carbon to carbon dioxide CO_2. If there is a lack of air or the combustion temperature and residence time in the flame are insufficient, part of the CO can escape along with the exhaust gases. If the hydrocarbons are not fully oxidized during combustion, in the combustor a decomposition process takes place, leading to the formation of new hydrocarbons via the separation of hydrogen atoms. In this way substances like aromatic and polycyclic aromatic hydrocarbons (PAH's) are formed, which also are emitted along with the exhaust plumes.
The formation of soot (carbon-containing particles) results from the application of high flame temperatures and from incomplete combustion due to lack of oxygen O_2. Also nitrogen oxides NO_x are formed during combustion processes with high flame temperatures through oxidation of the nitrogen N_2 in the air and through the combustion of the fuel-bound nitrogen. Primarily, nitrogen monoxide NO is produced, whereas nitrogen dioxide NO_2 is formed only after

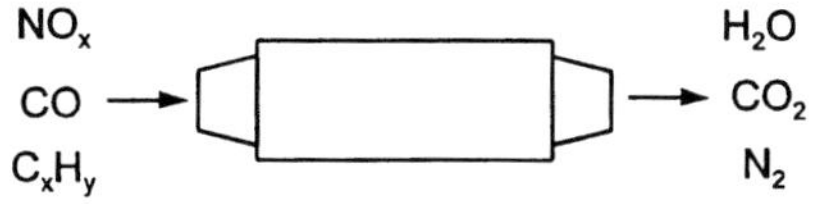

Figure 1.5-1. Motor vehicle catalytic converter.

the combustion when there is a sufficient oxygen content in the exhaust gases and finally in the atmosphere.

A feasible measure to reduce pollutant emissions from motorized vehicles is the use of a *three-way catalytic converter* (Figure 1.5-1). The term "three-way" indicates that the three pollutants UHC, CO, and NO_x are transformed, via the following gross reactions (Ref. 5):

1. Transformation of hydrocarbon:

$$C_xH_y + (x + y/4)\, O_2 \rightarrow x\, CO_2 + y/2\, H_2O\,,$$
$$CH_y + 2H_2O \rightarrow CO_2 + (2 + y/2)\, H_2\,.$$

2. Transformation of carbon monoxide:

$$CO + 1/2\, O_2 \rightarrow CO_2\,,$$
$$CO + H_2O \rightarrow CO_2 + H_2\,.$$

3. Transformation of nitrogen oxides and unburned hydrocarbons:

$$NO + CO \rightarrow 1/2\, N_2 + CO_2\,,$$
$$2\,(x + y/4)\, NO + C_xH_y \rightarrow (x + y/4)\, N_2 + (y/2)\, H_2O + x\, CO_2\,,$$
$$NO + H_2 \rightarrow 1/2\, N_2 + H_2O\,.$$

The body of the catalytic converter is placed in the engine exhaust pipe before the muffler. It should be noted that after disposal, the catalytic converter will have to be processed as well.

With the state of the art technology, the catalytic converter is only applicable to gasoline-powered and diesel-powered piston engines. Although diesel-powered vehicles emit lower amounts of carbon monoxide and unburned hydrocarbons than gasoline-powered vehicles, their emission of carbon particulates is considerably higher.

fuel	heating value, J/kg
hydrogen	11.85×10^7
methane	4.89×10^7
kerosene	4.34×10^7
coal	3.30×10^7
natural gas	3.16×10^7
brown coal (lignite)	2.70×10^7
fire wood	9.83×10^6

For remarks regarding the determination of the heating value, the reader is referred to Chapter 10, Section 10.3.

Table 1.5-1. Heating value of various fuels.

Other measures to reduce air pollution from road traffic are:

1. The utilization of unleaded petrol. However, to accomplish a better knock-resistance, aromatic compounds such as benzene are to be used as a fuel additive.

2. Speed reduction, i.e., a reduction of the combustion temperature.

3. The development of *clean* fuels and economic engines. Economic engines involve adequate combustion and a high thermal efficiency η_{th}, which is defined by:

$$\eta_{th} = \frac{P_{br}}{m_f H}, \tag{1.5-1}$$

where P_{br} is the power delivered to the shaft, m_f the fuel mass flow rate and H the *heating value* or *heat of combustion* of the fuel. For gasoline, $H \approx 4.3\times10^7$ J/kg (Table 1.5-1). The thermal efficiency of current piston engines, approximately, amounts to 30%.

1.6 AERO-ENGINE EMISSIONS

As shown in Figure 1.6-1, in an ideal (stoichiometric) combustion process, carbon dioxide CO_2 and water H_2O are formed as combustion products. Unfortunately, ideal combustion processes do not exist and also undesirable compounds are formed, mainly consisting of nitrogen oxides NO_x, carbon monoxide CO, unburned hydrocarbons (UHC), sulfur oxides SO_x, and soot

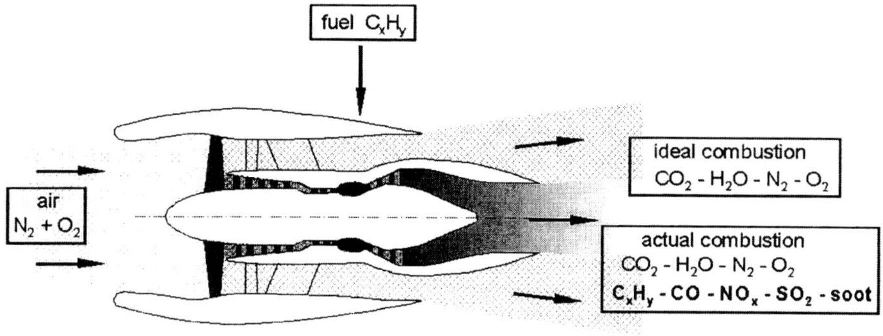

Figure 1.6-1. The combustion process in a turbofan engine.

particles. At present there is only a limited knowledge about the formation and behavior of minor, trace species and aerosols found in the exhaust gases of turbo-engines. Even less is known about how they are influenced by engine features and characteristics (Ref. 1).

Emissions of NO_x, CO, and UHC can be reduced by improved combustion techniques, while reduction of CO_2 and water H_2O requires reduced fuel consumption. A favorable feature of the kerosene fuel used for jet airplane propulsion is that it contains almost no sulfur. Therefore, the emission of sulfur dioxide SO_2 by airplanes is very low and usually negligible.

Although air transport uses just 5 percent of the global oil consumption per year and about 13 percent of the yearly consumption of fossil fuels taken by all forms of transport, control of pollutant emissions from the combustion of kerosene is of increasing importance in the design and operation of airplanes and airplane propulsion systems.

This is reflected by the fact that already for many years, standards for the pollutants produced by aviation engines, namely carbon monoxide CO, unburned hydrocarbons (UHC), nitrogen oxides NO_x, and smoke, are set by the International Civil Aviation Organization (ICAO) in Ref. 6. According to its definition by ICAO, smoke is formed by the carbonaceous materials in exhaust emissions which obscure the transmission of light.

In the past, man was merely concerned with the effects of pollutant emissions at ground level near the airports. In this respect, with success, attention has been given to the reduction of CO and UHC, which appear to be dominating at low thrust settings of the engine (Figure 1.6-2). Smoke emissions, which used to dominate at high thrust settings, have also been greatly reduced.

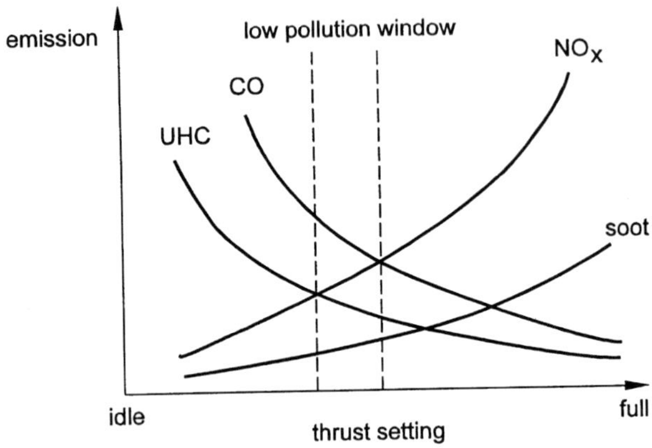

Figure 1.6-2. Emission level versus thrust setting.

Presently, much effort is devoted to the effects of emissions from aircraft upon the upper region of the troposphere and the lower region of the stratosphere, since at these heights the airplane is the only anthropogenic source of pollutants. Especially, the emissions of NO_X, the result of burning kerosine at high temperatures, have been increased during the past decades.
For long-range flights, naturally, the largest amount of NO_X will be emitted during the cruise part of the flight, where is flown at altitudes near the tropopause.

Today, therefore, considerable effort is being made to lower the NO_X emissions by improving the combustion process. Also, reducing the fuel consumption per flight will remain an effective way of decreasing the impact of aviation on the environment.

1.7 THE HYDROLOGICAL CYCLE

A great influence on the removal of pollutants from the atmosphere has the *hydrological cycle* by which water evaporates from oceans and land, is transported by air motions, forming clouds and precipitations, and returns from oceans and land to the atmosphere (Figure 1.7-1).

Wind blowing across the oceans and wave action produce a fine spray of water droplets containing salt. Upon these solid particles the water vapor in the atmosphere may condense.
Figure 1.7-2 illustrates the general, global pressure and surface wind

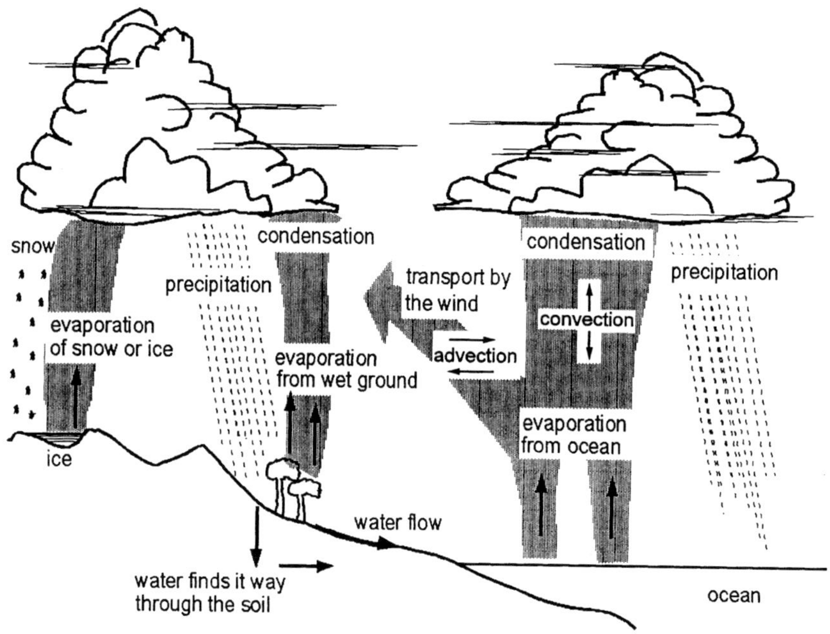

Figure 1.7-1. The hydrological cycle.

distributions. Surface winds are defined as winds occurring at very low heights. Close to the Equator the surface water, and so the air in contact with it, is heated intensively by its direct exposure to solar radiation. The warming of air produces a region of low atmospheric pressure into which air is drawn.

The low pressures near the Equator are the source of the well-known *trade winds*. Mariners exploited these winds, but dreaded the equatorial belt of light variable winds lying between them, which they called the *doldrums*.

In the mid-latitudes the familiar westerly winds (*westerlies*) prevail, while in both polar regions easterly winds (*easterlies*) occur.

Note that the direction of the wind is that direction from which it is blowing. If the wind is coming from the southwest, its direction is said to be southwest (S.W.). Usually wind directions are reported by directions on a 32 points compass or in degree (Figure 1.7-3). Thus, at the northern hemisphere the direction of the trade wind is N.E. (northeast) and at the southern hemisphere S.E. (southeast).

Like wind, also oceans and seas are important to the dispersion of heat energy over the Earth. Oceans and seas cover more than 70 percent of the total surface of the Earth. The average depth is 3.73 km and together they contain 1370 million km^3 of water. If the Earth would be covered by one vast ocean, the ocean would be 2.7 km deep.

The trade winds drive the surface currents toward the Equator, through which

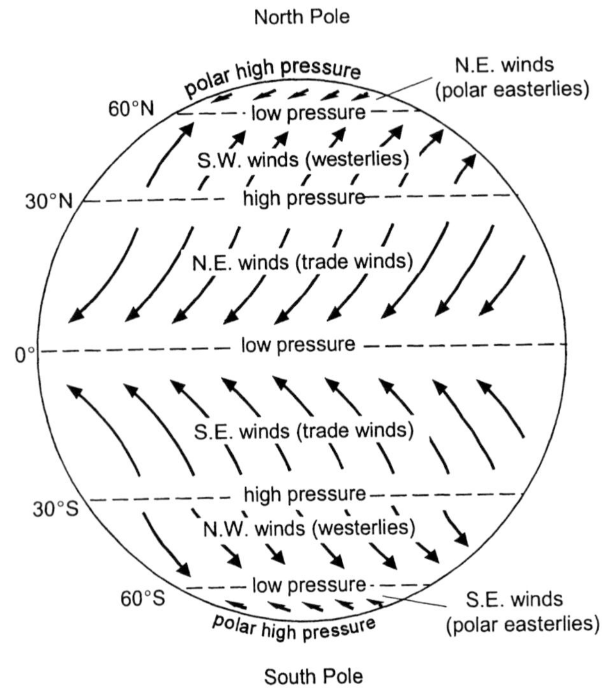

Figure 1.7-2. Global surface pressures and air currents.

cold water is heated and returns to warm the higher latitudes from which it came. Clearly, the transport of cool water to the Equator prevents the tropical regions from being as hot as they would be otherwise.

The atmosphere is warmed from underneath through which convection may ensue, with vertical ascents and descents of air, leading to a thorough mixing of the air. This situation is opposite to that at the surfaces of the oceans, where the water is heated from above by the Sun. This implies that there are layers of

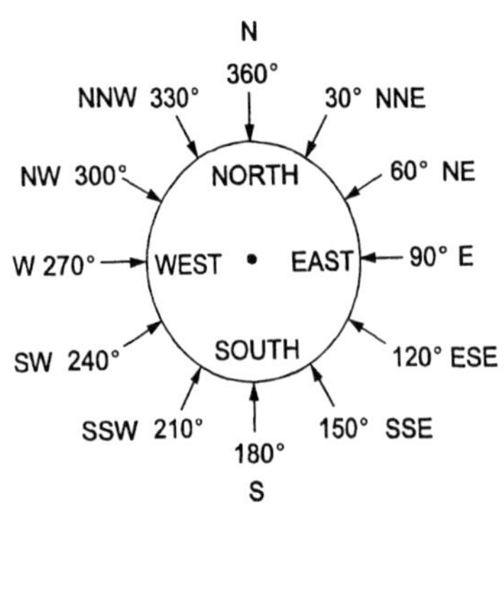

Figure 1.7-3. Wind direction scales.

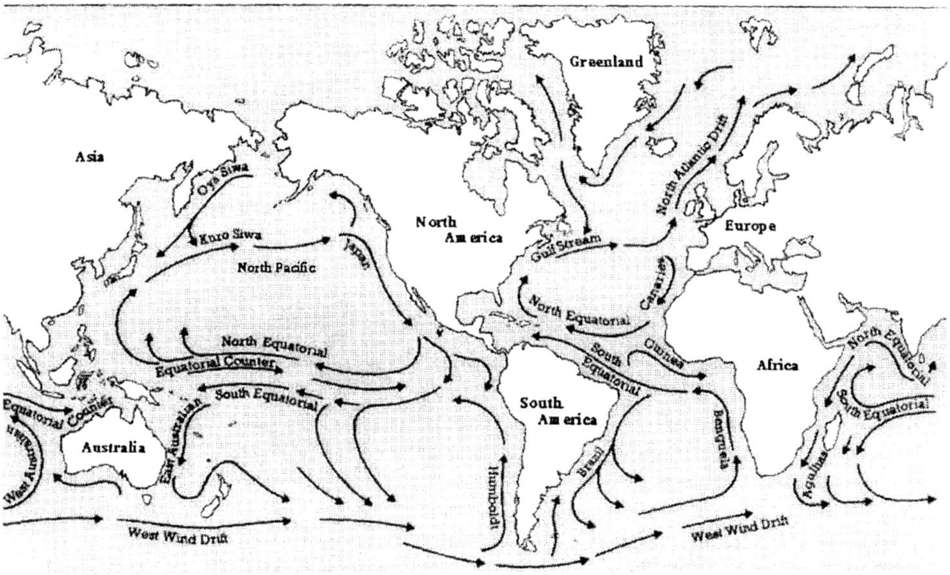

Figure 1.7-4. Major ocean currents.

relatively warm water floating above cooler and denser water masses. As little vertical mixing occurs, the oceans can be divided into approximately horizontal layers with different water temperatures.
At the same time, the water in the oceans is in constant horizontal motion. Friction between the water and the moving air keeps the water flowing in the direction of the wind. The interaction between wind and water has created the major global currents, which brings warm tropical water to the higher latitudes.

Figure 1.7-4 shows great surface motions of the oceans and seas, as caused by the prevailing surface winds. Due to the *Coriolis force*, which bends winds because of the rotation of the Earth about its own axis, also the flow of water is deflected toward the right north of the Equator and toward the left south of it (see also Chapter 2, Section 2.6). Summarizing, it may be said that by carrying warm water to cool latitudes, and cool water to warm latitudes, the ocean currents have a major effect on the local weather and the climates of the world's regions.

In addition to the surface winds, the influence of tidal currents, as caused by the combined effect of the gravitational pull of the Earth, Moon and Sun upon the water mass on the Earth, is of importance to the transfer of heat over the Earth. The essentials of *ebb and flow* of the *tides* are discussed separately in the next section.

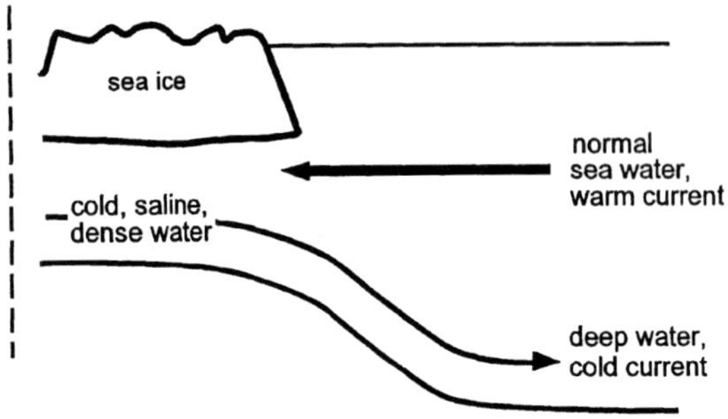

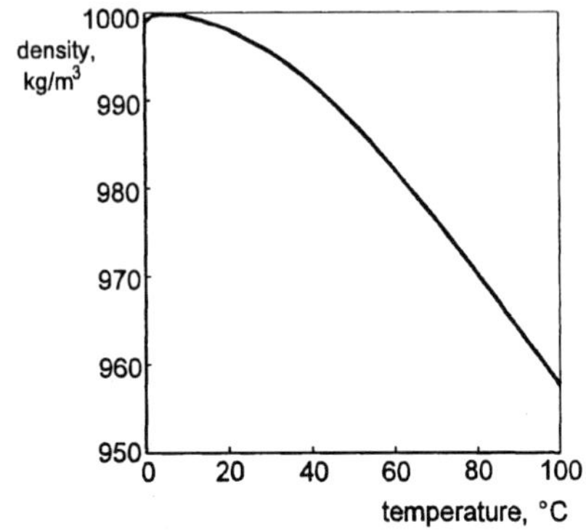

Figure 1.7-5. Formation of deep water current. *Figure 1.7-6. Density of water.*

Besides the movement of surface water, there are the deepwater currents (Figure 1.7-5). These are flowing from the polar regions, mainly by the formation of sea ice, which removes fresh (saltless) water from the sea surface.
Immediately below the sea ice, the water is therefore more saline. Moreover, the water has a temperature at which it has approximately its greatest density (Figure 1.7-6). These factors make the surface water more dense than the water below it. Accordingly, the water sinks below the less denser water and, at the same time, moves away from the sea ice. In this way a flow is established of very cold, saline water moving away from the Poles along the bottom of the ocean toward the Equator. Its place is taken by water flowing polewards beneath the surface water, and then is forced to rise as it encounters the denser water that is sinking. The above mechanisms are believed to drive the oceanic system of heat transport by ocean currents.

To conclude, it is important to emphasize that pollution not only implies the release of undesired substances into the atmosphere and soil, but also their introduction into the water. In this respect it should be noted that the oceans also are of great importance to the environment and the climate in the sense that they provide yearly the storage of an enormous amount of carbon dioxide from the atmosphere. The CO_2 leaving the atmosphere dissolves in the upper layers of the oceans and seas and, finally, becomes mixed into the deeper layers. Concerning this storage, it is worthy of mention that, as with oxygen, the solubility of water decreases as its temperature rises. In plain words, warm water cannot contain as much carbon dioxide as cool water.
Another consequence of a rise in the global average temperature at the Earth's surface, is that the waters of the oceans will warm and expand. Moreover, their volumes will increase as polar glaciers and other large land ice caps melt. This

would lead to a rise in sea level all over the world, and drown the land of low-lying countries and islands.
A temperature rise also will lead to increased evaporation and, therefore, to increased precipitation.

1.8 TIDAL CURRENTS

Tidal currents from the periodic rice and fall of the oceans are created by the combined gravitational action of the Earth, Sun and mainly the Moon upon the water mass on the Earth.
Figure 1.8-1 shows the orbital motions in the Sun-Earth-Moon system. Within this system, the Earth and Moon form a single system, mutually rotating around their common center of mass. The Earth-Moon barycenter is inside the Earth, about 1700 km below the surface (Figure 1.8-2).
Assuming circular orbits, the angular velocities and the radii of the paths traveled are the same everywhere. As a result of the eccentric motion, all points on and within the Earth will experience an equal centrifugal force. These forces are directed parallel to a line joining the centers of the Earth and the Moon.

On the contrary, the gravitational force exerted by the Moon on the Earth varies inversely with the square of the distance from the Moon. Points on the Earth nearest the Moon will experience a greater gravitational pull from the Moon than will points on the opposite side of the Earth. Because the component parts of the gravitational force are directed toward the Moon's center, the forces will not be exactly parallel to the direction of the centrifugal forces, except the force directed along the line joining the centers of the Earth and Moon.

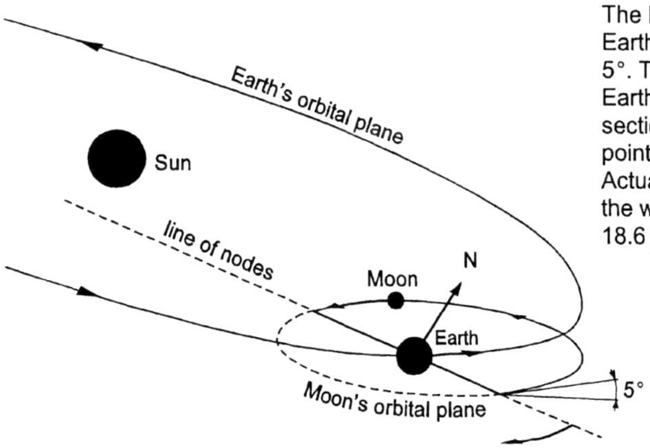

The Moon's orbital plane intersects the Earth's orbital plane at an angle of about 5°. The two planes are ellipses. As the Earth revolves around the Sun, the intersection line (*line of nodes*) continues to point in a fixed direction among the stars. Actually, the line of nodes rotates toward the west and completes a revolution in 18.6 years.

Figure 1.8.1. The orbital motions in the Sun-Earth-Moon system.

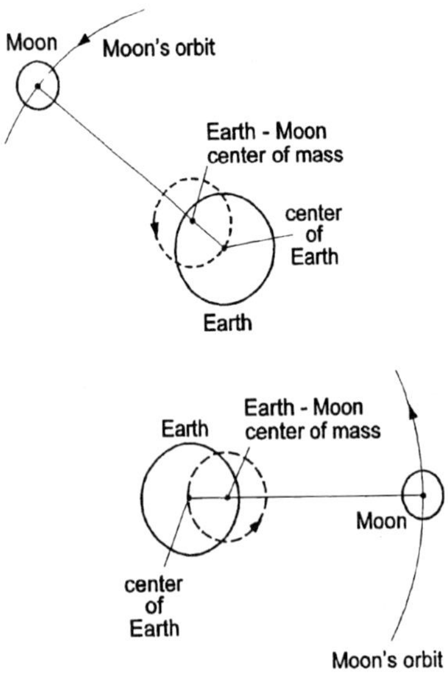

Figure 1.8-2. The rotation of Earth and Moon about the center of mass of the Earth-Moon system.

However, the total centrifugal force within the Earth-Moon system exactly balances the total force of gravitational attraction between the two bodies, so that the system as a whole is in equilibrium.

The resultant of the two forces at each point on the Earth's surface is known as the *tide-producing force*, and, depending upon its position on the Earth's surface with respect to the Moon, will be directed into, away from, or parallel to the

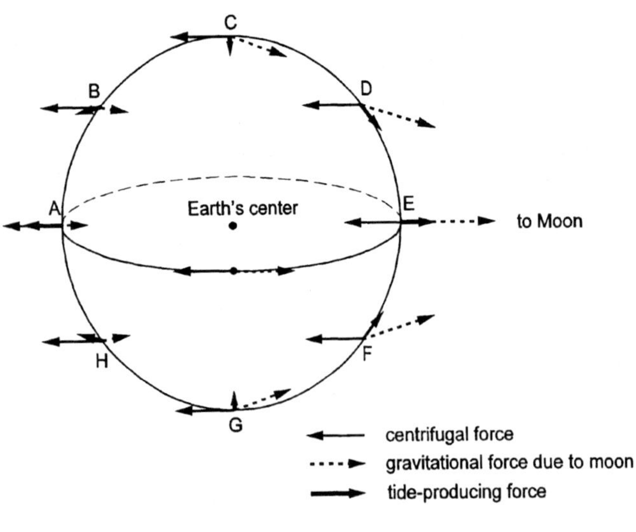

Figure 1.8-3. Tide-producing forces (not to scale).

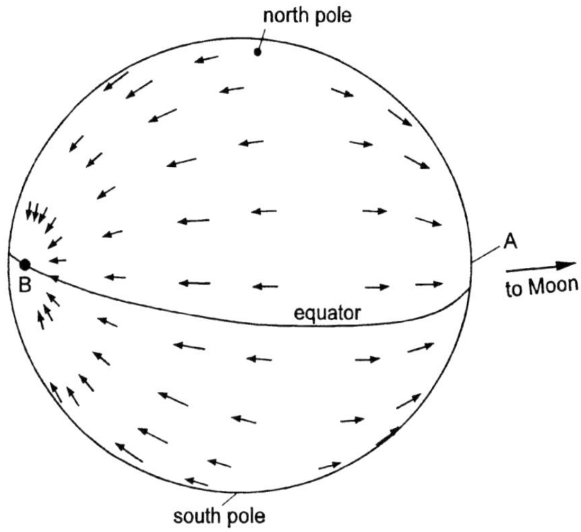

Figure 1.8-4. Tractive forces.

Earth's surface. The relative strengths and directions of the forces are illustrated in Figure 1.8-3.

Obviously, at points A and E in Figure 1.8-3 the difference between the centrifugal force and the Moon's gravitational force is greatest. Nevertheless, the local effects of these forces on ebb and flow of the tides are negligible as they are acting against the pull of the Earth's own gravity, which strongly exceeds the resulting tide-producing force.

At other points the effects of the tide-producing forces are well perceptible when there is a large horizontal component of this force. These so-called *tractive forces* cause the water to move, because, although small compared with the Earth's gravitational field, this horizontal component is not opposed by any other lateral force.

Figure 1.8-4 portrays where on the Earth's surface the tractive forces are at a maximum when the Moon is over the Equator. In this special situation, the tractive forces would result in a movement of water toward points A and B, producing an ellipsoid with its two bulges directed toward and away from the Moon (Figure 1.8-5a).

Because the Moon rotates about the Earth-Moon center of mass once every 27.3 days in the same direction as the Earth rotates upon its *polar axis* once every 24 hours, the period of the Earth's rotation with respect to the Moon is about 24 hours and 50 minutes (a *lunar day*), so that the lunar semi diurnal tide has a period of 12 hours and 25 minutes (Figure 1.8-6). Clearly, the times of high tides at many locations are almost an hour later each successive day.

Variations in the lunar-induced tides occur due to fact that the relative positions and orientations of the Earth and Moon vary according to the Moon's

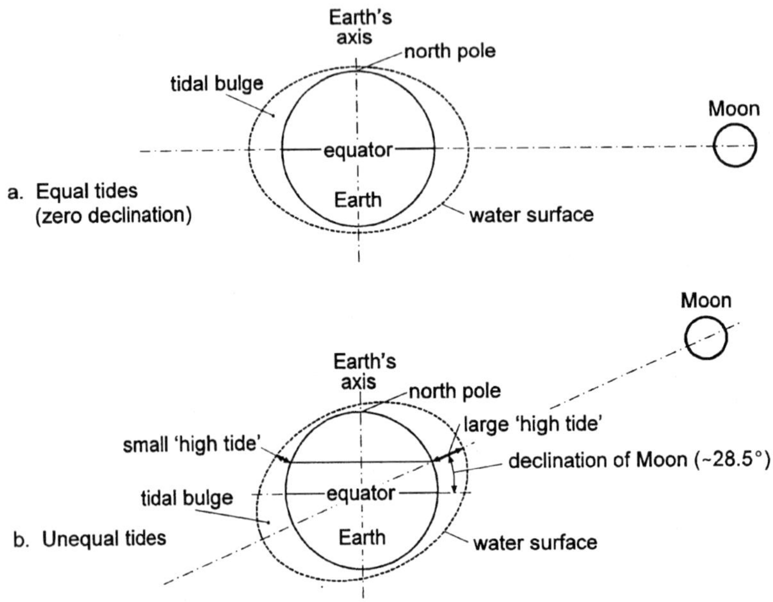

Figure 1.8-5. Effect of Moon's declination.

declination. As shown in Figure 1.8-5b, the Moon's orbital plane is inclined at an angle of 28.5° relative to the plane of the Earth's Equator. Consequently, when the Moon is at a large angle of declination, particularly, at mid-latitudes unequal tides will be produced. Hence, the heights reached by the two daily tides will show diurnal inequalities.

Moreover, the elliptical orbit of the Moon around the Earth-Moon center of mass results in variations in the tide-producing forces. When the Moon is closest to the Earth, the tide-producing force is increased above the average value, and decreased when the Moon is furthest from Earth.

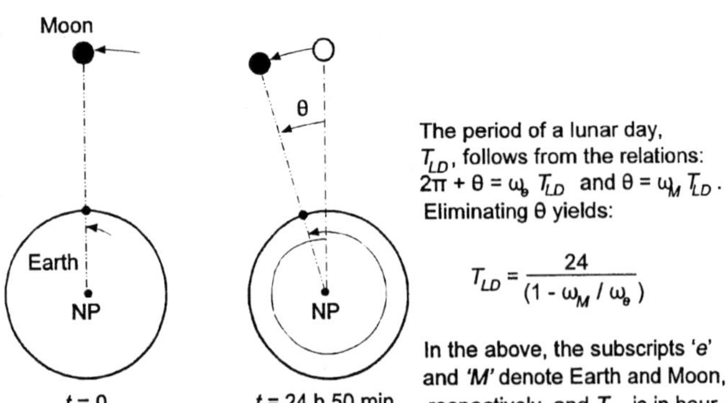

Figure 1.8-6. Solar and lunar day relationship.

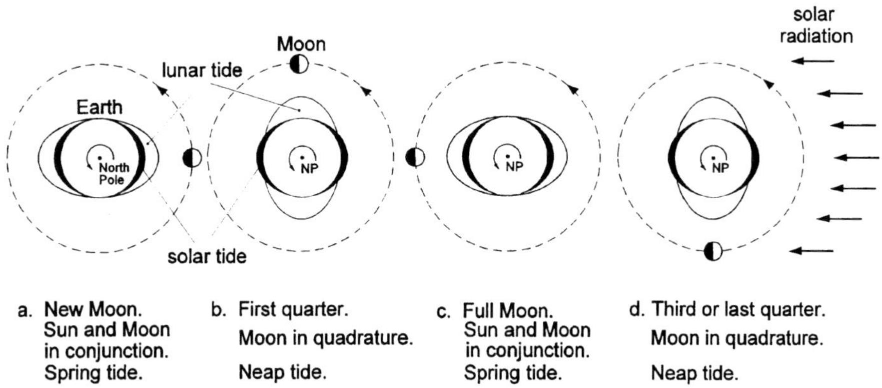

Figure 1.8-7. Interaction of lunar and solar tides.

Tides are highest when the Moon is full or new, and the times of high tide at any given location can be approximately related to the position of the Moon in the sky. The Sun also causes a tide-producing force, of which the magnitude is much smaller than that of the Moon, because, although enormously greater in mass than the Moon, the Sun is some 360 times further from the Earth. The resulting effect depends on the position of the Earth, Sun and Moon together, whether their position is in line or not. Figure 1.8-7 shows the simplest case, where the declinations of Moon and Sun are both zero.

When the tide-generating forces of the Sun and Moon are acting in conjunction, the high tide is higher and the low tide is lower than the average. Such tides are known as *spring tides* (Figures 1.8-7a and c. On the other hand, when the Sun and Moon act at right angles to each other, the tidal range is correspondingly smaller than average. These tides are known as *neap tides* (Figures 1.8-7b and d).

Of course, the force fields outlined above are strongly simplified. The regular changes in the declination of the Sun and Moon, and the elliptical orbits of the Earth around the Sun and the Moon around the Earth, will affect the tide at any particular time and place. Furthermore, the effects of topography, wind and atmospheric pressure distribution are also of significance. For practical use, tide-tables are available, indicating times of ebb and flow, and water levels. For further information and the theory on the dynamics of the tides, the reader may turn to Ref. 7.

1.9 THE SI-SYSTEM OF UNITS

Throughout this book the *International System of Units* (officially called *Système International d'Unités* in French) is used. This metric system has been adopted by many nations as the recommended system of units for weights and measures. In Table 1.9-1 are tabulated the seven basic SI units, and derived units that are relevant to this text.
It is of interest to note that, despite the legal adoption of the SI system of units, in engineering practice frequently the so-called *technical system of units* is used. In this system the quantity force, having also the name kilogram, is a basic unit instead of mass. In order to distinct these two kilograms, in the technical system the quantity force is (often) denoted as kilogram force, abbreviated as kgf. The following relationship is defined:

$$1\ \text{kgf} = 9.80665\ \text{N}.$$

quantity	name of unit	symbol	definition
	basic SI-units		
length	meter	m	
mass	kilogram	kg	
time	second	s	
temperature	kelvin	K	
electric current	ampère	A	
luminous intensity	candela	cd	
amount of substance	mole	mol	
	derived SI-units		
force	newton	N	kg m/s^2
pressure	pascal	Pa	N/m^2
work (energy)	joule	J	J = N m
power	watt	W	J/s
velocity	meter per second	V	m/s
acceleration	meter per second squared	a	m/s^2
density	kilogram per unit cubic meter	ρ	kg/m^3

Table 1.9-1. Basic and derived SI- units.

Also note that the technical system may also be expressed in English units, using foot (ft) and pound force (lbf) as basic units for length and force, respectively. The following relationships are defined:

1 ft = 0.3048 m,

1 lbf = 0.45359 kgf = 4.44822 N.

A number of technical units and corresponding SI units are given in Table 1.9-2. Although the SI unit of temperature is the kelvin (K), also the celsius or centigrade scale is employed. Since the unit degree celsius (°C) is exactly equal to the unit kelvin, temperature expressed in degree celsius can be readily converted to temperature in kelvin by the relationship:

°C = K - 273.15.

In nonscientific usage one may use the temperature scale fahrenheit. To convert degree celsius to degree fahrenheit (°F): multiply by 9/5 and add 32,

°F = °C ×9/5 + 32.

Similarly, to convert degree fahrenheit to degree celsius: subtract 32 and multiply by 5/9.

quantity	technical system		SI - system
	metric	English	
length	m	ft	m
time	s	s	s
force	kgf	lbf	kg m/s^2 (newton)
mass	kgf s^2/m	lbf s^2/ft (slug)	kg
pressure	kgf/m^2	lbf/ft^2	N/m^2
work (energy)	kgf m	lbf ft	kg m^2 /s^2 = N m (joule)
power	kgf m/s	lbf ft/s	kg m^2/s^3 = J/s (watt)
density	kgf s^2/m^4	lbf s^2/ft^4	kg/m^3

Table 1.9-2. System of units.

1.10 POLLUTANT CONCENTRATION

Naturally, the degree of harmfulness is different for the various pollutants. That's why central to the concept of air pollution is the concentration, since there is no substance so dangerous that there is no dose below which it is harmless to all organisms.

Important to the quantification of air pollution is the fact that atmospheric properties mostly involve large quantities, whereas pollution usually concerns very small concentrations.

Table 1.10-1 presents the prefixes, frequently used to avoid extremely large or small numbers.
When referring to trace gases, usually, the concentration is expressed in terms of ppm (parts per million, 10^6), ppb (parts per billion, 10^9), or ppt (parts per trillion, 10^{12}). These units indicate that the pollution particles differ from the other million, billion or trillion particles in the sample.

A distinction may be made by noting: ppm (v) for 'parts per unit volume' (ml/m^3) and ppm (m) for 'parts per unit mass' (mg/kg). Also, the ratio mg/m^3 may be used.
Finally, it should be noted that the denomination of some large numbers has different meanings in the USA and European countries (Table 1.10-2).

multiple	prefix	symbol	fraction	prefix	symbol
10^{24}	yotta	Y	10^{-1}	deci	d
10^{21}	zetta	Z	10^{-2}	centi	c
10^{18}	exa	E	10^{-3}	milli	m
10^{15}	peta	P	10^{-6}	micro	µ
10^{12}	tera	T	10^{-9}	nano	n
10^{9}	giga	G	10^{-12}	pico	p
10^{6}	mega	M	10^{-15}	femto	f
10^{3}	kilo	k	10^{-18}	atto	a
10^{2}	hecto	h	10^{-21}	zepto	z
10	deca	da	10^{-24}	yocto	y

Table 1.10-1. Standard multiples and decimal fractions.

number	denominator	
	European	American
10^{6}	million	million
10^{9}	milliard	billion
10^{12}	billion	trillion
10^{15}	thousand billion	qaudrillion
10^{18}	trillion	quintillion
10^{21}	quadrillion	septillion
10^{30}	quintillion	nonillion

Table 1.10-2. Denomination of numbers.

2 NATURE OF THE ATMOSPHERE

2.1 APPLICATION OF NEWTON'S LAW OF MOTION

According to Newton's second law of motion, the translational motion of a body with mass M is described by the well-known equation:

$$\boldsymbol{F} = M\boldsymbol{a} \ , \tag{2.1-1}$$

where $\boldsymbol{F}$ is the vector sum of all external forces acting on the body, and $\boldsymbol{a}$ is the (absolute) acceleration. Note that in this text boldface type letters are used to designate vector quantities.

Equation (2.1-1) requires that the motion is written down with respect to an *inertial frame of reference*, which is an axis system in a state of complete rest, or any coordinate system which translates with uniform velocity relative to the frame at rest.

Equation (2.1-1) can be applied in a coordinate system attached to the Earth if two apparent forces are added to the force $\boldsymbol{F}$ (see the analysis in Ref. 8),

$$\boldsymbol{F} - M\boldsymbol{\omega}_e \times (\boldsymbol{\omega}_e \times (\boldsymbol{R} + \boldsymbol{h})) - M(2\boldsymbol{\omega}_e \times \boldsymbol{V}) = M\boldsymbol{a}_r \ , \tag{2.1-2}$$

where $\boldsymbol{\omega}_e$ is the Earth's angular velocity, $\boldsymbol{R}$ is the Earth's radius vector, $\boldsymbol{h}$ is the height above the surface of the Earth, $\boldsymbol{V}$ is the velocity of the body with respect to the Earth, and $\boldsymbol{a}_r$ is the acceleration relative to the Earth.

The second term of the left-hand side of Equation (2.1-2) is the *centrifugal force*,

$$\boldsymbol{F}_t = -M\boldsymbol{\omega}_e \times (\boldsymbol{\omega}_e \times (\boldsymbol{R} + \boldsymbol{h})) = -M\boldsymbol{a}_t \ , \tag{2.1-3}$$

where $\boldsymbol{a}_t$ is the centripetal acceleration.

The third term of the left-hand side of Equation (2.1-2) is the *Coriolis force*, named after the French physicist Gaspard de Coriolis (1792 - 1843),

$$\boldsymbol{F}_c = - M(2\boldsymbol{\omega}_e \times \boldsymbol{V}) = -M\boldsymbol{a}_c \ , \tag{2.1-4}$$

where $\boldsymbol{a}_c$ is the Coriolis acceleration.

The derivation of Equation (2.1-2) is based on the assumption that the Earth translates with constant velocity along a straight line. This idealization of the Earth's motion around the Sun may be correct in view of the large distance that the Earth is removed from the Sun (see Section 2.2).

By expressing the Equations (2.1-3) and (2.1-4) in trigonometric form, the magnitude of the centrifugal force and Coriolis force due to the Earth's rotation about its polar axis read:

$$F_t = M\omega_e^2 (R + h) \cos\theta \ , \tag{2.1-5}$$

and

$$F_c = M 2 \omega_e V \sin\varphi \ . \tag{2.1-6}$$

In Equation (2.1-5) the angle θ is *geographic latitude* (Figure 2.1-1). Latitude is measured north and south from the Equator through 90° in each direction. The Equator is the 0° latitude parallel. Perpendicular to the parallels are drawn the *meridians of longitude*. By international agreement the 0° meridian passes through Greenwich, near London. Longitude is measured east and west from the 0° meridian through 180° in each direction.

In Figure 2.1-1 also is depicted a *great circle* which, marked on a sphere, divides it into two equal halves. The Equator, the 0° latitude line, is the only parallel that is a great circle, whereas all meridians of longitude are great circles.

A great circle route is of importance to aviation, because it always yields the shortest distance between any two points on Earth.

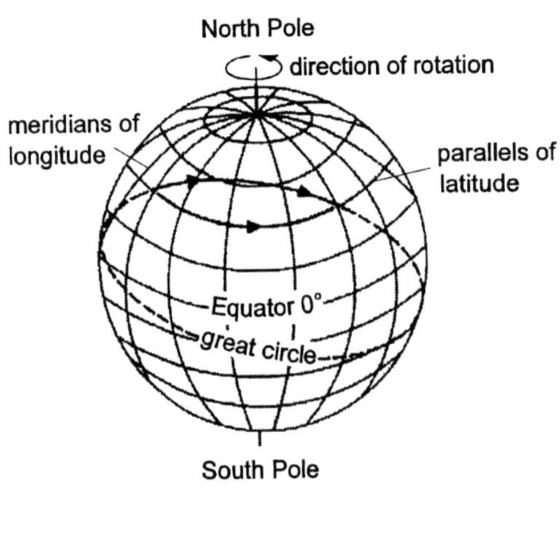

Figure 2.1-1. Parallels and meridians.

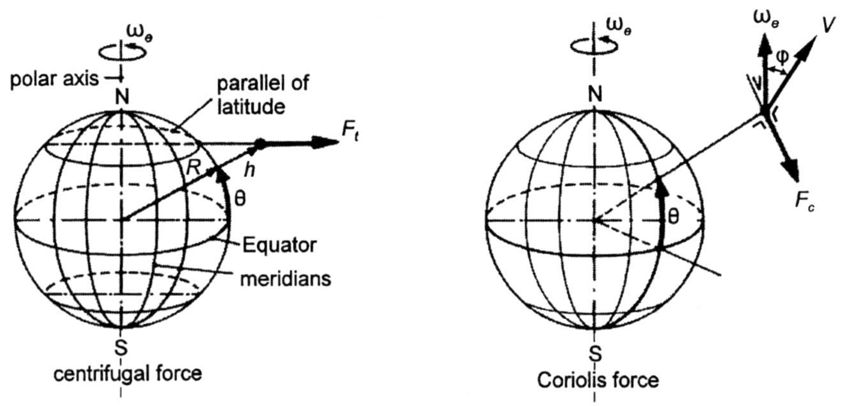

Figure 2.1-2. Forces due to the rotation of the Earth.

The forces and geometry used in Equations (2.1-5) and (2.1-6) are depicted in Figure 2.1-2. The angle φ defines the direction of the velocity *V* relative to the polar axis. The centrifugal force is directed perpendicular to the Earth's axis and points out from the Earth along a line intersecting the axis of rotation.

Equation (2.1-6) shows that at a given velocity, the Coriolis acceleration, a_c, has its maximum value when the velocity of the body is directed perpendicular to the polar axis (φ = 90°).

Figure 2.1-3 illustrates the Coriolis force in case of a northward velocity, where the body is at ground level. Then it appears that the body is subject to an eastward force:

$$F_c = M 2 \omega_e V \sin\theta \ . \tag{2.1-7}$$

Similarly, for a southward velocity the Coriolis force acts in westward direction.

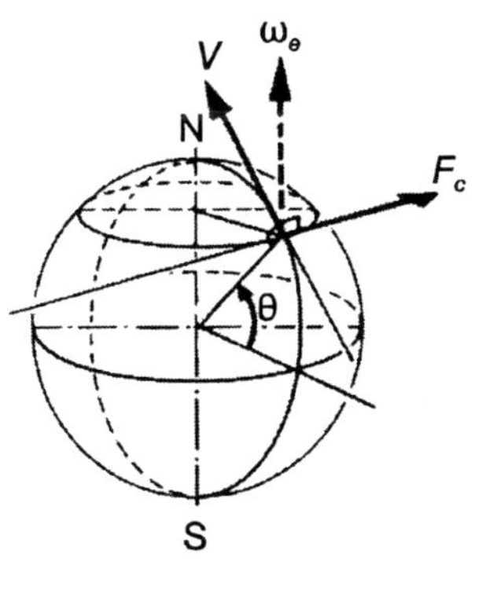

Figure 2.1-3. Eastward Coriolis force due to north velocity.

2.2 THE SEASONS

The seasons are produced by the ellipticity of the Earth's orbit around the Sun, and the fact that the Earth's axis of rotation is inclined at an angle of 23.5 degrees with respect to the direction normal to its orbital plane. This plane is called the *plane of the ecliptic* (Figure 2.2-1).

The distance between Earth and the Sun varies between 147.2×10^6 km and 152.1×10^6 km. As a consequence, the Earth's orbit around the Sun makes the northern hemisphere summer about five days longer than the southern hemisphere summer.
The Earth is closest to the Sun in early January, and most distant in early July. The closest and farthest points along the Earth's orbit are called *perihelion* and *aphelion*, where the distances between the Earth and the Sun differ from the mean distance by about 2.5×10^6 km.
From March 21 to September 23 the northern hemisphere is tilted toward the Sun, and so receives a greater input of heat than the southern hemisphere. On June 21, which is called the *summer solstice*, northern regions are tilted most directly toward the Sun, giving the longest day and shortest night of the year. At around that time there the Sun appears high in the sky at noon.
During the other half of the year, the northern hemisphere is tilted away from the Sun, and receives less heat than the southern hemisphere. At December 21, the *winter solstice*, the Sun rises late in the northern temperate zone, passes low to the south and sets early, giving the shortest day and the longest night.
In the southern hemisphere the daylight patterns are reversed, and the winter and summer solstices are at June 21 and December 21, respectively.

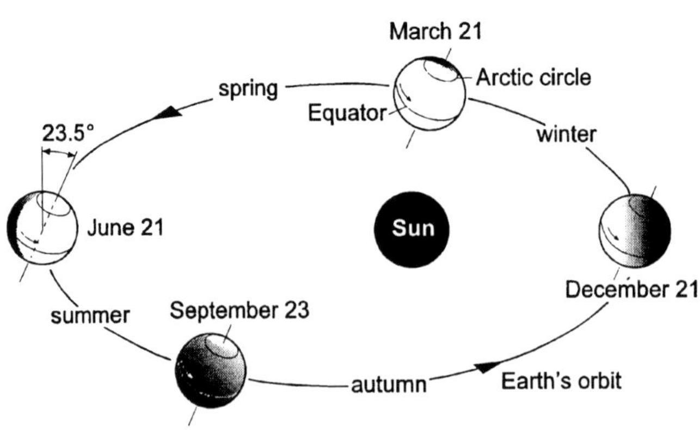

Figure 2.2-1. Orbit of the Earth around the Sun.

Regions near the Equator, face the Sun at about the same angle throughout the year. Hence, in the tropics the seasons are less marked, so that the Sun always climbs high in the sky and the number of hours of daylight varies only slightly during the year.
The effect on the lengths of the day and night is maximum near the Poles. The arctic and antarctic circles are defined as the limits, pollards of which there is at least one period of 24 hours during the year when the Sun does not appear above the horizon, and one 24-hour period when it does not sink below the horizon. The circles are therefore at latitudes 90° - 23.5°= 66.5°N. and S.
Halfway between the solstices, on days called *equinoxes*, the Sun crosses the Equator plane, so that daylight and night are equally long and the Sun is overhead at midday on the Equator. In the northern hemisphere the spring equinox falls at March 21 and the autumn equinox at September 23. These data introduce spring and autumn, which, of course, are reversed in the southern hemisphere.

It will be clear that the amount of solar energy falling on the surface of the Earth reaches a maximum when the Sun is overhead, and becomes insignificant at or near sunrise or sunset. In addition to the diurnal variation of the Sun's altitude from dawn to dusk, a pronounced seasonal variation is present in the insolation, at least for high and intermediate geographical latitudes. The variation is modest at the Equator and greatest at high latitudes as a result of the extreme seasonal variation in those regions. It is also important to recognize that most of the southern hemisphere is covered by water, unlike the northern hemisphere, which is dominated by land. Therefore, the southern hemisphere has fewer extreme temperatures because oceans heat and cool more slowly than the land.
At this point, it is interesting to note that the Earth is not precisely spherical, but rather pear-shaped. Since the introduction of artificial satellites we know that the Poles are closer to the center of the Earth than a point on the Equator, and that the radius to the North Pole is somewhat larger than the radius to the South Pole. The gravitational influence of Sun and Moon on the equatorial bulge causes a very slow gyroscopic motion of the Earth's polar axis, which type of motion is known as precession (Ref. 9).

The complete revolution of the Earth about its polar axis in 24 hours, allows a convenient division of the Earth surface into 24 standard time zones, each of about 15° of longitude wide. In general each zone is one hour ahead of the next zone to the west, except at the international date line in the Pacific Ocean (the 180° meridian), where a westbound traveler will go forward by 24 hours. At some places the zones have been modified for convenience. In certain parts of the world daylight-saving time is used, i.e., during summer clocks are put one or two hours ahead of the appropriate setting for their time zone.

2.3 ACCELERATION OF GRAVITY

The acceleration of gravity results from Newton's law of gravity. According to this law, the *gravitational force* F_g acting upon a particle with mass M located at a height h above the Earth's surface, is given by (Figure 2.3-1):

$$F_g = \mu \frac{M_e M}{(R_e + h)^2} , \tag{2.3-1}$$

where M_e is the mass of the Earth ($M_e = 5.98\times10^{24}$ kg), R_e is the mean radius of the Earth ($R_e = 6371$ km). The factor μ is a proportionality factor, known as the *universal gravitational constant* ($\mu = 6.673\times10^{-11}$ m^3/kg s^2).
To derive Equation (2.3-1) the assumption must be made that the Earth can be considered a sphere, of which the density is a function of the distance to its center only. It is worth noting that, as the mass of the Earth is 5.98×10^{24} kg, its mean density amounts to 5520 kg/m^3. Because the mean density of the material near the Earth's surface is about 2700 kg/m^3, the interior of the Earth must be much denser!
As depicted in Figure 2.3-1, there is also a centrifugal force F_t acting due to the rotation of the Earth about its polar axis (see also the previous Figure 2.1-2). With respect to an earthbound, rotating frame of reference this force is given by:

$$F_t = M\omega_e^2 (R_e + h) \cos\theta , \tag{2.3-2}$$

where ω_e is the angular velocity of the Earth ($\omega_e = 2\pi/(24\times3600) = 7.29\times10^{-5}$

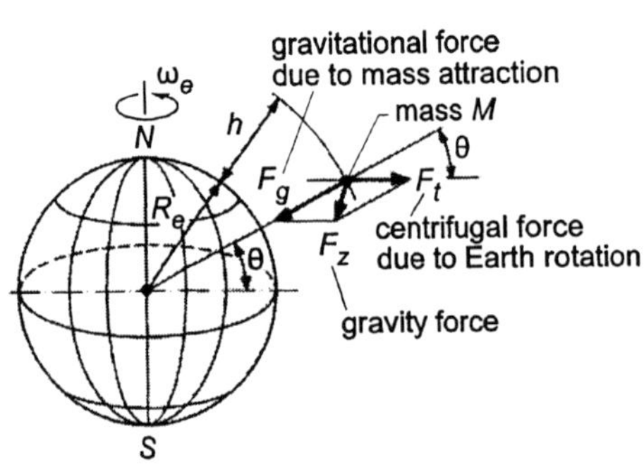

Figure 2.3-1. Components of gravity force.

rad/s), and angle θ is the geographic latitude.
The *gravity force* F_z stems from the vector sum of the gravitational force F_g and the centrifugal force F_t:

$$\boldsymbol{F}_z = \boldsymbol{F}_g + \boldsymbol{F}_t \ . \tag{2.3-3}$$

Figure 2.3-1 indicates that the gravity force F_z is smaller than the gravitational force F_g. Also the gravity force varies with geographical latitude, due to the variation of the centrifugal force. At the Equator the gravity force is minimum, while at the Poles it reaches a maximum value, and $F_z = F_g$. The acceleration g that a mass M experiences in the direction of the center of the Earth during free fall from a height h above the Earth's surface, follows from $F_z = M\,g$ and Equations (2.3-1) and (2.3-2):

$$g = \mu \frac{M_e}{(R_e + h)^2} - \omega_e^2 (R_e + h) \cos^2\theta \ . \tag{2.3-4}$$

At sea level, where $h = 0$:

$$g_0 = \mu \frac{M_e}{R_e^2} - \omega_e^2 R_e \cos^2\theta \ . \tag{2.3-5}$$

Apparently, g_0 increases with geographical latitude. At the Equator, where cos θ equals one, g_0 is minimum:

$$g_0 = \mu \frac{M_e}{R_e^2} - \omega_e^2 R_e = 9.827 - 0.034 = 9.793 \text{ m/s}^2 \ . \tag{2.3-6}$$

At the Poles, it has a maximum value of:

$$g_0 = \mu \frac{M_e}{R_e^2} = 9.827 \text{ m/s}^2 \ . \tag{2.3-7}$$

The Earth not only has a pear shape. Also, its mass distribution is not homogeneous. As a consequence, the value of g_0 will differ from the values predicted with Equation (2.3-5).
On the northern hemisphere, an adequate prediction of the sea-level acceleration of gravity can be obtained from the following equation, named after the famous cartographer Johann Heinrich Lambert (1728-1777), see Ref 16:

$$g_0 = 9.780356\,(1 + 0.0052885 \sin^2\theta - 0.0000059 \sin^2 2\theta)\,,\ \text{m/s}^2 \ . \tag{2.3-8}$$

The values of g_0 at the Equator, at 45°N., and at the Pole from Lambert's formula are 9.78036, 9.80616, and 9.83208 m/s^2, respectively.

Ignoring the effect of the centrifugal force in Equation (2.3-4), the acceleration of gravity varies inversely with the square of the distance to the center of the Earth:

$$g = g_0 \frac{R_e^2}{(R_e + h)^2} = \frac{g_0}{\left(1 + \frac{h}{R_e}\right)^2} . \tag{2.3-9}$$

For example, at an altitude of 11 km follows: $g/g_0 = 0.9966$.

2.4 VARIATION OF AIR PRESSURE AND DENSITY WITH ALTITUDE

The weight of the column of air at rest above a unit area will produce a certain pressure, p, at that surface. The higher one rises in the atmosphere, the smaller will be the weight of the air above the unit area, and the smaller will be the occurring pressure. The change in pressure dp with altitude dh is described by the *aerostatic equation* (Figure 2.4-1):

$$\mathrm{d}p = -\rho g \,\mathrm{d}h , \tag{2.4-1}$$

where ρ is the density, and g is the acceleration of gravity.
In adopting the aerostatic equation it is assumed that the atmosphere is at rest with respect to the Earth.
Atmospheric air can be assumed to satisfy the equation of state for an ideal gas,

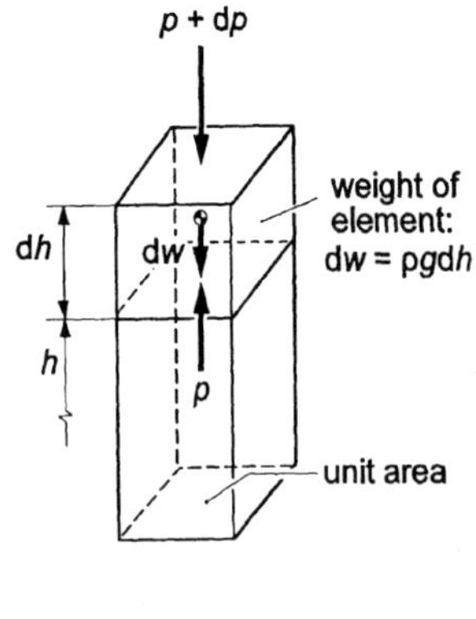

Figure 2.4-1. Forces acting on an element of air.

i.e., the *perfect gas law*:

$$\frac{p}{\rho} = \frac{R_a}{M} T \; . \tag{2.4-2}$$

In this equation, M is the molecular mass of the gas, T is the absolute temperature (kelvin), and R_a = 8314.32 J/K kmol, is the universal gas constant per kilomole (kmol) of the gas, where one kilomole is defined as that number of kilograms numerically equal to the molecular mass of the substance.
Substitution of Equation (2.4-2) into Equation (2.4-1) gives:

$$\frac{\mathrm{d}p}{p} = -\frac{Mg}{R_a T} \mathrm{d}h \; . \tag{2.4-3}$$

Integration from $h = 0$ (sea level) to height h yields:

$$\ln \frac{p}{p_0} = -\int_0^h \frac{Mg}{R_a T} \mathrm{d}h \; , \tag{2.4-4}$$

where the subscript "0" indicates sea level.

Equation (2.4-4) indicates that the calculation of atmospheric pressure requires knowledge of:

- The variation of molecular mass with height.
- The variation of acceleration of gravity with height.
- The variation of temperature with height.
- The sea-level atmospheric pressure.

In order to get round the difficulty of the varying acceleration of gravity it is customary to use the quantity *geopotential altitude*, H, defined by:

$$H = \int_0^H \mathrm{d}H = \frac{1}{g_0} \int_0^h g \, \mathrm{d}h \; . \tag{2.4-5}$$

In this equation, g_0 is the local sea-level value of the acceleration of gravity. Since both h and H are taken equal to zero at sea level, the geopotential altitude H is the height in a uniform gravity field at which the potential energy per unit mass is the same as at a *geometric height h* in a variable gravity field.
Using geopotential altitude, Equation (2.4-4) can be expressed as:

$$\ln \frac{p}{p_0} = - \int_0^H \frac{M g_0}{R_a T} \, \mathrm{d}H \, . \tag{2.4-6}$$

As discussed in Chapter 1, when calculating pressure changes in the atmospheric region up to 90 km, the variation of molecular mass with height may be neglected (M = constant).

Then, with $R = R_a / M$ is the specific gas constant of air, the perfect gas law (Equation (2.4-2)) can be written as:

$$\frac{p}{\rho} = R T \, . \tag{2.4-7}$$

As observed earlier in Figure 1.3-1, the temperature in the troposphere may be assumed to vary linearly with height:

$$T = T_0 + \lambda h \approx T_0 + \lambda H \, , \tag{2.4-8}$$

where $\lambda = \mathrm{d}T/\mathrm{d}H$ is the temperature gradient, which quantity has a negative value.

The values of p_0 and T_0 vary considerably with time of day and year, and geographical location. Table 2.4-1 presents extreme values of pressure and temperature at sea level, measured in the Netherlands, and elsewhere.

At mid-latitudes, the average values for pressure and temperature at sea level can be taken as: $p_0 = 101325$ N/m^2, and $T_0 = 288.15$ K = 15 °C (see Chapter 3).

Substitution of Equation (2.4-8) and the relationship $R = R_a / M$ into Equation (2.4-6) yields, after integration, the following equation for the variation of

	The Netherlands	Elsewhere
maximum pressure (hPa)	1050	1086 - Western Siberia
minimum pressure (hPa)	964.2	887.0 - Saparoea
maximum temperature (°C)	38.6	59 - various places
minimum temperature (°C)	- 27.4	- 88.3 Vostoc, Antarctica

Table 2.4-1. Extreme values of pressure and temperature at sea level.

pressure with height in the troposphere:

$$\ln\frac{p}{p_0} = -\frac{g_0}{\lambda R}\ln\left[1+\frac{\lambda H}{T_0}\right]. \qquad (2.4\text{-}9)$$

Equation (2.4-9) can be written as:

$$\frac{p}{p_0} = \left[1+\frac{\lambda H}{T_0}\right]^{-\frac{g_0}{\lambda R}} = \left[\frac{T}{T_0}\right]^{-\frac{g_0}{\lambda R}}. \qquad (2.4\text{-}10)$$

With the perfect gas law the corresponding relation between density and altitude can be established:

$$\frac{\rho}{\rho_0} = \frac{p}{p_0}\frac{T_0}{T} = \left[1+\frac{\lambda H}{T_0}\right]^{-\left(\frac{g_0}{\lambda R}+1\right)}. \qquad (2.4\text{-}11)$$

In general, the occurring temperature lapse rate in the troposphere is such that also the air density decreases with increasing height.
Obviously, the adoption of a particular temperature gradient dT/dH allows the computation of pressure and density as functions of altitude, provided that also the molecular mass of the air is known.

2.5 AIR MASSES AND FRONTS

As discussed in Section 2.2, there are diurnal and seasonal variations of the solar energy falling on the surface of the Earth. Also, the presence of clouds has a tremendous climatological impact. Cloudiness diminishes the incoming solar radiation, and the emitted infrared nocturnal radiation. It also causes all forms of precipitation, and consequently it plays an important part in life on Earth. In addition to the uneven heating of the Earth by the Sun, the transport of heat by the motion of air and water masses over the Earth produces a variation of the air temperature from Equator to Poles. Surfaces near the Equator heat most, leading to high tropical temperatures. On the contrary, near the Poles low temperatures exist. Consequently, the atmospheric temperature has a strong dependence on latitude, whereby the horizontal temperature gradient is much larger in winter than in summer. Most decrease of the temperature between

Equator and Pole occurs in the mid-latitude regions between 30° and 60° (Ref. 10).
The motion of air masses over the Earth is caused by pressure differences in the atmosphere. Important to this is the fact that the locations on Earth at which high and low pressure zones occur, more or less remain unchanged.
Figure 2.5-1 shows the main meridional circulations in the atmosphere, along with the average surface pressures and wind distributions identical to the previous Figure 1.7-2. Concerning the usual terminology, wind is the horizontal, or nearly horizontal movement of air. Vertical streams of air are not designated wind, but as convective currents.
The high surface temperatures near the Equator cause the air to expand, leading to a low pressure zone near the equator surface. Its place near the surface is taken by cooler air, drawn in from higher latitudes. The rising hot tropical air moves away to higher latitudes of about 30° to 40°, where the cooled denser air is pulled to the Earth's surface and high pressure zones are found. Once there, the air splashes north and south. The portion moving to the Equator becomes the *trade winds*. The air mass that moves northward on the northern hemisphere becomes the prevailing *westerlies*, blowing from west to east. At 50° to 70° latitude, there are low pressure zones, toward also the polar *easterlies* blow. In the polar regions, finally, high pressure zones occur.

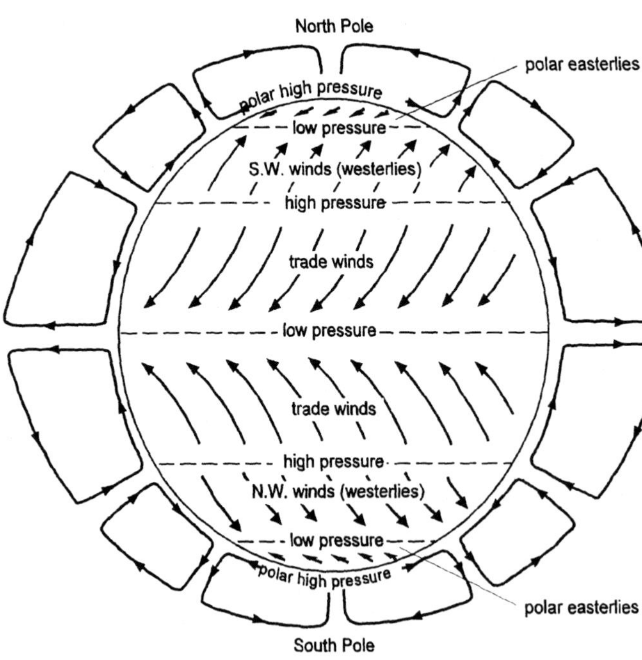

Figure 2.5-1. Global atmospheric circulation.

The spinning of the Earth affects the surface winds so that the major global circulation systems flow toward the east or west rather than directly north or south. This bending of the wind is caused by the Coriolis force. Its effect on wind direction is discussed in the next section.

With respect to the general circulation of the air, it is to be noted that each air mass has its own, distinct characteristics that produce the climate typical of the geographical area from which the air mass originates.
When the air mass moves away from its place of origin, the associated weather may be modified by the type of ground surface over which it travels. E.g., if the air mass moves across irregular land surfaces with many and large obstacles, friction forces will reduce the wind velocity.
Although the determining distinction of air masses is that they are in motion, all have a different velocity, so collisions frequently come about. However, once an air mass has acquired a certain level of temperature and humidity, it tends to retain that level and not to mix with other air masses. Collisions of air masses of different origins, usually, are coupled with a strong change for the worse of the weather. Especially at mid-latitudes, where the masses of tropical and polar air often meet, the most frequent and extreme changes in weather occur.

If a high pressure, cold air mass meets a warm air mass, the cold, heavier air pushes forwards beneath the warmer and lighter air mass. The boundary between the two air masses moving at different velocities is called the frontal surface. The intersection of the frontal surface with the ground surface is named a *front*. A front is called a warm or cold front according to the nature of the air mass behind it. So the arrival of a cold air mass, pushing its way into a warm air mass, is marked by the apparency of a *cold front* (Figure 2.5-2a). In this case the warm, less dense air rides up the frontal surface, with only a limited amount of mixing. This causes the rising air to cool, and its water vapor may condense to form clouds and create precipitation.
The frontal surface of a cold front slopes back from its direction of movement. When it passes a point on the ground, the cloud formation and accompanying

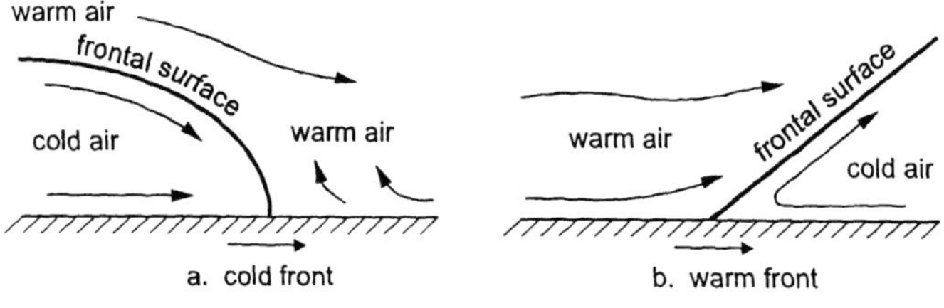

Figure 2.5-2. Vertical cross sections of atmospheric fronts.

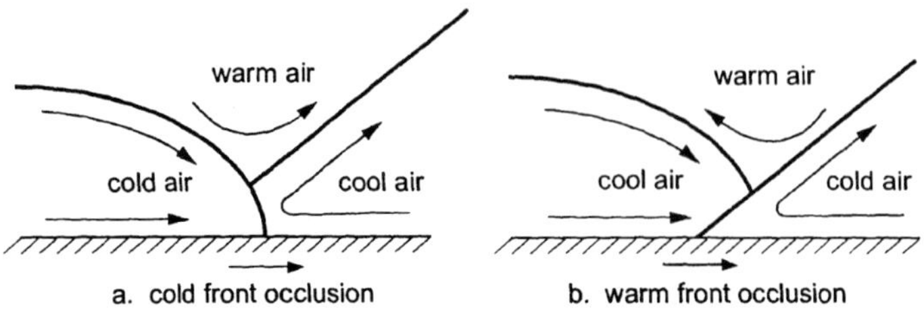

Figure 2.5-3. Occluded fronts.

precipitation start as the front arrives, and continue behind it, because warm air is still being lofted up what is an overhanging slope when observed from below. When the whole of the front has passed, the weather will that of the cold air mass itself.

If a warm air mass is displacing cold air, it is called a *warm front* (Figure 2.5-2b). A warm front slopes forwards and the cloud formation and precipitation begin already some time before the front arrives, and may continue inside the warm air mass.

In the case that a cold front meets a warm front, an *occluded ("closed in") front* may arise. The way in which the occlusion works out depends on the relative temperature of the cold air masses underlying the two frontal surfaces. In a so-called *cold front occlusion*, the air behind the cold front is the colder, and so displaces the cool air behind the warm front (Figure 2.5-3a). In a *warm front occlusion* the air behind the cold front is less cold than the air below the warm front, so that the cold front climbs over the wedge of the warm front (Figure 2.5-3b).

A discussion on the weather changes that characterize the passage of air mass fronts will be held in Chapter 4.

There are two regions in each hemisphere where major fronts are more or less constantly existent. On the northern hemisphere the more northerly is called the *polar front* and marks the boundary between the polar and tropical air masses. The more southerly marks the northern boundary of the subtropical high-pressure zone.

In these regions, the prevailing wind velocities reach their maximum values at an altitude between 9 and 13 km, due to the strong north-south temperature gradients across the fronts. These very strong winds are known as *jet streams*. They follow meandering paths, reach velocities as great as 400 km/h, and are blowing in fairly narrow bands, which form sharp boundaries between cold and warm air. The highest wind velocities occur in winter, because then the

horizontal temperature profiles are at their steepest as the tropical temperatures remain almost unchanged while the high-latitude temperatures drop.
On the southern hemisphere a comparable situation is found. There, the jet streams are even more powerful as the larger ice masses near the South Pole create sharper temperature and pressure differences between the polar regions and the tropics.
In civil aviation, jet streams may be employed by high-flying airplanes to shorten flight time, and so to decrease their fuel consumption.

2.6 GEOSTROPHIC WIND

Primarily, the wind direction is determined by the pressure forces from the pressure zones indicated in the previous Figure 2.5-1. In addition, two other horizontal forces are involved. These are the friction force arising from the relative motion between air and ground surface, and the Coriolis force, acting on the moving air due to the rotation of the Earth around its polar axis.

To explain the occurring *pressure gradient force*, we consider a rectangular axis system *X Y Z*, of which the origin is at a convenient point on the Earth's surface. The *X*-, *Y*-, and *Z* axes are pointing horizontally eastward, northward, and vertically upward, respectively (Figure 2.6-1).
Since isobars in Figure 2.5-1 are parallel to the Equator, there is a horizontal gradient of air pressure along the *Y*-axis. Then the resultant force acting on an element of air of volume $dxdydz$, is: $-(\partial p/\partial y)dxdydz$. Since the mass of air contained within the element is $\rho dxdydz$, the magnitude of the pressure gradient force per unit mass, f_p, is:

$$f_p = -\frac{1}{\rho}\frac{\partial p}{\partial y}, \tag{2.6-1}$$

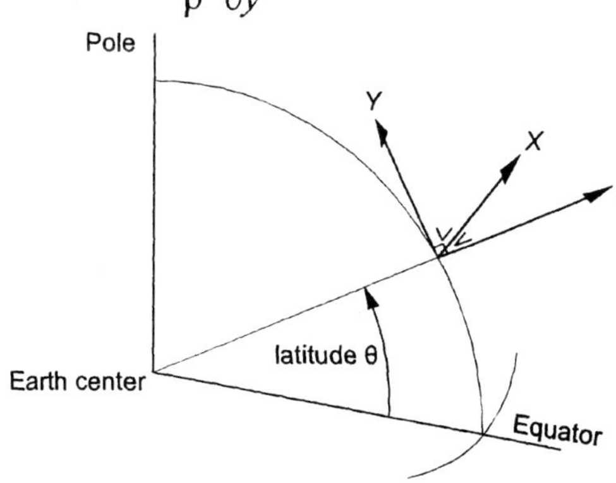

Figure 2.6-1. Axis system with origin on the ground surface.

where $\partial p/\partial y$ is the horizontal pressure gradient.
The *friction force* is pointed in opposite direction of the wind, hence it tends to decrease the wind velocity. Through the decreasing effect of ground friction, the wind velocity increases with increasing height. So the Coriolis force grows and as a result the direction of the wind at some height above ground level no longer coincides with the normal to the isobar (Figure 2.6-2a).
Obviously, the friction force depends on the type and roughness of the Earth's surface. For example, above the oceans, the angle that the wind makes with the isobars will be much smaller due to the water waves. A discussion on the effect of ground friction on the variation of the wind speed with height will be deferred until Section 2.8.
Referring to the previous Equation (2.1-4), the Coriolis force equals the mass of the air times twice the vector product of the angular velocity of the Earth $\boldsymbol{\omega}_e$ and the wind velocity $\boldsymbol{V}_w$.
The Coriolis force acts in a direction perpendicular to the plane defined by the Polar axis and the wind velocity (see the previous Figure 2.1-2). According to the previous Equation (2.1-7), the magnitude of the specific Coriolis force (the Coriolis force acting on a unit mass), f_c, is given by:

$$f_c = 2\,\omega_c V_w \sin\theta \,, \tag{2.6-2}$$

where θ is geographic latitude.
As stated before, starting from zero at ground level, the wind velocity will increase with increasing height. In the so-called *surface boundary layer*, which reaches to a height of about 50 meters, the Coriolis force is negligible small in comparison with the friction force, and the air moves at right angles to the isobars near the Earth's surface. Above the surface boundary layer, between, say, 50 and 500 m, the velocity of the wind not only increases due to the decreasing influence of ground friction, but also its direction changes, due to

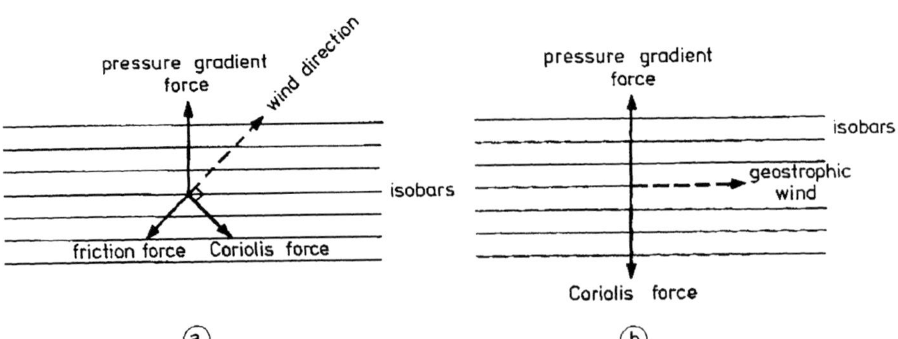

Figure 2.6-2. Change of wind direction due to Coriolis force.

the deflecting action of the increasing Coriolis force.
When the endpoints of the wind speed vectors at several altitudes are interconnected and projected on a horizontal plane, a so-called *Ekmann-spiral* is obtained (Ref. 11). This curvature of the course of the wind is named after the Swedish scientist Vagn Walfrid Ekmann (1874-1954). The layer where this occurs is called the *Ekmann layer*.
The layer formed by the surface boundary layer and Ekmann layer is termed the *planetary boundary layer*.
As shown before in Figures 1.7-2 and 2.5-1, due to the rotation of the Earth from west to east, the air flow on the northern hemisphere moving toward the Equator is deflected toward the west, whereas the air flow moving away from the equator is deflected to the east by the Coriolis force. Therefore, at northern latitudes the trade winds are blowing from northeastern direction, and the westerlies are blowing toward the Pole from southwestern direction. The air moving southward from the North Pole is deflected into a northeasterly flow.

The layer above the planetary boundary layer is known as the *free atmosphere*. Here virtually no influence of the friction with the ground surface can be found. Above this level only the horizontal pressure gradient force and the Coriolis force are acting on a moving air mass. As a result, the air flow is parallel to the isobars, and the associated wind is named *geostrophic wind*, V_{wg} (Figure 2.6-2b).

This phenomenon was verified and first published in 1857 by the Dutch scientist Christophorus Buys Ballot (1817-1890) and is known as the *law of Buys-Ballot*, which may be expressed as: “If you face the wind in the northern hemisphere at a sufficient height, the atmospheric pressure decreases toward your right and increases toward your left”. In the southern hemisphere the reverse is true. The suggestion that moving air masses tend to be deflected to the right in the northern hemisphere and to the left in the southern, was first stated in 1856 by the US meteorologist William Ferrel (1817-1891).
By definition, the velocity of geostrophic wind is obtained from equating the force due to the horizontal pressure gradient and the Coriolis force:

$$\frac{1}{\rho}\frac{\partial p}{\partial y} = 2\,\omega_e\,V_{wg}\sin\theta\ . \qquad (2.6\text{-}3)$$

The significance of (quasi) geostrophic flow is that it reflects the actual behavior of wind in the absence of frictional forces.
On the average, the wind speed increases up to the tropopause. For various geographical locations this wind profile has been determined. Figure 2.6-3

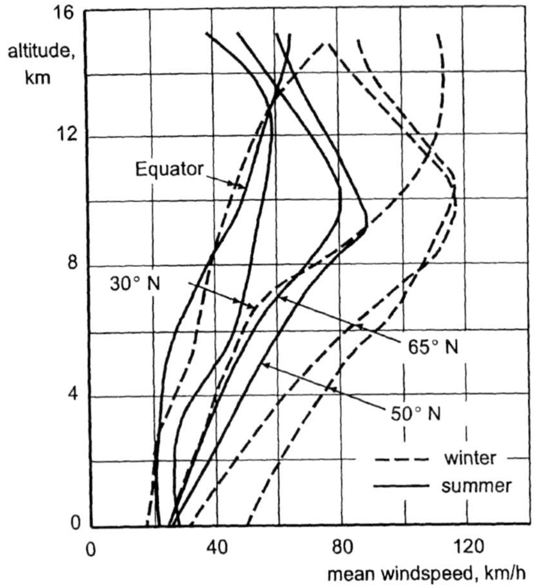

Figure 2.6-3. Wind velocity profiles at several latitudes.

represents the average wind speeds for several latitudes on the northern hemisphere (Ref. 12). Apparently, besides the aforementioned jet streams, in the stratosphere very high wind speeds may occur.

To explain the variation of geostrophic wind with height, we consider an atmosphere in aerostatic equilibrium. Then, from the discussion in Section 2.4, it is apparent that the decrease of pressure between two isobaric surfaces with heights h_2 and h_1 ($h_2 > h_1$) is given by:

$$\ln \frac{p_2}{p_1} = - \int_{h_1}^{h_2} \frac{g}{RT} \, \mathrm{d}h \ , \tag{2.6-4}$$

where R is the specific gas constant of air, and p_2 and p_1 are the pressures at height h_2 and h_1, respectively.
Using a mean temperature of the layer, $\overline{T}$, and taking the acceleration of gravity g constant, the thickness of the layer yields:

$$h_2 - h_1 = - \frac{R\overline{T}}{g} \ln \frac{p_2}{p_1} \ . \tag{2.6-5}$$

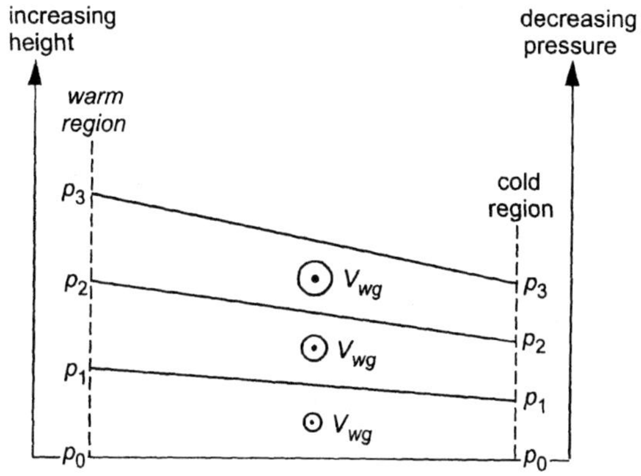

Figure 2.6-4: Variation of geostrophic wind velocity with height.

This expression indicates that in an atmosphere at rest, the distance between two isobaric surfaces with pressures p_2 and p_1 is proportional to the mean air temperature $\overline{T}$ of the layer. In other words, the warmer the air the greater the distance between the isobaric surfaces at heights h_2 and h_1. This behavior is pictured in Figure 2.6-4, showing that the isobaric planes are sloping upwards due to the growing thickness of the layer when going from a cold to a warm region.

It can also be seen that this slope increases with height because sequential layers of air are lighter in the warm region. Assuming that the horizontal temperature and pressure gradient are in the same direction and that the pressure at ground level is uniform, we observe an enlargement in the horizontal pressure gradient with height.
Hence, according to Equation (2.6-3), the geostrophic wind will increase with height. Clearly, the magnitude of the vertical shears of the geostrophic wind depends on the occurring horizontal temperature gradient. In meteorology, the relationship between the horizontal temperature gradient and vertical shear of geostrophic wind is known as the *thermal wind relation*.
It is evident that the geostrophic wind velocities are at right angles to the horizontal temperature gradient, with the warm region to the right-hand side and the cold region to the left-hand side of the wind vector (Figure 2.6-4).

2.7 SMALL-SCALE AIR MOTIONS

Besides the generally prevailing mayor wind systems, also local air flows occur, involving smaller bodies of air, of which the motions are influenced by local conditions.

From the discussion in Section 2.6, we know that the geostrophic wind vector is parallel to the isobars with the low pressure region to the left-hand side of the wind vector. Likewise, it will be clear that around a low pressure center in the northern hemisphere, the geostrophic wind will circulate in a counterclockwise direction (Figure 2.7-1). This type of motion is called a *cyclonic motion*, and the related low pressure weather system may be called a *cyclonic depression*, often called simply a *low*. Naturally, areas of low pressure result from upward atmospheric motion of warm, less dense, air. An example is the presence of a warm front. If an observer is standing at the front and facing the warm air mass, the wind will blow from his left in the warm air and from his right in the cold air. In other words, the wind will shift cyclonically across a front.
Similarly, around a high pressure center in the northern hemisphere, the geostrophic wind will circulate in a clockwise direction. This type of motion is called an *anticyclonic motion*, and results from downward motion of atmospheric air. The related high pressure weather system is usually referred to as a *high* (Figure 2.7-1).
Of course, in the southern hemisphere a depression will cause a motion in clockwise direction, and a high pressure area a motion in anticlockwise direction.

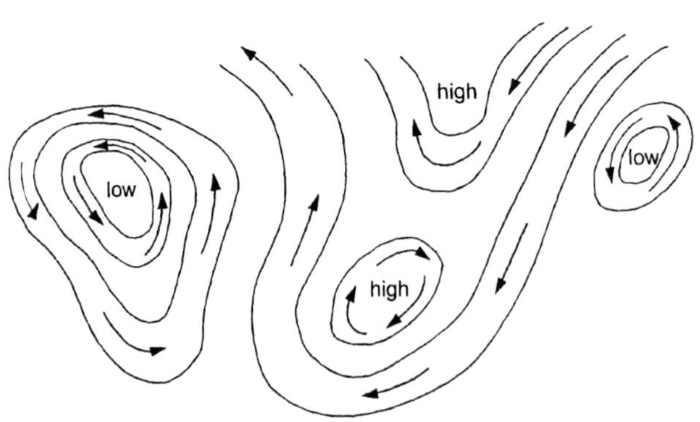

Figure 2.7-1. Wind flow around low and high pressure centers in the northern hemisphere.

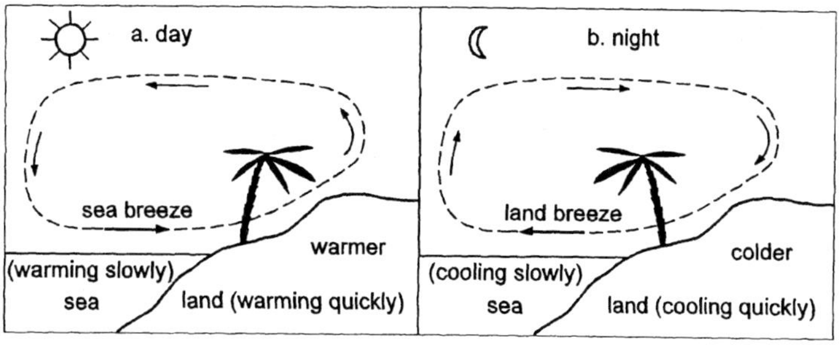

Figure 2.7-2. Local thermal circulations.

Tropical cyclones originate in tropical and subtropical regions over the warm oceans. They involve low pressure areas of very high intensity, around which extreme high wind velocities occur in combination with heavy rainfall (see Chapter 4). Dependent on the part of the world where they arise, tropical cyclones are also designated as hurricanes or as typhoons. The most violent parts of thunderstorms are tornadoes, often accompanying cold fronts in the form of a band of stormy weather with squally winds.

An example of slow motion of small air masses is the cooling wind that blows from the sea on a hot day. This surface wind is induced when during the day the land surface temperature is higher than that above the sea surface. Hot air rises, creating a region of low surface pressure above land. Due to the uneven distribution of heating and cooling rates, cooler air from the region of high pressure over the sea is drawn in to equalize the pressure differences (Figure 2.7-2a).
The resulting *sea breeze* is the tangible presence of the sea observed along coastal areas at warm days.

In the evening and at night, the distinction between land and sea surface

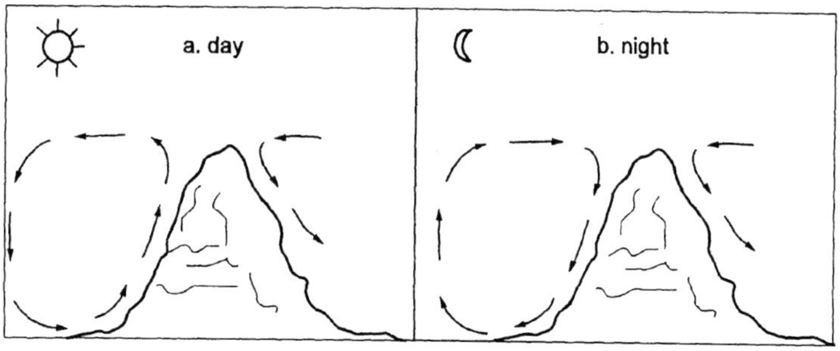

Figure 2.7-3. Valley breeze.

temperature is reversed. Then a small-scale circulation in the opposite direction is developed and the surface wind is now blowing from land to sea. This wind is designated as a *land breeze* (Figure 2.7-2b).

A similar local wind system that is exclusively based on thermal effects is found in hilly or mountainous regions during warm, clear days with little or no wind (Figure 2.7-3). If the mountain slope is heated by the Sun, in turn, the air near it is heated. This means that this air becomes warmer than air at a same height lying above the valleys. The air near the mountain therefore moves up as a breeze from the valley. On the other hand, the air over the valleys slowly sinks as a mountain breeze. At night, mountainside air cools, becomes heavier, and flows down the hillsides and rises above the valleys.

2.8 THE VERTICAL WIND GRADIENT NEAR THE GROUND

The friction force arising from the relative motion between air and ground surface, causes vertical wind gradients or *wind shears* (Figure 2.8-1).

The increase in wind velocity with height near the ground, up to several meters of height may be considered a flow in a turbulent boundary layer. Here the average wind speed increases with increasing height, its direction being independent of height. The variation with height of the Coriolis force and the pressure gradient force are negligible. The turbulent shear force τ, of which the magnitude may be assumed to be independent of height, follows from:

$$\tau = \rho\varepsilon \frac{\partial V_w}{\partial h} , \qquad (2.8\text{-}1)$$

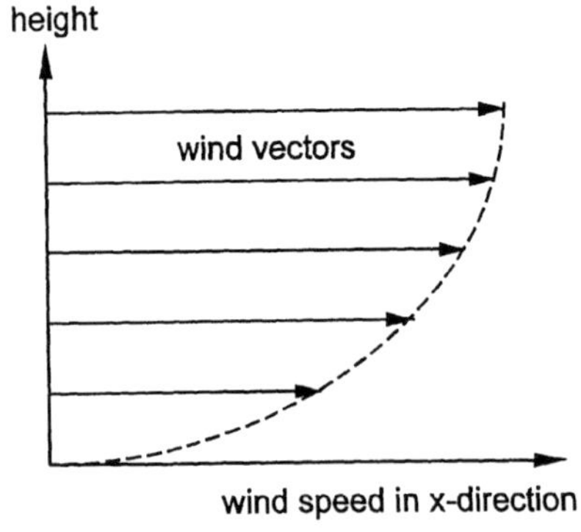

Figure 2.8-1. Vertical wind shear near the ground surface.

where ε is the turbulent diffusion coefficient.
For the ratio τ/ρ may be written:

$$\frac{\tau}{\rho} = U_x^2 , \qquad (2.8\text{-}2)$$

where U_x is termed the *friction velocity*.

If the temperature gradient is relatively small, the air density ρ and hence the friction velocity U_x will be virtually independent of height. Near the ground the velocity in vertical direction is approximately zero, hence ε will be very small here, and increases with height.
Introducing:

$$\varepsilon = K U_x h , \qquad (2.8\text{-}3)$$

into Equation (2.8-2) furnishes:

$$\frac{\tau}{\rho} = U_x^2 = K U_x h \frac{\partial V_w}{\partial h} , \qquad (2.8\text{-}4)$$

where K is the *Von Kármán constant*, named after the Hungarian-born American scientist Theodore von Karman (1881-1963). Usually, K is adopted as $K = 0.4$.
Integration of Equation (2.8-4) yields:

$$V_w = \frac{U_x}{K} \ln \frac{h + h_0}{h_0} , \qquad (2.8\text{-}5)$$

where h_0 represents the *roughness height*, which depends on the dimension of the prevailing roughness of the surface. In Table 2.8-1 are given values of the

surface condition	h_0, m	U_x, m/s
flat and smooth	0	0,2
short grass up to 0.01 m	0	0,3
thin grass up to 0.10 m	0,01	0,4
thick grass up to 0.10 m	0,023	0,5
thin grass up to 0.50 m	0,05	0,6
thick grass up to 0.50 m	0,09	0,6

Table 2.8-1. Friction velocity.

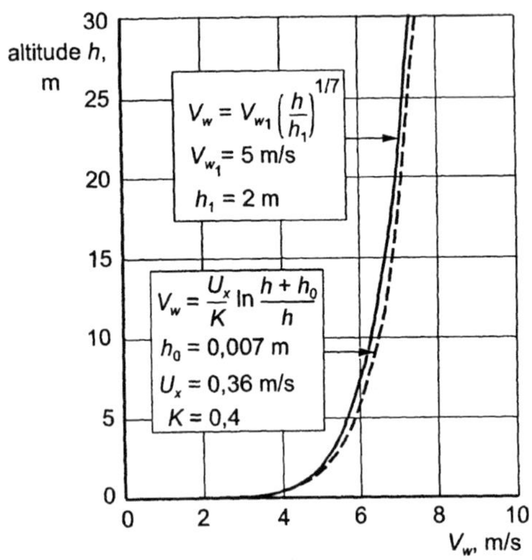

Figure 2.8-2. Vertical wind velocity profiles near the ground.

roughness height and the friction velocity for a temperature gradient of $dT/dh = -1$ K/100 m and a wind velocity of $V_w = 5$ m/s at a height of 2 m (Ref. 13).

The variation of wind velocity with height in the surface boundary layer often is represented by the relationship:

$$V_w = V_{w_1}\left[\frac{h}{h_1}\right]^{\frac{1}{n}} \; ; \; n > 0 \; , \tag{2.8-6}$$

where V_{w_1} is the average wind velocity at a fixed reference height h_1. For blowing over a relatively smooth surface and under the conditions of a normal lapse rate, n is approximately equal to 7. Figure 2.8-2 displays the shape of the wind profile according to equation (2.8-6), with $V_{w_1} = 5$ m/s at $h_1 = 2$ m. Also the wind profile according to Equation (2.8-5), for $h_0 = 7$ mm and $U_x = 0.36$ m/s is portrayed.
Of course, local wind velocities and wind shear are affected by surface features, such as buildings, vegetation, forests, rivers, lakes, hills, and mountains (see Ref. 5).

Surface winds may also be expressed in terms of the Beaufort scale of wind force. Table 2.8-2 gives the relationship between the 'equivalent ten-minutes average wind speed' and the Beaufort number as defined for wind at a height of 10 meters above the ground. This scale, devised in 1806 by Admiral Sir Francis

Beaufort (1774-1857) to estimate the wind velocity on the basis of sea state, presently assesses the effects of wind strength on land surface features such as trees and structures.

In wintertime, air temperature and wind velocity introduce the so-called *wind chill effect*, indicating the extra unpleasant feeling of coldness by living beings due to the combined effect of low temperatures and cold winds (Ref. 14). Evidently, the same discomfort level can occur when the temperature is very low but the wind is light, and also when the temperature is higher but the wind is stronger. The polar scientist Paul Siple (1908-1968), was the first to measure the heat loss from warm objects exposed to various combinations of temperature and wind velocity. Although others have since attempted to improve on Siple's experiments, his data remain today the most widely used to estimate the wind chill effect. The way to quantify this effect for a given

Baufort number	description of wind	wind velocity, m/s	effect
0	calm	0.0 - 0.5	Smoke rises vertically
1	light air	0.6 - 1.7	Smoke shows direction of the wind but wind vanes do not move
2	light breeze	1.8 - 3.3	Leaves rustle; wind vanes move; wind felt on face
3	gentle breeze	3.4 - 5.2	Leaves and twigs move constantly; light flags move
4	moderate breeze	5.3 - 7.4	Dust and loose paper blow about; small branches move
5	fresh breeze	7.5 - 9.8	Small trees with leaves sway
6	strong breeze	9.9 - 12.4	Large branches move; difficult to use umbrellas
7	moderate gale	12.5 - 15.2	Whole trees move; pressure when walking into the wind
8	fresh gale	15.3 - 18.2	Twigs break from trees; difficult to walk
9	strong gale	18.3 - 21.5	Chimney pots and slates blown from roofs; slight damage to structures
10	whole gale	21.6 - 25.1	Considerable damage to buildings; trees uprooted
11	storm	25.2 - 29.0	Widespread damage
12	hurricane	> 29.0	Devastation

Table 2.8-2. Beaufort wind scale.

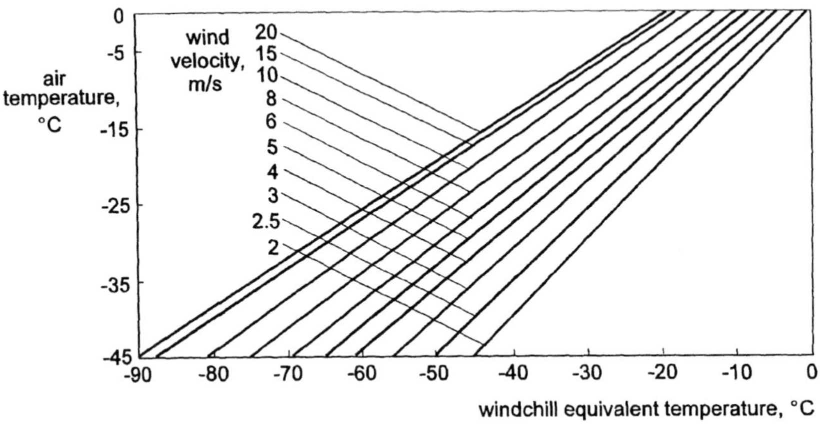

Figure 2.8-3. Windchill effect.

combination of temperature and wind speed is to determine what (lower) temperature, together with a wind speed of 1.8 m/s (4 mph), would result in the same level of discomfort. The temperature determined in this way is known as *wind chill equivalent temperature* (*WET*). A plot of Siple's data is shown in Figure 2.8-3, which gives values of *WET* for any temperature lower than 4.5 °C, and for any wind velocity in the range between 2 and 20 m/s.

2.9 TURBULENCE AND GUSTS

Apart from large-scale and small-scale motion systems, there are also local air motions in a very small scale, in horizontal as well as in vertical direction. In this respect, the designation *turbulence* and *gust* are employed. Turbulence may be described as a chaotic atmospheric motion in both space and time. It is especially characterized by strong currents of rising and falling air. When the wind gusts, the air blows with short and strong rushes, and shows rapid changes in direction. More precisely expressed, a gust is a short positive departure from the ten-minute average wind speed.

In contrast to the quasi-stationary nature of wind, turbulence and gusts are characterized by a state of irregularity.

Turbulence is mainly caused by atmospheric instability and *wind shear*, i.e., the variation of wind velocity with height. The wind shear required to maintain a certain degree of turbulence depends to some extent on the stability of the atmosphere, which is determined by the local vertical temperature gradient (see Chapter 4).

On an unclouded day the attendant turbulence is denoted as *clear air turbulence* (CAT). It gives the crew and passengers of an airplane the typical "cobblestone" feel. CAT usually arises from local wind shear and mainly occurs at altitudes just below the tropopause.

Up to 3000 m, gust speeds increase due to the convective motion. Above 3000 m they decrease again. Just below the tropopause gust speeds may reach again high values, due to the jet streams at this altitude, inducing a large wind shear (see the previous Section 2.5).

The horizontal distance in which the gust velocity grows from zero to aG maximum value, varies considerably. However, there is no correlation to the maximum value of the prevailing gust velocity. The distance may vary between several meters and more than 100 m.

At lower heights, turbulence and gusts, especially, occur in the vicinity of storm clouds. In these altitude regions strong convective air currents with horizontal and vertical wind shear are found (see Figure 4.4-2 in Chapter 4).

Other causes of airflow that will lead to turbulence and gust velocities are shearing stresses within the atmospheric boundary layer and wind blowing over a rough or irregular surfaces. Of course, these effects of friction on air motion, which increase with increasing wind velocities and roughness of the ground, are of importance to the dispersion of air pollutants near ground level.

Generally, gust speeds are measured indirectly through recording of the accelerations experienced by airplanes in vertical and horizontal air motions. The correlation of these accelerations and the maximum gust velocity proceeds from assumptions made on the way in which the gust speed increases to its maximum value, and the response of the airplane to this.

3 STANDARD ATMOSPHERES

3.1 THE INTERNATIONAL STANDARD ATMOSPHERE

In the atmosphere there are always large alterations of pressure, temperature and density at any time, height and place on Earth. In this respect, the *International Standard Atmosphere* (*ISA*) was developed to satisfy the need for standardization of the performance of aircraft and aircraft engines (Ref. 15).
The International Standard Atmosphere is considered to be a satisfactory reflection of the average atmospheric conditions at mid-latitude regions on the northern hemisphere.

Fundamentally, the International Standard Atmosphere is defined in terms of an ideal air that is assumed to be dry, clean, and at rest relative to the Earth, and obeying the perfect gas law. It is based upon standard values for atmospheric temperature, pressure, and density at sea level, and on the adoption of three connected linear segments involving variations of temperature with geopotential altitude up to 32 km (Figure 3.1-1).

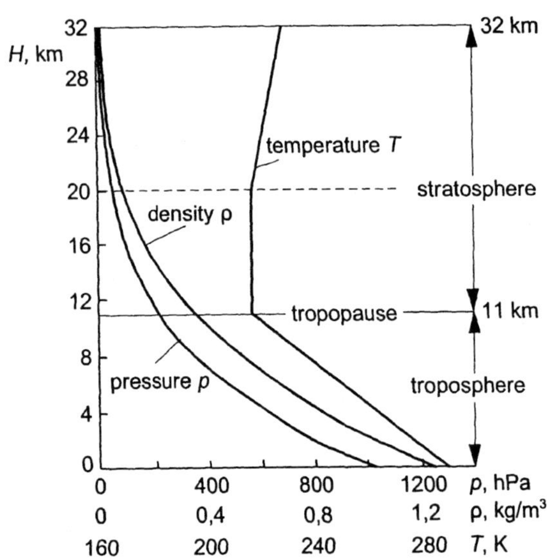

Figure 3.1-1. Variation of temperature, pressure and density in the International Standard Atmosphere (ISA).

sea-level pressure	$p_0 = 101325\ \mathrm{N/m^2}$
sea-level temperature	$T_0 = 288.15$ K (15 °C)
sea-level density	$\rho_0 = 1.225\ \mathrm{kg/m^3}$
acceleration of gravity at sea level	$g_0 = 9.80665\ \mathrm{m/s^2}$
universal gas constant	$R_a = 8314.32$ J/K kmol
ratio of specific heats of air	$\gamma = c_p/c_v = 1.4$
temperature gradient in troposphere (H = 0 - 11 km)	$\lambda_t = dT/dH = -\ 0.0065$ K/m
temperature gradient in (lower) stratosphere (H = 11 - 20 km)	$\lambda_s = 0$
temperature gradient in (upper) stratosphere (H = 20 - 32 km)	$\lambda_m = 0.001$ K/m

Table 3.1-1. Adopted primary constants.

The numerical values adopted as exact for the computations of the atmospheric properties are given in Table 3.1-1, where the value for the sea-level acceleration, g_0, is found at a northern latitude of about 45° (see Chapter 2, Section 2.3).
Using the numbers given in Table 3.1-1, the molecular mass of clean, dry air at sea level follows from the equation of state for a perfect gas as:

$$M_0 = \frac{\rho_0 R_a T_0}{p_0} = 28.96442\ \mathrm{kg/kmol}\ .$$

The specific gas constant of air is determined from the relationship:

$$R = \frac{R_a}{M_0} = 287.05\ \mathrm{m^2/s^2K}\ .$$

The specific heat of air at constant pressure is given by (see Chapter 7, Section 7.3):

$$c_p = R\frac{\gamma}{\gamma - 1}\ . \tag{3.1-1}$$

This yields with $\gamma = 1.4$: $c_p = 287.05 \times 1.4 / 0.4 = 1004.68\ \mathrm{m^2/s^2\ K}$.
The specific heat of air at constant volume follows from the expression:

$$c_v = \frac{R}{\gamma - 1}\ , \tag{3.1-2}$$

yielding: $c_v = 287.05/\ 0.4 = 717.63\ \mathrm{m^2/s^2\ K}$.

For the International Standard Atmosphere the value of the acceleration due to gravity at sea level, g_0, is taken as 9.80665 m/s^2. According to Equation (2.3-8) this value applies to a latitude of 45°32' 33".

As indicated in Figure 3.1-1, the International Standard Atmosphere is divided into the altitude regions *troposphere* and *stratosphere*, defined by three linear temperature segments (see also the nomenclature of the atmosphere in the previous Figures 1.3-1 and 1.3-2).

In the troposphere (H = 0 - 11 km), pressure and density of the air can be computed from the previous Equations (2.4-10) and (2.4-11). For $\lambda = \lambda_t$, these equations read:

$$\frac{p}{p_0} = \left[1 + \frac{\lambda_t H}{T_0}\right]^{-\frac{g_0}{\lambda_t R}} = \left[\frac{T}{T_0}\right]^{-\frac{g_0}{\lambda_t R}} \tag{3.1-3}$$

and

$$\frac{\rho}{\rho_0} = \frac{p}{p_0}\frac{T_0}{T} = \left[1 + \frac{\lambda_t H}{T_0}\right]^{-\left(\frac{g_0}{\lambda_t R}+1\right)} . \tag{3.1-4}$$

In the stratospheric layer between H = 11 km and H = 20 km, the *lower stratosphere*, the temperature is constant ($\lambda_s = 0$). In this case the expression for the pressure ratio is obtained by direct integration of Equation (2.4-6) between the altitude of the tropopause H_s and altitude H:

$$\ln\frac{p}{p_s} = -\int_{H_s}^{H} \frac{g_0}{R T_s}\, \mathrm{d}H = -\frac{g_0}{R T_s}(H - H_s) \tag{3.1-5}$$

or

$$\frac{p}{p_s} = e^{-\frac{g_0}{R T_s}(H - H_s)} . \tag{3.1-6}$$

In these equations, the subscript '*s*' denotes the condition at the tropopause (H = 11 km).
From the perfect gas law, we find the relation:

$$\frac{\rho}{\rho_s} = \frac{p}{p_s} \; . \qquad (3.1\text{-}7)$$

Thus:

$$\frac{\rho}{\rho_s} = e^{-\frac{g_0}{RT_s}(H - H_s)} \; . \qquad (3.1\text{-}8)$$

In the upper part of the stratosphere (H = 20 - 32 km), the pressure and temperature ratio variations are easily derived as:

$$\frac{p}{p_m} = \left[1 + \frac{\lambda_m(H - H_m)}{T_s}\right]^{-\frac{g_0}{\lambda_m R}} \qquad (3.1\text{-}9)$$

and

$$\frac{\rho}{\rho_m} = \left[1 + \frac{\lambda_m(H - H_m)}{T_s}\right]^{-\left(\frac{g_0}{\lambda_m R} + 1\right)} \; . \qquad (3.1\text{-}10)$$

In these equations the subscript '*m*' indicates the conditions at H = 20 km.

The computed variations of temperature, pressure, and density with geopotential altitude are presented in Figure 3.1-1. Numerical values of T, p, and ρ are listed in References 4, 8, and 15.

Before closing this section, we note that, with respect to the nomenclature of the atmosphere, in aeronautical engineering the layer between 11 and 32 km ISA, alternatively, may be subdivided into the stratosphere (11 - 20 km) and the mesosphere (20 - 32 km).

3.2 OFF-STANDARD ATMOSPHERIC CONDITIONS

Obviously, in real live, certain standard atmospheric conditions never happen. For example, vertical temperature gradients near the Equator will differ from those near the Poles. As shown earlier in Figure 1.3-3, in the polar regions also the tropopause is located at a much lower level than near the Equator.
However, although not throughout the whole year, the temperature lapse rate and tropopause level adopted for the International Standard Atmosphere will appreciably correspond to the temperature profiles at mid-latitude regions on the northern hemisphere.

With regard to air pollution, the tropopause level is of great importance because at this height, in fact, the temperature inversion in the stratosphere begins (see the previous Figure 1.3-2). This behavior of the vertical temperature gradient is closely related to the presence of the stratospheric ozone layer. In this layer the absorption of short-wave solar radiation takes place, causing a maximum temperature to occur at the stratopause.
Certainly, airplanes flying at their cruise altitudes emit gases and particles directly into the upper troposphere and lower stratosphere where they have an impact on the composition of the local atmosphere, and so may participate in stratospheric ozone depletion and global warming.

The magnitude and variations in ambient temperature with the altitude are an

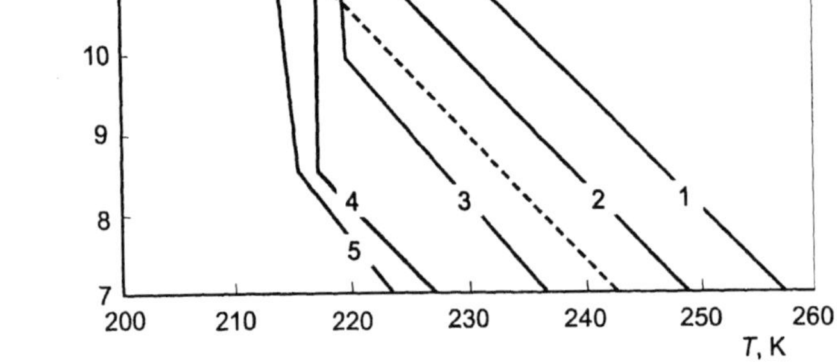

Figure 3.2-1. Temperature profiles for several latitudes.

Geographic latitude	approx. tropopause altitude, km	dT/dH, K/m	g_0, m/s^2
15°N, annual	17	- 0.0071	9.78381
30°N, January	12	- 0.0065	9.79324
45°N, January	10	- 0.0060	9.80665
60°N, January	8.5	- 0.0070	9.81911
75°N, January	8.5	- 0.0055	9.82860
ISA	11	- 0.0065	9.80665

Table 3.2-1. Tropopause altitude, temperature gradient, and sea-level acceleration of gravity at various geographic latitudes.

important factor in the performance of airplanes. To be able to investigate the effect of flight technique on pollutant emissions when cruising at geographic locations different from mid-latitude regions, mean January temperature versus altitude profiles at latitudes 30°, 45°, 60° and 75°N. are given in Figure 3.2-1 (Ref. 16). In Figure 3.2-1 also is shown the mean annual atmosphere at 15°N., and the temperature profile according to the International Standard Atmosphere.

In Table 3.2-1 is given the altitude of the tropopause at the various latitudes considered in Figure 3.2-1, the corresponding numerical value of the tropospheric temperature gradient, and the sea-level acceleration of gravity.

Off-standard atmospheres, for the performance prediction of aircraft flying at mid-latitudes, are defined by a variation of air temperature with the altitude, which is achieved by changing the temperature according to the International

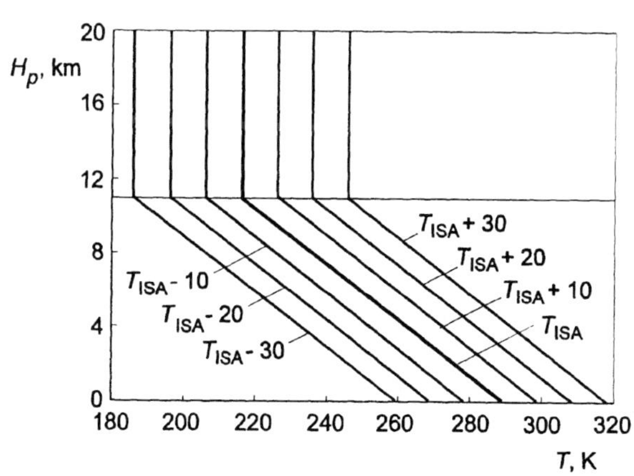

Figure 3.2-2. Off-standard atmosphere temperatures versus geopotential pressure altitude.

Standard Atmosphere with increments ΔT (Figure 3.2-2). The resulting temperatures then are expressed as:

$$T = T_{ISA} + \Delta T, \text{ K}, \qquad (3.2\text{-}1)$$

where the subscript 'ISA' indicates temperature in kelvin in the International Standard Atmosphere.

The constant increments in temperature, ΔT, are related to the so-called *geopotential pressure altitude*, H_p.
The latter altitude agrees with the reading of an airplane altimeter. This flight instrument is simply a pressure measuring device as it is driven by the air pressure.
As shown in Figure 3.2-3, the altimeter is made up of an airtight case, where a differential-pressure capsule is installed. The capsule consists of a number of diaphragms which are connected and sealed to make hollow rooms.
The instrument chamber is connected to a pressure tapping on the airplane so that the interior of the case is at the ambient pressure p. The diaphragm unit is evacuated, through which it will expand when the air pressure decreases. A linkage structure conveys the displacement of the capsule to the indicator pointer.
To provide an indication of height in place of pressure, the instrument is calibrated by exposing its differential-pressure capsule to various pressures and marking on the dial the values of geopotential altitudes that in the International Standard Atmosphere correspond to these air pressures.
Therefore, when at a given air pressure the temperature of the air is different from that of the International Standard Atmosphere, the altitude displayed by the altimeter is the geopotential pressure altitude.

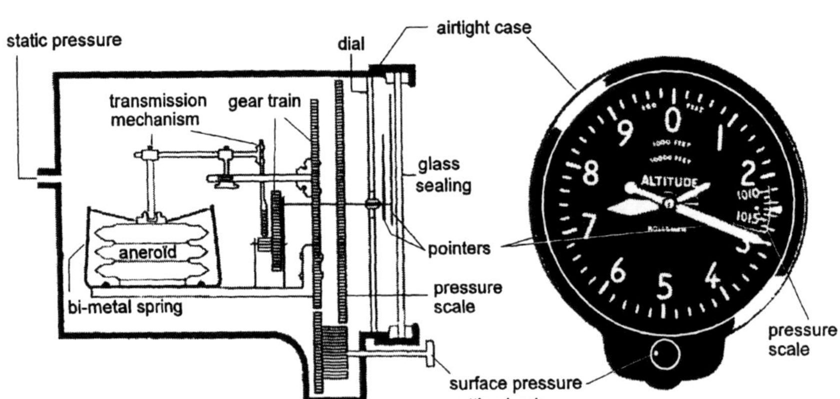

Figure 3.2-3. Principle of altimeter.

Apparently, the geopotential pressure altitude is the geopotential altitude in the International Standard Atmosphere at which the air pressure is equal to the actual pressure. For example, when the altimeter is calibrated according to Equation (3.1-3), we have in the troposphere the following relationship between pressure and geopotential pressure altitude ($H_p \leq 11{,}000$ m):

$$H_p = \left[\left(\frac{p}{p_0} \right)^{-\frac{\lambda_t R}{g_0}} - 1 \right] \frac{T_0}{\lambda_t} . \tag{3.2-2}$$

Likewise, for the isothermal part of the stratosphere, where $\lambda_s = 0$, it follows from Equation (3.1-6) that the indicated geopotential pressure altitude is given by:

$$H_p = H_s - \frac{R T_s}{g_0} \ln \frac{p}{p_s} , \tag{3.2-3}$$

where the subscript '*s*' denotes the value at the tropopause.

Since the altimeter is a pressure measuring device, the vertical position of an airplane is expressed in terms of pressure, usually called *Flight Level* (FL), defined by: FL = $H_p/100$. E.g., FL 200 means that the altimeter indicates a geopotential pressure altitude of 20,000 ft.

If, besides altitude H_p, also the actual temperature T and hence ΔT is available, the air density can be calculated from the perfect gas law:

$$\frac{p}{R} = \rho_{\text{ISA}}\, T_{\text{ISA}} = \rho (T_{\text{ISA}} + \Delta T) , \tag{3.2-4}$$

or

$$\rho = \frac{\rho_{\text{ISA}}}{1 + \dfrac{\Delta T}{T_{\text{ISA}}}} . \tag{3.2-5}$$

Normally, airplane performance data are given as functions of geopotential pressure altitude H_p and air temperature $T = T_{\text{ISA}} + \Delta T$.

Clearly, when the atmospheric conditions depart from those in International Standard Atmosphere, H_p differs from H and the altimeter furnishes only an approximation of the geometric height above the ground surface. The greatest error is caused by the influence of the alteration of the prevailing sea-level pressure from the standard value $p_0 = 101325$ N/m^2. In order to correct the altimeter reading for its deviation from the true height, the altimeter is provided with a subsidiary scale. Now the datum pressure at which the altimeter will read zero height can be set on this sub-scale. This may be accomplished by means of a manually operated adjustment knob on the instrument (see Figure 3.2-3). The adjustment knob can be tuned until the altimeter reads the elevation of the airport *above mean sea level* (*AMSL*), or zero height when the airplane is at the airport level. The latter air pressures, which are determined at the airport, are reported to the pilot at regular times.

The three settings of the datum pressure on the sub-scale are designated by the following codes (Figure 3.2-4):

QNE - setting the standard sea-level pressure $p_0 = 101325$ N/m^2.

QNH - pressure setting such that the altimeter reads the altitude above mean sea level at the airport level.

QFE - pressure setting such that the altimeter reads zero height at the airport level.

The QNE setting is used for normal flight operations above a prescribed *transition level*. Then the altimeter is a pressure measuring device, and a means to secure the vertical separation of airplanes.

The QNH and QFE settings are applied for takeoff and landing maneuvers and

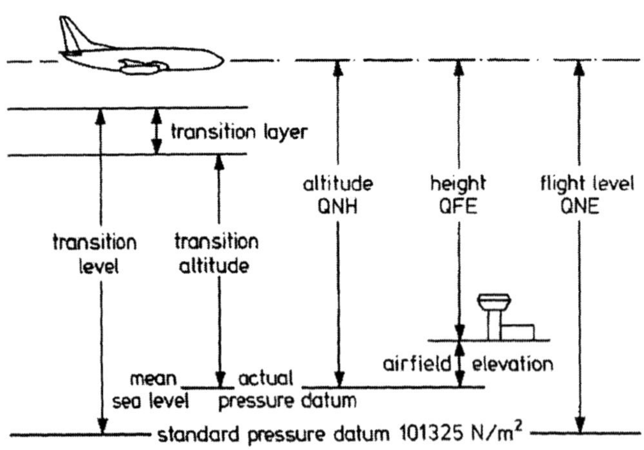

Figure 3.2-4. Definition of altimeter settings.

for low-level flights below a certain *transition altitude*.
Between the transition altitude and the transition level there is a transition layer with a thickness of at least 1000 ft.
It can be seen from Figure 3.2-4 that the vertical position of an airplane during the climb after takeoff is expressed in terms of altitude (in feet) until it arrives at the transition altitude above which the vertical position is given in terms of flight level (FL).

Finally, it is of interest to note that also the so-called *geopotential density altitude*, H_ρ may be a valuable quantity. This is the geopotential altitude in the International Standard Atmosphere at which the air density is equal to the actual density.
By solving Equation (3.1-4) for height, we obtain the geopotential density altitude in the troposphere as:

$$H_\rho = \left[\left(\frac{\rho}{\rho_0}\right)^{-\frac{\lambda_t R}{g_0 + \lambda_t R}} - 1\right]\frac{T_0}{\lambda_t} . \qquad (3.2\text{-}6)$$

In the lower part of the stratosphere, we have from Equation (3.1-8),

$$H_\rho = H_s - \frac{R T_s}{g_0} \ln \frac{\rho}{\rho_s} . \qquad (3.2\text{-}7)$$

Figure 3.2-5 portrays the variation of $H\rho$ with T for given values of H_p. Applying this chart, an altimeter reading can be converted into geopotential density altitude, provided that the prevailing air temperature is known.

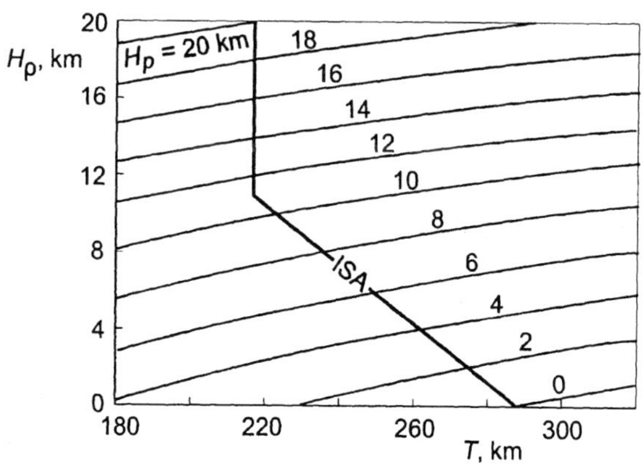

Figure 3.2-5. H_ρ *as functions of* T *and* H_p.

3.3 U.S. STANDARD ATMOSPHERE, 1962

The *U.S. Standard Atmosphere, 1962* (Ref. 4) provides an extension of numerical values of basic atmospheric properties up to an altitude of 700 km, and is for all practical purposes in agreement with the International Standard Atmosphere (ISA) over the geopotential altitude region from 0 to 32 km.
Like the International Standard Atmosphere, the U.S. Standard Atmosphere, 1962 is defined in terms of clean, dry, and still air, which also is assumed to be a perfect gas.
Table 3.3-1 gives the numerical values of the primary constants adopted for the computations, showing that the standard sea-level values of pressure, temperature, density, and acceleration of gravity are according to those used in ISA (see Table 3.1-1).

As depicted in the previous Figure 2.3-1, the gravity force is the vector sum of the centrifugal force F_t, resulting from the use of an earthbound, rotating axis system, and the gravitational force F_g, on the basis of Newton's law of gravity. Therefore, the gravity field, being a conservative force field, is derived from the geopotential $\mathbf{\Phi}$, given by:

$$\mathbf{\Phi} = \mathbf{\Phi}_t + \mathbf{\Phi}_g \,, \tag{3.3-1}$$

where $\mathbf{\Phi}_t$ is the potential energy per unit mass, coupled with the centrifugal force. The term $\mathbf{\Phi}_g$ in Equation (3.3-1) is the potential energy per unit mass of the gravitational attraction.

sea-level pressure	$p_0 = 101325$ N/m^2
sea-level temperature	$T_0 = 288.15$ K (15 °C)
sea-level density	$\rho_0 = 1.225$ kg/m^3
acceleration of gravity at sea level	$g_0 = 9.80665$ m/s^2
universal gas constant	$R_a = 8314.32$ J/K kmol
ratio of specific heats of air	$\gamma = c_p/c_v = 1.4$
Sutherland's constant	$S = 110.4$ K; $\beta=1.458\times10^{-6}$ kg s^{-1} K$^{-1/2}$
diameter or air molecule	$\delta = 3.65 \times 10^{10}$ m
Avogadro's number	$N_A = 6.02257 \times 10^{26}$ kmol^{-1}

Table 3.3-1. Primary constants.

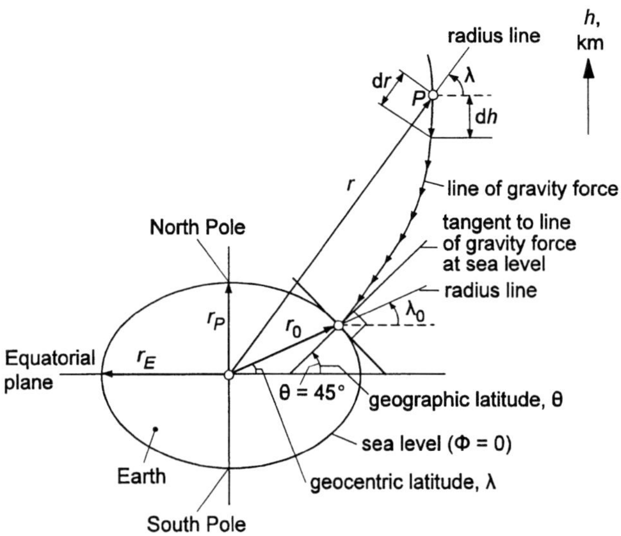

Figure 3.3-1. The geopotential Φ.

The gravity force per unit mass is given by:

$$\boldsymbol{g} = -\nabla\boldsymbol{\Phi}\ , \tag{3.3-2}$$

where $\nabla\boldsymbol{\Phi}$ is the gradient of the geopotential.

The acceleration of gravity is the magnitude of $\boldsymbol{g}$:

$$g = |\nabla\boldsymbol{\Phi}|\ . \tag{3.3-3}$$

The gravity field is represented by fictitious equipotential surfaces, on each of which the geopotential $\boldsymbol{\Phi}$ is constant, the surfaces being pierced orthogonally by curves called *the lines of gravity force* (Figure 3.3-1). At each point on a line of force, the tangent has the direction of the corresponding gravity vector $\boldsymbol{g}$. The geometric altitude h of a point is defined as the distance, measured along the line of force through the point, from the equipotential surface for which $\boldsymbol{\Phi}$ = 0 (sea level) and $g = g_0$ to the point in question. The slight differences between geometric altitudes as defined here, and radial distances r shown schematically in Figure 3.3-1, are assumed to be negligible.

On the basis of this definition, the differential relation of the geometric altitude h and the geopotential $\boldsymbol{\Phi}$ is given by:

$$\mathrm{d}\boldsymbol{\Phi} = g\,\mathrm{d}h. \tag{3.3-4}$$

This is the incremental work done in elevating a unit mass a distance dh against the gravity force. Therefore, the geopotential at a point whose altitude is h is given by:

$$\mathbf{\Phi} = \int_0^h g\,\mathrm{d}h \ , \qquad (3.3\text{-}5)$$

where the integration is performed along the line of gravity force, passing through the point in question.

In Figure 3.3-1 the shape of the Earth, that is, the equipotential surface $\mathbf{\Phi} = 0$, is treated as an ellipsoid of revolution, which coincides with the Earth's polar axis. In Ref. 4 the values used for its semi-major axis (equatorial radius, r_E) and semi-minor axis (polar radius, r_P) are 6.378178×10^6 m and 6.356798×10^6 m, respectively.
From Figure 3.3-1 it can be seen that the deviation of the ellipsoid from a pure sphere affects the value of the latitude of a place on Earth, because the local vertical on the ellipsoid will not pass through the Earth's center. It strikes the equatorial plane at an angle called the *geographic latitude*, θ. This geographic latitude differs from the *geocentric latitude*, λ, of that place.

The centrifugal potential per unit mass $\mathbf{\Phi}_t$ can be written as:

$$\mathbf{\Phi}_t = \frac{1}{2}\,\omega_e^2\,[\,r_E^2 - (r\cos\lambda)^2\,] \ , \qquad (3.3\text{-}6)$$

where ω_e is the angular velocity of the Earth and $r \cos \lambda$ is the distance, measured normally to the Earth's axis, from this axis to the point P.
An expression for the gravitational potential per unit mass $\mathbf{\Phi}_g$ is given in Ref. 4, in which $\mathbf{\Phi}_g$ is expanded in an infinite series of spherical harmonics. We will not give a derivation here but merely quote the result as obtained by applying only the most important terms:

$$\mathbf{\Phi}_g = -\frac{\mu M_e}{r} + \frac{\mu M_e}{r}\frac{J}{3}\left(\frac{r_E}{r}\right)^2 (3\sin^2\lambda - 1) + \frac{\mu M_e}{r_E}\left(1 + \frac{J}{3}\right) \ , \qquad (3.3\text{-}7)$$

where μ is the universal gravitational constant, M_e is the mass of the Earth, and J is the second harmonic coefficient.

Equations (3.3-1) to (3.3-3), (3.3-6), and (3.3-7), define the gravity field of the Earth. The hypothetical surface for which $\Phi = 0$, corresponds closely to mean sea-level.

Since the surface of the Earth is considered to be an ellipsoid, the adopted geopotential Φ is not dependent on longitude. This means that the gravity field is axially symmetric, so that each line of force lies in a meridian half-plane, and obeys the differential equation:

$$r \; = \; \frac{\mathrm{d}\lambda}{\mathrm{d}r} \; = \; \frac{\dfrac{1}{r}\dfrac{\partial \Phi}{\partial \lambda}}{\dfrac{\partial \Phi}{\partial r}} \; . \tag{3.3-8}$$

The gravity line of force passing through the point, for which $\Phi = 0$ and $g = g_0$ is the curve along which the present standard atmosphere is postulated. In order that the point for which $\Phi = 0$ and $g = g_0$, corresponds precisely to the geographic latitude $\theta = 45°$, the following average value for J is used:

$$J \; = \; 1.623495 \times 10^{-3} \; \text{(dimensionless)} \; .$$

The corresponding value of $\mu \, M_e$ equals 3.9862216×10^{14} $\mathrm{m^3/s^2}$.

In terms of r_E and r_P, the values λ_0 and r_0, which correspond to sea level and $\theta = 45°$, are found from the analytic geometry of the ellipse to be:

$$\tan \lambda_0 \; = \; \frac{r_P^2}{r_E^2} \; = \; (1 - f)^2$$

and

$$r_0^2 \; = \; \frac{r_E^4 + r_P^4}{r_E^2 + r_P^2} \; = \; \frac{r_E^4 + r_E^4 (1 - f)^4}{r_E^2 + r_E^2 (1 - f)^2} \; ,$$

where f is the flattening of the ellipsoid:

$$f \; = \; \frac{r_E - r_P}{r_E} \; = \; \frac{1}{298.32} \; .$$

From the above values of f and r_E we find: $\lambda_0 = 44.808°$ and $r_0 = 6.367533 \times 10^6$ m.

Since the geopotential given by Equations (3.3-6) and (3.3-7) is only dependent on latitude, the binomial series is applied to obtain the following expanded form of Equation (3.3-3):

$$g = |\nabla \Phi| = \sqrt{\left(\frac{\partial \Phi}{\partial r}\right)^2 + \left(\frac{1}{r}\frac{\partial \Phi}{\partial \lambda}\right)^2}$$

$$= \frac{\partial \Phi}{\partial r}\left[1 + \frac{1}{2}\left(\frac{1}{r}\frac{\partial \Phi / \partial \lambda}{\partial \Phi / \partial r}\right)^2 + \ldots\right] . \tag{3.3-9}$$

By retaining only the first two terms in the series in the parenthesis and replacing $(\partial \Phi / \partial r)^2$ in the denominator of the second term by $(\mu M_e / r_E^2)^2$, Equation (3.3-9) reduces to:

$$g = \left[\frac{\mu M_e}{r^2}\left[1 - J\left(\frac{r_E}{r}\right)^2 (3\sin^2\lambda - 1)\right] - \omega_e^2 r \cos^2\lambda\right] \times$$

$$\times \left[1 + \frac{1}{2}\frac{1}{r^2}\left[\frac{r_E^2}{\mu M_e}\right]^2 \left[\frac{(2 M_e J r_E)^2}{r^3} + \omega_e^2 r^2\right]^2 (\sin\lambda \cos\lambda)^2\right] . \tag{3.3-10}$$

Simultaneous numerical integrations of Equation (3.3-8) and the relation:

$$\mathrm{d}h = \sqrt{(\mathrm{d}r)^2 + (r\,\mathrm{d}\lambda)^2} , \tag{3.3-11}$$

have revealed that the following linear approximations are adequate for the present purposes:

$$\lambda = \lambda_0 + \left(\frac{\mathrm{d}\lambda}{\mathrm{d}r}\right)_0 (r - r_0) \tag{3.3-12}$$

and

$$r = r_0 + \frac{h}{\left(\frac{\mathrm{d}h}{\mathrm{d}r}\right)_0} = r_0 + \frac{h}{\sqrt{1 + r_0^2\left(\frac{\mathrm{d}\lambda}{\mathrm{d}r}\right)_0^2}} . \tag{3.3-13}$$

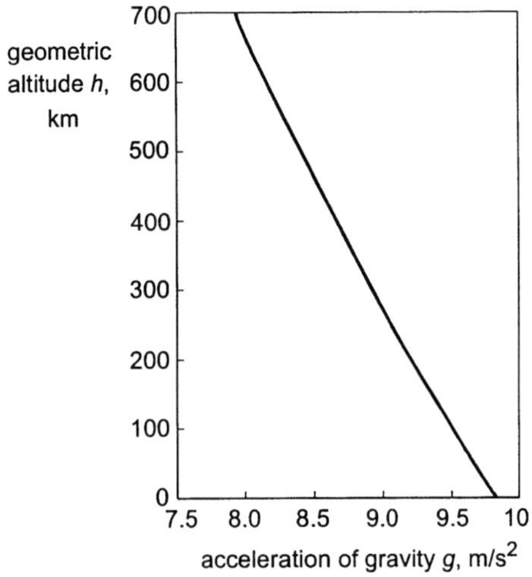

Figure 3.3-2. Acceleration of gravity versus geometric altitude.

With the use of Equations (3.3-12) and (3.3-13) to replace r and λ by linear functions of h, Equation (3.3-10) yields an approximate equation for the acceleration of gravity as a function of the geometric altitude. Computational results are shown in Figure 3.3-2.

It is obvious, to draw in Figure 3.3-2 also the simplified variation of g with geometric altitude h, in the manner of the previous Equation (2.3-9). However, this curve has been omitted because its difference from the curve according to Equations (3.3-10) to (3.3-13) would not be discernible on the graph!

In accordance with Equation (3.3-5), the previous Equation (2.4-5) can be employed to define the corresponding geopotential altitude H (see the tables in Ref. 4).

As discussed before in Chapter 1, the molecular mass of the air changes with the altitude. The variation of M assumed in Ref. 4 is depicted graphically in Figure 3.3-3, showing that up to 90 km the molecular mass is taken as constant at M_0 = 28.9644 joule/K kmol. Above 90 km, mainly because of molecular dissociation and diffuse separation, the molecular mass decreases.
In order to allow for the change of the molecular mass, use is made of the *molecular-scale temperature,* T_M, defined by:

$$T_M = T\frac{M_0}{M} \quad . \qquad (3.3\text{-}14)$$

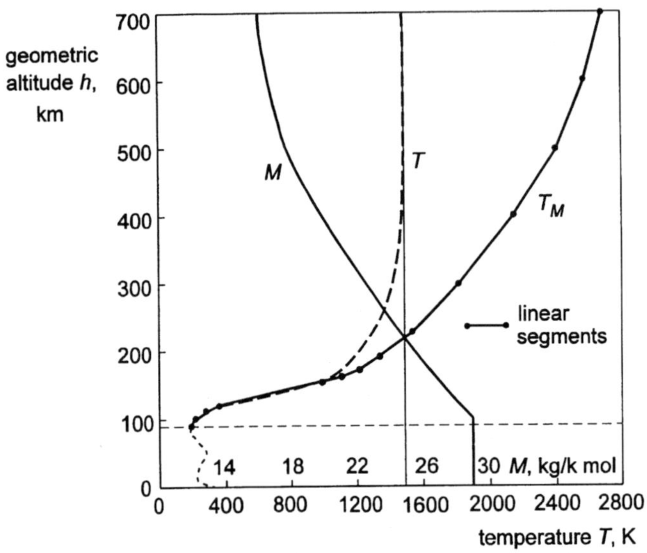

Figure 3.3-3. Graphic representation of molecular mass, temperature, and molecular-scale temperature versus geometric altitude.

where M_0 is the molecular mass at geometric heights lower than 90 km. Molecular-scale temperature is the defining property in the U.S. Standard Atmosphere, 1962. Up to the geometric altitude $h = 90$ km, the molecular-mass ratio M/M_0 is unity and hence $T_M = T$. Above $h = 90$ km, $M < M_0$ and hence $T_M > T$.

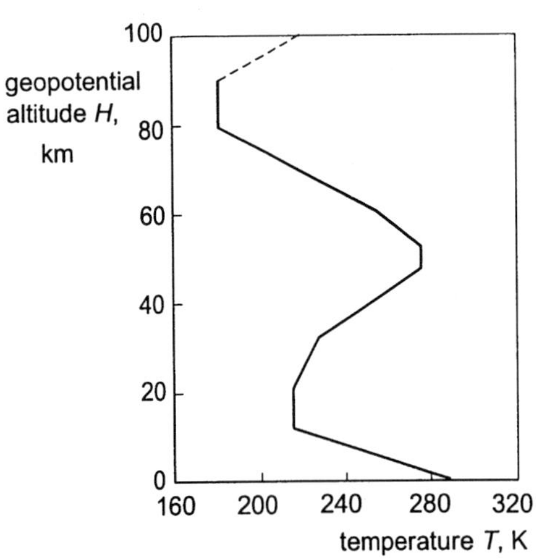

Figure 3.3-4. Temperature versus geopotential altitude up to 90 km.

Also, the variations of T_M and T with geometric altitude h are shown in Figure 3.3-3. See also the variation of $T_M = T$ with geopotential altitude H up to 90 km in Figure 3.3-4. Apparently, the molecular-scale temperature variation is defined as a series of connected segments linear in geopotential altitude to $h =$ 90 km, and linear in geometric altitude above $h = 90$ km.

The general form of each linear segment is:

$$T_M = T_{Mb} + \lambda_b (H - H_b) \tag{3.3-15}$$

to $h = 90$ km, and

$$T_M = T_{Mb} + \lambda_b (h - h_b) \tag{3.3-16}$$

above $h = 90$ km.
In these equations the subscript 'b' denotes the condition at the base of each layer. It is to be noted that in Equation (3.3-15) λ_b is the gradient of the molecular-scale temperature with geopotential altitude $\mathrm{d}T_M/\mathrm{d}H$, and that in Equation (3.3-16) λ_b is the gradient of molecular-scale temperature with geometric altitude $\mathrm{d}T_M/\mathrm{d}h$. The quantity H_b is the geopotential altitude at the base of a particular layer characterized by a specific value of λ_b, while T_{Mb} is the value of T_M at altitude H_b or h_b as appropriate.
At and above 90 km, h_b, the geometric altitude at the base of a particular layer, characterized by a specific value of λ_b, is used instead of H_b. The defining base values and gradients are given in Table 3.3-2.

Within an atmospheric layer throughout which T_M is a linear function of H, the aerostatic equation and the perfect gas law furnish the following expression for the pressure variation with geopotential altitude (see the derivations in Section 3.2):

$$\frac{p}{p_b} = \left[\frac{T_{Mb} + \lambda_b (H - H_b)}{T_{Mb}} \right]^{-\frac{M_0 g_0}{\lambda_b R_a}} . \tag{3.3-17}$$

The density is obtained from the pressure and the molecular-scale temperature by means of the perfect gas law:

$$\frac{p}{\rho} = \frac{R_a}{M} T = \frac{R_a}{M_0} \frac{M_0}{M} T = \frac{R_a}{M_0} T_M . \tag{3.3-18}$$

The variation of density with altitude H is given by:

geopotential altitude H, km	molecular-scale temperature T_M, K	temperature gradient λ_b, K/km	molecular mass M_0, kg/kmol	temperature T, K
0	288.15		28.9644	288.15
		- 6.5		
11	216.65		28.9644	216.65
		0.0		
20	216.65		28.9644	216.65
		+ 1.0		
32	228.65		28.9644	228.65
		+ 2.8		
47	270.65		28.9644	270.65
		0.0		
52	270.65		28.9644	270.65
		- 2.0		
61	252.65		28.9644	252.65
		- 4.0		
79	180.65		28.9644	180.65
		0.0		
88.743	180.65		28.9644	180.65
geometric altitude h, km	**molecular-scale temperature T_M, K**	**temperature gradient λ_b, K/km**	**molecular mass M, kg/kmol**	**temperature T, K**
090	180.65		28.9644	180.65
		+ 3.0		
100	210.65		28.88	210.02
		+ 5.0		
110	260.65		28.56	257.00
		+ 10.0		
120	360.65		28.07	349.49
		+ 20.0		
150	960.65		26.92	892.79
		+ 15.0		
160	1110.65		26.66	1022.2
		+ 10.0		
170	1210.65		26.45	1105.5
		+ 7.0		
190	1350.65		25.85	1205.5
		+ 5.0		
230	1550.65		24.69	1321.7
		+ 4.0		
300	1830.65		22.66	1432.1
		+ 3.3		
400	2160.65		19.94	1487.4
		+ 2.6		
500	2420.65		17.94	1499.2
		+ 1.7		
600	2590.65		16.84	1506.1
		+ 1.1		
700	2700.65		16.17	1507.6

Table 3.3-2. Defining properties of the U.S. Standard Atmosphere, 1962.

H = 0 - 11000 m	**H = 47000 - 52000 m**
$T = 288.15 + \lambda_b H$, K $\lambda_b = -\ 0.0065$ K/m $p = 1013.25\,(288.15/T)^{\frac{M_0 g_0}{R_a \lambda_b}}$, hPa $\rho = 1.2250\,(288.15/T)^{\frac{M_0 g_0}{R_a \lambda_b}+1}$, kg/m^3	$T = 270.65$ K $p = 1.10905\ e^{-\frac{M_0 g_0}{R_a T}(H-47000)}$, hPa $\rho = 0.0014275\ e^{-\frac{M_0 g_0}{R_a T}(H-47000)}$, kg/m^3
H = 11000 - 20000 m	**H = 52000 - 61000 m**
$T = 216.65$ K $p = 226.320\ e^{-\frac{M_0 g_0}{R_a T}(H-11000)}$, hPa $\rho = 0.36392\ e^{-\frac{M_0 g_0}{R_a T}(H-11000)}$, kg/m^3	$T = 270.65 + \lambda_b\,(H - 52000)$ K $\lambda_b = -\ 0.002$ K/m $p = 0.590005\,(270.65/T)^{\frac{M_0 g_0}{R_a \lambda_b}}$, hPa $\rho = 0.00075943\,(270.65/T)^{\frac{M_0 g_0}{R_a \lambda_b}+1}$, kg/m^3
H = 20000 - 32000 m	**H = 61000 - 79000 m**
$T = 216.65 + \lambda_b\,(H - 20000)$ K $\lambda_b = +\,0.001$ K/m $p = 54.7487\,(216.65/T)^{\frac{M_0 g_0}{R_a \lambda_b}}$, hPa $\rho = 0.088035\,(216.65/T)^{\frac{M_0 g_0}{R_a \lambda_b}+1}$, kg/m^3	$T = 252.65 + \lambda_b\,(H - 61000)$ K $\lambda_b = -\ 0.004$ K/m $p = 0.182099\,(252.65/T)^{\frac{M_0 g_0}{R_a \lambda_b}}$, hPa $\rho = 0.00025109\,(252.65/T)^{\frac{M_0 g_0}{R_a \lambda_b}+1}$, kg/m^3
H = 32000 - 47000 m	**H = 79000 - 88743 m**
$T = 228.65 + \lambda_b\,(H - 32000)$ K $\lambda_b = +\,0.0028$ K/m $p = 8.68014\,(228.65/T)^{\frac{M_0 g_0}{R_a \lambda_b}}$, hPa $\rho = 0.013225\,(228.65/T)^{\frac{M_0 g_0}{R_a \lambda_b}+1}$, kg/m^3	$T = 180.65$ K $p = 0.010377\ e^{-\frac{M_0 g_0}{R_a T}(H-79000)}$, hPa $\rho = 0.00002001\ e^{-\frac{M_0 g_0}{R_a T}(H-79000)}$, kg/m^3

$M_0 = 28.9644$ kg/kmol; $g_0 = 9.80665$ m/s^2; $R_a = 8314.32$ J/K kmol

Table 3.3-3. Expressions for temperature, pressure, and density, in the U.S. Standard Atmosphere, 1962.

$$\frac{\rho}{\rho_b} = \left[\frac{T_{Mb} + \lambda_b (H - H_b)}{T_{Mb}}\right]^{-\left[\frac{M_0 g_0}{\lambda_b R_a} + 1\right]} . \tag{3.3-19}$$

For layers where T_M = constant ($\lambda_b = 0$), we have:

$$\frac{p}{p_b} = \frac{\rho}{\rho_b} = e^{-\frac{M_0 g_0}{R_a T_{Mb}}(H - H_b)} . \tag{3.3-20}$$

These expressions, which are in terms of geopotential altitude H, are used up to a geometric altitude of 90 km. Table 3.3-3 gives an outline of the various expressions for the computation of pressure and density in this altitude region.

Seeing that in the altitude region above h = 90 km, geometric altitude h is employed and λ_b is always greater than zero, the pressure variation in this altitude region is given by:

$$\ln \frac{p}{p_b} = -\frac{1}{\lambda_b} \frac{M_0}{R_a} \int_{h_b}^{h} \frac{g\,dh}{\frac{T_{Mb}}{\lambda_b} + (h - h_b)} . \tag{3.3-21}$$

At each geometric altitude h the air density can be computed from the known values of pressure and molecular-scale temperature by using Equation (3.3-18).

The computed values of pressure and density in the U.S. Standard Atmosphere, 1962 are shown in the Figures 3.3-5a and 3.3-5b.

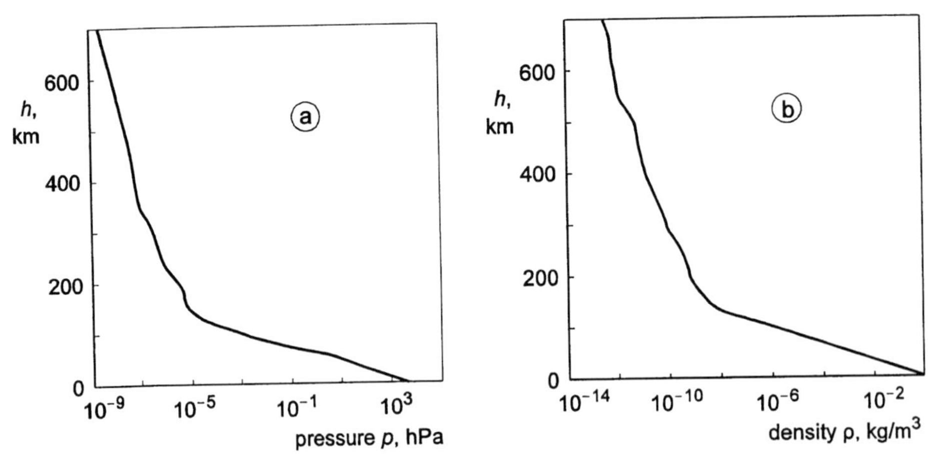

Figure 3.3-5. Pressure and density as function of geometric altitude.

Since sea level is the zero-altitude reference point in the computations, it is convenient to present expressions and numerical values of some derived atmospheric quantities for this reference condition.

1. Number density, n.

   The *number density of air* is defined as the number of atmospheric particles per unit volume:

$$n_0 = \frac{N_A p_0}{R_a T_0} = 25.47 \times 10^{24}\ \mathrm{m}^{-3} ,$$

   where N_A denotes Avogadro's number (see Table 3.3-1).

2. Mean air-particle speed, $\bar{\nu}$.

   The *mean air-particle speed* is the arithmetic average of the velocity of all air particles in the volume element being considered:

$$\bar{\nu}_0 = \sqrt{\frac{8}{\pi}\frac{R_a}{M_0} T_0} = 458.94\ \mathrm{m/s} .$$

3. Mean free path length, λ.

   The *mean free path length* is the mean value of the distances traveled by 'each' of the particles in a selected volume:

$$\lambda_0 = \frac{1}{\pi \delta^2 \sqrt{2}} \frac{R_a}{N_A} \frac{T_0}{p_0} = 6.6328 \times 10^{-8}\ \mathrm{m} ,$$

   where δ denotes the diameter of an air molecule; $\delta = 3.65 \times 10^{-10}$ m.

4. Average collision frequency, η.

   The *average collision frequency* is the average velocity of the air particles within a selected volume divided by the mean free path length of the particles within that volume:

$$\eta_0 = \frac{\bar{\nu}_0}{\lambda_0} = 4 \delta^2 N_A \sqrt{\frac{\pi}{M_0 R_a}} \frac{p_0}{\sqrt{T_0}} = 6.9193 \times 10^{9}\ \mathrm{s}^{-1} .$$

5. Mole volume, υ.

Mole volume is defined as the volume per mole of air:

$$\upsilon_0 = \frac{M_0}{\rho_0} = \frac{R_a T_0}{p_0} = 23.64 \ \mathrm{m^3/mol} \ .$$

6. Coefficient of dynamic viscosity, μ.

The *coefficient of dynamic viscosity* is defined as a coefficient of internal friction developed where gas regions move adjacently to each other at different velocities:

$$\mu_0 = \beta \frac{T_0^{3/2}}{T_0 + S} = 1.7894 \times 10^{-5} \ \mathrm{kg/m\,s} \ ,$$

where β and S are Sutherland's constants (see Table 3.3-1).
It should be noted that the above expression fails at extremely low and high temperatures and under conditions occurring at great altitudes.

7. Kinematic viscosity, η.

Kinematic viscosity is defined as the ratio of the coefficient of dynamic viscosity of a gas to the density of that gas:

$$\eta_0 = \frac{\mu_0}{\rho_0} = 1.4607 \times 10^{-5} \ \mathrm{m^2/s} \ .$$

8. Coefficient of thermal conductivity, k.

For the *coefficient of thermal conductivity* the empirical expression given in Ref. 17 may be used:

$$k_0 = \frac{2.64638 \times 10^{-3} \times T^{\frac{3}{2}}}{T_0 + 245.4 \times 10^{-\frac{12}{T_0}}} = 2.5362 \times 10^{-5} \ \mathrm{J/msK} \, .$$

9. Speed of sound, c.

The *speed of sound* is the rate at which a sound wave propagates through the air:

$$c_0 = \sqrt{\gamma R T_0} = 340.294 \text{ m/s} .$$

This expression for the speed of sound will be derived in Chapter 7.

Graphic representations as well as tabulated values of the various atmospheric properties as a function of the altitude are given in Ref. 4.

3.4 U.S. STANDARD ATMOSPHERE, 1976

The *U.S. Standard Atmosphere, 1976*, with tables and graphs extending to the geometric height h = 1000 km, is identical with the U.S. Standard Atmosphere, 1962 below 51 geopotential km, but replaces this standard atmosphere at higher altitudes (Ref. 17). A much greater availability of experimental data at that time was the basis for the revision and extension of the U.S. Standard Atmosphere, 1962.
The portion of the 1976 U.S. Standard Atmosphere up to H = 32 km, like the 1962 U.S. Standard Atmosphere, is equal to the International Standard Atmosphere.

The 1976 Standard Atmosphere is completely defined in terms of (kinetic) temperature, T. In the 1962 U.S. Standard Atmosphere this was accomplished by defining a molecular-scale temperature, T_M, and the vertical profile of molecular mass of air, M. Kinetic temperature was derived from T_M and M,

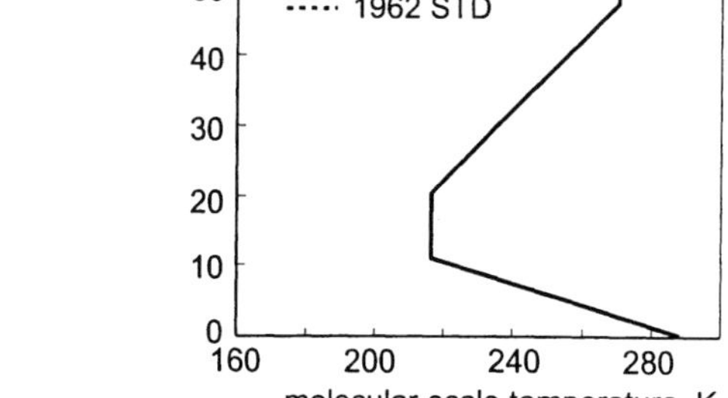

Figure 3.4-1. Molecular-scale temperature versus geopotential altitude.

using Equation (3.3-14); see the previous Figure 3.3-3.
Traditionally, standard atmospheres have defined temperature by a number of linear functions of height to eliminate the need for numerical integration in the computation of air pressure versus height. The 1976 U.S. Standard Atmosphere follows this approach to altitudes up to 86 km, and the function T_M versus geopotential altitude H is expressed as a series of seven successive linear equations. The general form of each linear segment is given by the previous Equation (3.3-15), repeated below:

$$T_M = T_{Mb} + \lambda_b(H - H_b) \ . \tag{3.3-15}$$

The value of subscript b in this equation ranges from 0 to 6, in accordance with each of the seven successive layers.
A graph of this function is compared with the similar function of the 1962 U.S. Standard Atmosphere in Figure 3.4-1. Note that in this region, where M is taken at 28.9644 kg / kmol, T_M is equal to T in accordance with Equation (3.3-14).
For the development of the 1976 U.S. Standard Atmosphere sufficient observational data of different kinds were available to estimate kinetic temperature at various high levels and to fit these with an analytical expression of temperature versus geometric height, h, which is continuous from the mesopause through the thermosphere. This analytical temperature-height profile joins smoothly with the temperature-altitude profile for lower altitudes, made up of linear segments.
At heights above 86 km the variation of temperature with geometric height is defined in terms of four successive functions. A graph of T versus h from sea-level to 500 km is shown in Figure 3.4-2.

The variation of the molecular mass of the air with geometric height adopted in

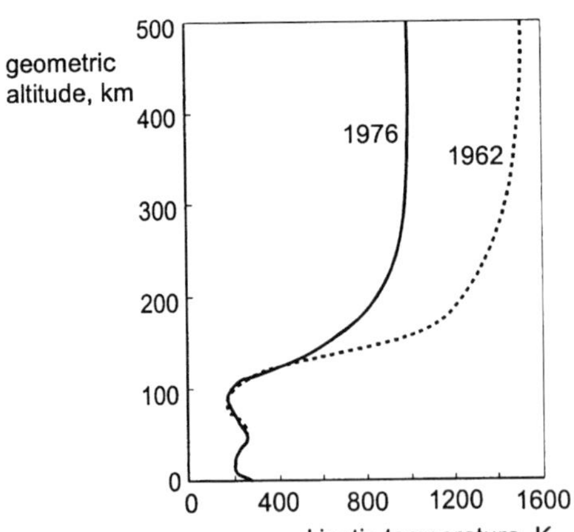

Figure 3.4-2. Kinetic temperature versus geometric altitude.

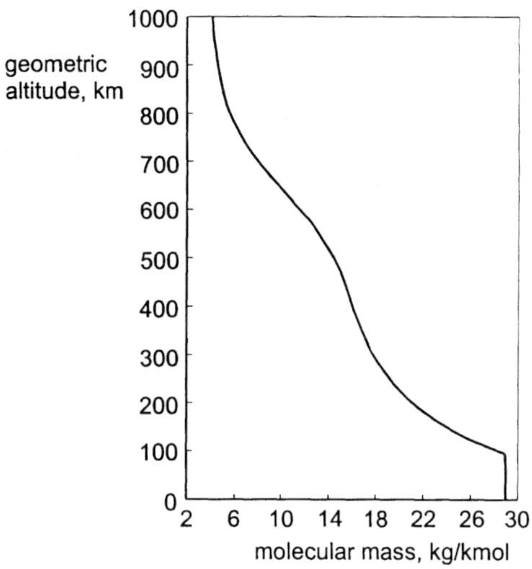

Figure 3.4-3. Mean molecular mass versus geometric altitude.

the 1976 U.S. Standard Atmosphere is depicted in Figure 3.4-3, showing that up to 86 km the molecular mass remains constant at its sea-level value M_0 = 28.9644 kg / kmol. Above 86 km, the molecular mass, decreases gradually to a value of M = 3.94 kg / kmol at height h = 1000 km.

A suitable function to account for the height variation in acceleration of gravity was obtained in Ref. 17 by applying the inverse-square law of gravitation (Equation (2.3-9)). As found before, this equation predicts values of g with sufficient accuracy for model-atmosphere computations.

4 MOIST AIR

4.1 SATURATION VAPOR PRESSURE

The partial pressure caused by the presence of water molecules in the air is called *vapor pressure*, indicated by the symbol e. When the actual vapor pressure in moist air equals the saturation vapor pressure, the air is said to be saturated. The *saturation vapor pressure*, e_{max}, defined as the maximum vapor pressure to which water vapor can be compressed before its phase changes into water or ice, increases rapidly with temperature (Table 4.1-1).
A schematic representation of the variations of the saturation vapor pressures with temperature is shown in Figure 4.1-1.

As the temperature rises, a greater number of water molecules have sufficient kinetic energy to break from the liquid surface into the vapor phase. When the temperature is raised to the point where the saturated vapor pressure at that temperature equals the air pressure, boiling takes place. Also notice that ice has measurable saturated vapor pressures which are somewhat higher than those for water, so that at air temperatures below 0°C water vapor can change directly into ice without passing through the liquid phase, or can condense into supercooled water.
The saturation vapor pressure, in principle, does not depend on the attendance and pressure of the air.

The curves in Figure 4.1-1 can be approximated by:

temperature, °C	saturated vapor pressure, N/m^2	temperature, °C	saturated vapor pressure, N/m^2
-50	4.00×10^0	40	7.37×10^3
-10	2.60×10^2	50	1.23×10^4
0	6.11×10^2	60	1.99×10^4
5	8.72×10^2	70	3.12×10^4
10	1.23×10^3	80	4.73×10^4
15	1.71×10^3	90	7.01×10^4
20	2.33×10^3	100	1.01×10^5
25	3.17×10^3	120	1.99×10^5
30	4.24×10^3	150	4.76×10^5

Table 4.1-1. Saturated vapor pressures of water.

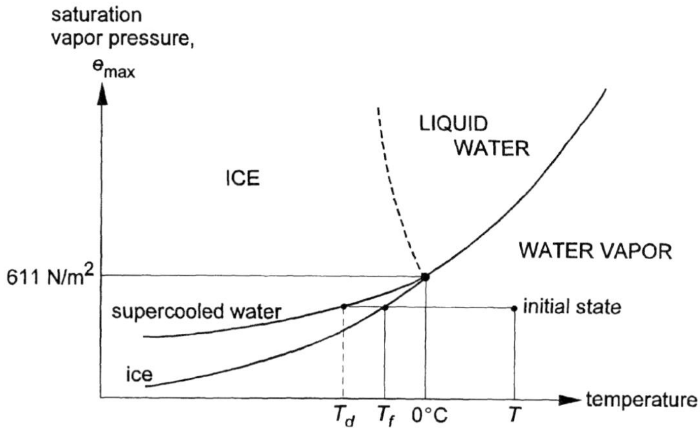

Figure 4.1-1. Saturation vapor pressure versus temperature.

$$e_{max_{water}} = 611 \times 10^{\frac{7.5T}{237.3 + T}} \qquad (\text{water vapor} \leftarrow \rightarrow \text{water}) \tag{4.1-1a}$$

and

$$e_{max_{water}} = 611 \times 10^{\frac{9.5T}{265.5 + T}} \qquad (\text{water vapor} \leftarrow \rightarrow \text{ice}) \; . \tag{4.1-1b}$$

In these expressions, the temperature T is in degree celsius, and the saturation vapor pressure e_{max} in N/m^2.

To explain the forms of the Equations (4.1-1), it may be recalled that for the isentropic condensation process, according to thermodynamics, the variation of saturation vapor pressure with temperature is given by the *Clausius-Clapeyron equation* (Ref.7):

$$\frac{\mathrm{d}e_{max}}{\mathrm{d}T} = \frac{r_v}{T\left(\dfrac{1}{\rho_v} - \dfrac{1}{\rho_w}\right)} \; , \tag{4.1-2}$$

where T is the temperature in kelvin, ρ_v and ρ_w are the densities of water vapor and water, respectively, and r_v is the temperature dependent *heat of condensation*, which is the heat released when converting 1.0 kg of water vapor into the liquid phase. Since heat of condensation changes only slightly with

temperature, a constant value for r_v often is used.
Furthermore, in Equation (4.1-2) $1/\rho_w << 1/\rho_v$, so the former quantity may be ignored. Then, using the perfect gas law for water vapor, we obtain:

$$\frac{e_{max}}{\rho_v} = \frac{R_a}{M_w} T = R_w T , \qquad (4.1\text{-}3)$$

where M_w and R_w are the molecular mass and the specific gas constant of water, respectively, Equation (4.1-2) can be written as follows:

$$\frac{de_{max}}{dT} = \frac{r_v e_{max}}{R_w T^2} . \qquad (4.1\text{-}4)$$

When the reverse process takes place we speak of the *heat of vaporization,* which is the heat required to change 1.0 kg of water from the liquid into the vapor phase.
If at temperatures below 0°C, the water vapor changes directly into ice, the heat released in this process is named the *heat of sublimation.*
The heat required to change the phase of 1.0 kg of ice into water is called the *heat of fusion.*
Values for the heats of condensation, vaporization, sublimation, and fusion also are named *latent heats.*

Using the above-mentioned approximation that r_v is practically independent of temperature, Equation (4.1-4) can be integrated to yield:

$$e_{max} = \text{constant} \times e^{-\frac{r_v}{R_w T}} , \qquad (4.1\text{-}5)$$

where the letter e on the right-hand side of Equation (4.1-5) is the base of the natural logarithm ($e = 2.7182\ldots$).
From the curves the Table 4.1-1, it may also be clear that due to the low values of the saturation vapor pressure at low temperatures, at greater altitudes and in cold regions near the Earth's surface, the air can only contain very small concentrations of water vapor.

4.2 MEASURES FOR THE WATER CONTENT

The following four quantities are used to express the amount of water vapor in the atmosphere:

1. The *absolute humidity*, a, which is the mass of water vapor per cubic meter of moist air. Hence, it is equal to the density of water vapor, ρ_v:

$$a = \rho_v = \frac{eM_w}{R_a T} = \frac{e}{R_w T} , \tag{4.2-1}$$

where M_w is the molecular mass of water. Knowing from the previous Table 1.2-1 that $M_w = 18.0$ kg/kmol, we get:

$$R_w = \frac{R_a}{M_w} = \frac{8314.32}{18} = 461.90 \text{ m}^2/\text{s}^2\text{K} .$$

The absolute humidity is maximum if at the actual temperature the air is saturated ($e = e_{max}$). According to Table 4.1-1 the saturation vapor pressure e_{max} at 15 °C equals 1710 N/m². Then for the absolute humidity we find:

$$a = \rho_v = \frac{1710}{461.90 \times 288.15} = 0.00128 \text{ kg/m}^3 .$$

2. The *relative humidity*, H, which is defined as the ratio of the actual vapor pressure e and the saturation vapor pressure e_{max} at that temperature:

$$H = \frac{e}{e_{max}} (\%) , \tag{4.2-2}$$

where H is expressed in percent.
As clarified in Figure 4.2-1, lines of constant values of relative humidity can be drawn. Evidently, the curve for the saturation vapor pressure corresponds to a relative humidity of 100%.

If moist air is isobarically cooled, the vapor pressure remains unchanged. The relative humidity however increases because the saturation vapor pressure diminishes with decreasing temperature. The temperature at which the saturation point is reached, is called the *dew point*, T_d (Figure 4.2-1). When during isobaric cooling the saturation pressure for ice is reached, the corresponding temperature is called the *frost point*, T_f (see Figure 4.1-1).

Dew point and relative humidity can be measured by use of a *psychrometer*, which consists of a *dry-bulb* and a *wet-bulb thermometer*. The wet bulb is moistened by covering it with a water-soaked gauze wick. When air is passed rapidly over both thermometers, cooling by evaporation from the wick causes the thermometers to read different temperatures. In principle, the dry

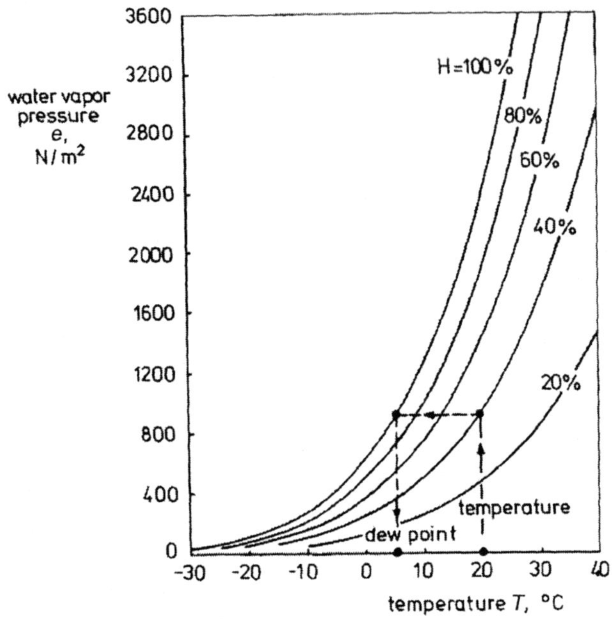

Figure 4.2-1. Relative humidity as functions of water vapor pressure and temperature.

bulb reads the prevailing air temperature and the wet bulb the dew point. For unsaturated air, naturally, the dew point is always lower than the actual air temperature. Air temperature and dew point values given in surface weather observations indicate the conditions near the ground under which fog may be produced. The possibility of fog formation should always be considered when the air temperature and dew point are close together, especially if air temperature is dropping.

3. The *specific humidity*, q, which is the ratio of the density of water vapor ρ_v to the density of the moist air:

$$q = \frac{\rho_v}{\rho_d + \rho_v} . \tag{4.2-3}$$

Here ρ_d is the density of dry air. As the pressure of moist air, p, is the sum of the partial pressure of dry air and the partial pressure of water vapor, the density of dry air is given by:

$$\rho_d = \frac{p - e}{RT} . \tag{4.2-4}$$

Substitution of Equations (4.2-1) and (4.2-4) into Equation (4.2-3) yields:

$$q = \frac{\dfrac{e}{R_w T}}{\dfrac{e}{R_w T} + \dfrac{p-e}{RT}} = \frac{\dfrac{R}{R_w} e}{p - e\left(1 - \dfrac{R}{R_w}\right)} \; . \tag{4.2-5}$$

Using $R = 287.05$ m^2/s^2 K and $R_w = 461.90$ m^2/s^2 K, we obtain:

$$q = \frac{0.622\, e}{p - 0.278\, e} \; . \tag{4.2-6}$$

Because $e << p$, Equations (4.2-5) and (4.2-6) may be approximated by:

$$q = \frac{R}{R_w}\frac{e}{p} = 0.622\,\frac{e}{p} \; . \tag{4.2-7}$$

4. The *mixing ratio*, x, which is defined as the ratio of the density of water vapor to the density of dry air:

$$x = \frac{\rho_v}{\rho_d} \; . \tag{4.2-8}$$

Substitution of Equations (4.2-1) and (4.2-4) into Equation (4.2-8) furnishes:

$$x = \frac{\dfrac{e}{R_w T}}{\dfrac{p-e}{RT}} = \frac{\dfrac{R}{R_w} e}{p-e} = \frac{0.622\, e}{p-e} \approx 0.622\,\frac{e}{p} \; . \tag{4.2-9}$$

For example, at $T = 15$ °C and $p = 101325$ N/m^2 (0 m ISA) the maximum mixing ratio (= maximum specific humidity) follows by substitution of the saturation vapor pressure at this condition into Equation (4.2-9):

$$x_{\max} = 0.622\,\frac{e_{\max}}{p} = 0.622\,\frac{1710}{101325} = 0.0105 \; .$$

Figure 4.2-2 shows, for constant values of relative humidity, the variation of the mixing ratio (= specific humidity) with the altitude in ISA. The interesting feature of this graph is the extremely low values of humidity in the upper troposphere due to the very low saturation vapor pressures.

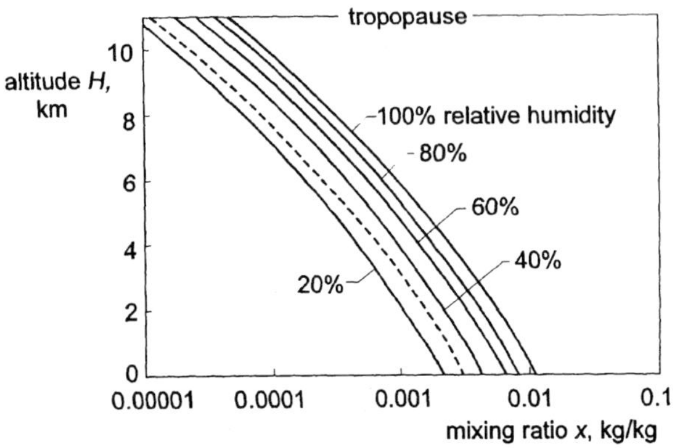

The dashed line represents a measured average vertical distribution of humidity in the troposphere at mid-latitude, taken from Ref. 10.
The curve indicates the occurrence of an almost constant relative humidity with altitude.

Figure 4.2-2. Mixing ratio (specific humidity) versus geopotential altitude.

The influence of water vapor on the density of moist air, ρ, can be analyzed by writing:

$$\rho = \rho_d + \rho_v = \frac{p-e}{RT} + \frac{e}{R_w T} = \frac{p}{RT}\left[1 - \frac{e}{p}\left(1 - \frac{R}{R_w}\right)\right] . \qquad (4.2\text{-}10)$$

By inserting $R/R_w = 0.622$ we obtain:

$$\rho = \frac{p}{RT}\left(1 - 0.378\,\frac{e}{p}\right) = \frac{p}{RT_v} . \qquad (4.2\text{-}11)$$

Here T_v is an enlarged temperature, called the *virtual temperature*, which is the temperature that dry air would need to have at the given pressure in order to have the same density as moist air.
From Equation (4.2-11) it follows that at a given pressure, moist air is somewhat less dense than dry air.

4.3 THE VERTICAL STABILITY OF THE ATMOSPHERE

Generally, the temperature in the troposphere decreases with increasing altitude (approximately with 6.5 K/km; see Table 3.1-1 and Figure 3.2-1). However, as portrayed in Figure 4.3-1, in the troposphere layers may occur in which the temperature increases with the altitude. The layers with a temperature inversion, for instance near the ground, are found in the evening and at night, due to the

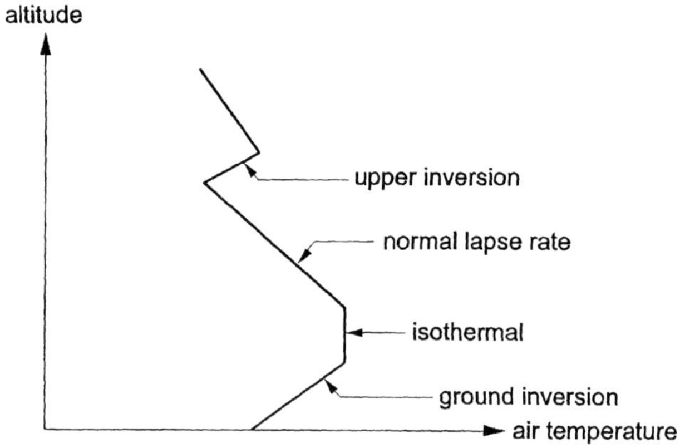

Figure 4.3-1. Possible lapse rates in the troposphere.

heat radiation from the Earth. If the temperature remains constant in a particular layer, it is called an isothermal layer.

From the aerostatic equation (2.4-1) it follows that the weight of the column of air at rest above a unit area produces the pressure at that surface. Hence the atmospheric pressure decreases with increasing height. In general, the change in pressure and temperature with increasing height results in a decrease in density with increasing height.
If for a particular layer ρ is constant, we have $\mathrm{d}\rho/\mathrm{d}h = 0$. Then with $R = R_a / M$ and Equations (2.4-1) and (2.4-2) we obtain:

$$\frac{\mathrm{d}\rho}{\mathrm{d}h} = \frac{p}{R^2 T^2} g - \frac{p}{R T^2} \frac{\mathrm{d}T}{\mathrm{d}h} = 0 \tag{4.3-1}$$

or:

$$\frac{\mathrm{d}T}{\mathrm{d}h} = -\frac{g}{R} = -\frac{9.80665}{287.05} = -0.0342\ \mathrm{K/m}\ .$$

Apparently, if in a particular layer the temperature decreases with 0.0342 K/m, the density of the air is constant, hence it is called a *homogeneous layer*. If the temperature decrease is even smaller, the air density will actually increase with height. The temperature gradient $(\mathrm{d}T/\mathrm{d}h)_a = -3.42$ K/100 m is called the *auto-convection gradient*. Homogeneous layers virtually never occur, because their instability makes them liable to vertical atmospheric disturbances.

To illustrate the static stability of unsaturated atmospheric air, the *first law of thermodynamics* is applied on a parcel of air moving in vertical direction:

$$\mathrm{d}q = \mathrm{d}u + \mathrm{d}w. \tag{4.3-2}$$

Here $\mathrm{d}q$ is the amount of energy added to the parcel, $\mathrm{d}u$ is the change in internal energy, and $\mathrm{d}w$ is the work done by the parcel, all per unit mass of the air parcel.
Assuming an isentropic process ($\mathrm{d}q = 0$), from Equation (4.3-2) we can obtain the expression (see the derivations in Chapter 7, Section 7.3):

$$0 = c_p \, \mathrm{d}T - \frac{1}{\rho} \mathrm{d}p. \tag{4.3-3}$$

Since atmospheric pressure decreases with increasing height, the pressure within the isolated mass of air will decrease when it rises in the surrounding air. We may assume that the ascending mass of air has nearly uniform properties and that its pressure $\bar{p}$ constantly equals the ambient pressure p. Furthermore we assume that the air expands adiabatically (no heat transportation from or to the air parcel), and no change of state occurs. In consequence, the volume of the rising air parcel will increase, doing the work required for expansion. This decreases the heat energy of the air parcel with the result that the temperature of the rising air drops.

Using ($\mathrm{d}p = \mathrm{d}\bar{p}$) and substitution of the aerostatic equation $\mathrm{d}p/\mathrm{d}h = -\varrho\, g$ (Equation (2.4-1)) into Equation (4.3-3), the state of the ascending air parcel can be expressed in the form:

$$c_p \, \mathrm{d}\bar{T} + \frac{\rho}{\bar{\rho}} g \, \mathrm{d}h = 0 \,. \tag{4.3-4}$$

We can apply again the perfect gas law to write the vertical temperature gradient within the rising mass of air as:

$$\left(\frac{\mathrm{d}\bar{T}}{\mathrm{d}h} \right) = - \frac{g}{c_p} \frac{\bar{T}}{T} \,. \tag{4.3-5}$$

Substituting $g = 9.80665$ m/s^2 and $c_p = 1004.68$ m^2/s^2 K, we get:

$$\left(\frac{\mathrm{d}\bar{T}}{\mathrm{d}h} \right)_d = - 0.00975 \frac{\bar{T}}{T} \,. \tag{4.3-6}$$

As T almost equals $\bar{\mathrm{T}}$, we find:

$$\left(\frac{\mathrm{d}\bar{T}}{\mathrm{d}h}\right)_d \cong -0.01\ \mathrm{K/m}\ .$$

Apparently, at low altitudes the temperature decrease is about 1K for a 100 m rise. This value of the temperature lapse rate only holds for dry or unsaturated air and is known as the *dry adiabatic rate.*

Water vapor in a moving body of air also cools as it ascends. When it reaches its saturation temperature, condensation or freezing will take place. The heat released by condensation or freezing is added to the air, through which the lapse rate is lower than the dry adiabatic rate. The temperature lapse rate for saturated air is therefore denoted as the *saturation adiabatic rate.*

As saturated air ascends, the first law of thermodynamics (4.3-2) is again the basis upon which we can investigate the effect of condensation on the variation of temperature.

Obviously, the heat given off in the change of the phase from water vapor into liquid water depends on the latent heat of condensation, r_v, and the mass of the water. Therefore, a third term must be added to Equation (4.3-4):

$$c_p \mathrm{d}\bar{T} - \frac{1}{\rho}\mathrm{d}p + r_v \mathrm{d}\left(\frac{\rho_v}{\rho_d + \rho_v}\right) = 0\ . \qquad (4.3\text{-}7)$$

The extra term represents the heat per kg moist air released into the ascending air mass due to the latent heat.

The differential $\mathrm{d}\left(\frac{\rho_v}{\rho_d + \rho_v}\right)$ is related to moist air that is saturated, so that the quantity behind the differential sign is equal to the maximum specific humidity, q_{max}. Then, according to Equation (4.2-7) we have:

$$r_v \mathrm{d}\left(\frac{\rho_v}{\rho_d + \rho_v}\right) = r_v \mathrm{d}q_{max} = r_v \frac{R}{R_w} \mathrm{d}\left(\frac{e_{max}}{p}\right)\ , \qquad (4.3\text{-}8)$$

where p is the pressure of the moist air.

Differentiation of e_{max}/p yields:

$$\mathrm{d}\left(\frac{e_{max}}{p}\right) = \frac{\mathrm{d}e_{max}}{p} - \frac{e_{max}}{p^2}\,\mathrm{d}p \ , \tag{4.3-9}$$

where $\mathrm{d}e_{max}$ is related to the temperature variation given by Equation (4.1-4). Introducing Equations (4.3-9) and (4.1-4) in Equation (4.3-8) gives:

$$r_v\,\mathrm{d}q_{max} = r_v\,\frac{R}{R_w}\left[\frac{r_v e_{max}}{R_w \overline{T}^2 p}\,\mathrm{d}\overline{T} - \frac{e_{max}}{p^2}\,\mathrm{d}p\right] . \tag{4.3-10}$$

Substitution of this result into Equation (4.3-7), delivers an expression in terms of the variations of p and T:

$$c_p\,\mathrm{d}\overline{T} - \frac{1}{\overline{\rho}}\,\mathrm{d}p + r_v\,\frac{R}{R_w}\left[\frac{r_v e_{max}}{R_w \overline{T}^2 p}\,\mathrm{d}\overline{T} - \frac{e_{max}}{p^2}\,\mathrm{d}p\right] = 0 \ . \tag{4.3-11}$$

Dividing by T and introducing the perfect gas law, $\frac{p}{\overline{\rho}} = R\overline{T}$, furnishes:

$$\left(c_p + \frac{R}{R_w^2}\,\frac{r_v^2 e_{max}}{\overline{T}^2 p}\right)\frac{\mathrm{d}\overline{T}}{\overline{T}} - \left(R + \frac{r_v R\, e_{max}}{R_w p\,\overline{T}}\right)\frac{\mathrm{d}p}{p} = 0 \ . \tag{4.3-12}$$

Integration leads to:

$$\frac{\overline{T}}{p^x} = \text{constant} \ , \tag{4.3-13}$$

where

$$x = \frac{R + \dfrac{r_v R\, e_{max}}{R_w p\,\overline{T}}}{c_p + \dfrac{R}{R_w^2}\,\dfrac{r_v^2 e_{max}}{\overline{T}^2 p}} \ .$$

It is appropriate at this point to consider again Equation (4.3-3), describing the isentropic process of a perfect gas.

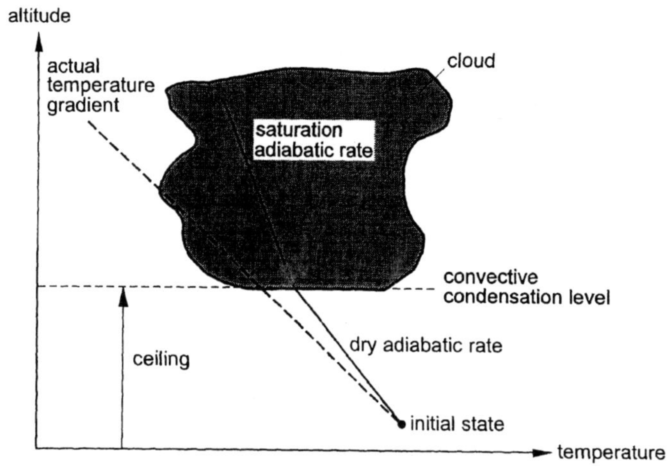

Figure 4.3-2. Convective condensation level and cloud formation.

Dividing this equation by T, and using the relation: $p/\rho = RT$, we obtain:

$$c_p \frac{dT}{T} - R\frac{dp}{p} = 0 \,. \tag{4.3-14}$$

Equation (4.3-14) can be integrated, with the result that:

$$\frac{T}{p^{R/c_p}} = \text{constant.} \tag{4.3-15}$$

This is the well-known *Poisson relation*, which describes the isentropic expansion process of an ideal gas (see Chapter 7, Section 7.3).

Comparing Equation (4.3-13) with Equation (4.3-15) manifests the influence of the released latent heat of condensation on the variation of temperature with pressure.

As depicted in Figure 4.3-2, from the altitude whereupon condensation starts, the air temperature falls off more slowly.
Putting in Equation (4.3-13) the suitable numbers given below, we find that for moist air of 25°C at sea level, the saturation adiabatic rate will be equal to $dT/dh = -0.0045$ K/m.
This result indicates that the temperature lapse rate is about half the value of the dry adiabatic rate.

The numbers used in the above prediction of the saturated adiabatic rate are:

- air temperature : $T_0 = 298.15$ K
- air pressure : $p_0 = 101325$ N/m^2
- specific gas constant of air : $R = 287.05$ m^2/s^{2}K
- specific gas constant of water : $R_w = 461.90$ m^2/s^{2}K
- specific heat of air at constant pressure : $c_p = 1004.68$ m^2/s^{2}K
- latent heat of condensation of water vapor : $r_v = 22.6 \times 10^5$ J/kg.

To examine the static stability in a general way, we consider an unsaturated air mass at a height h_1, where the ambient air temperature amounts to T_1 (Figure 4.3-3). If the air mass is moving in downward direction, e.g., due to a local disturbance, its temperature will increase with 1 K/100 m. The stirring air mass will tend to move back in upward direction, as it is warmer than the surrounding air (assuming the indicated temperature lapse rate). When the air mass is moving in upward direction, its temperature will decrease. Hence, it will be colder than the surrounding air, and therefore will return to its initial position (stable atmosphere).

If the air parcel is saturated with water vapor, it will be colder respectively warmer than the surrounding air, when sinking or rising, and hence it will move away from its initial position (unstable atmosphere).

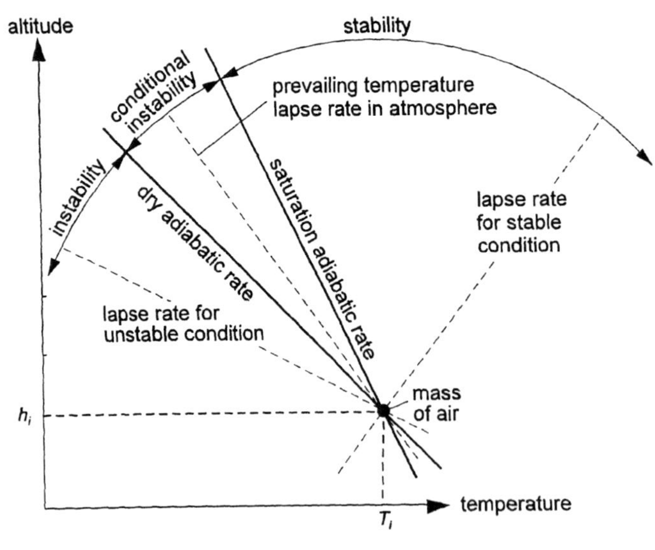

Figure 4.3-3. Lapse rate conditions for vertical stability.

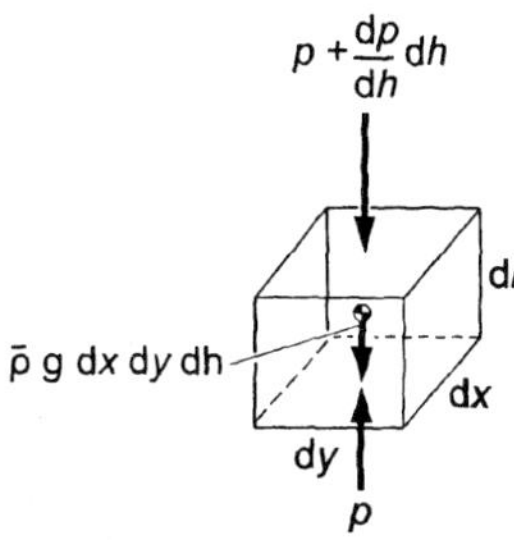

Figure 4.3-4. Force diagram.

The region between stable and unstable air is confined by the dry adiabatic and saturation adiabatic rate. When the actual vertical temperature lapse rate is greater than the saturation adiabatic rate, so-called *absolute stability* exists. In case the actual vertical lapse rate is less than the dry adiabatic rate, it is called *absolute instability*. When the actual temperature lapse rate lies in between the saturation and the dry adiabatic rate, *conditional stability* prevails. This term expresses that there is a stable condition for unsaturated air and an unstable condition for saturated air.

In order to explain the dynamics of lifting and sinking air, Newton's second law is applied on a moving infinitesimal element of air with dimensions dx, dy and dh (Figure 4.3-4). If we assume a quasi-static motion (dp = d$\bar{p}$) and the vertical motion involves no friction forces between the element and the environment, we may write:

$$\bar{\rho}\,\mathrm{d}x\,\mathrm{d}y\,\mathrm{d}h\,\frac{\mathrm{d}w}{\mathrm{d}t} = p\,\mathrm{d}x\,\mathrm{d}y - \left(p + \frac{\mathrm{d}p}{\mathrm{d}h}\,\mathrm{d}h\right)\mathrm{d}x\,\mathrm{d}y - \bar{\rho}g\,\mathrm{d}x\,\mathrm{d}y\,\mathrm{d}h \tag{4.3-16}$$

or

$$\frac{\mathrm{d}w}{\mathrm{d}t} = -\frac{1}{\bar{\rho}}\frac{\mathrm{d}p}{\mathrm{d}h} - g\,, \tag{4.3-17}$$

where w is the velocity in vertical direction.

Note that in Equation (4.3-17) the vertical component of the Coriolis force is neglected.

Using the aerostatic equation dp/dh = $-\,\varrho\, g$ yields:

$$\frac{\mathrm{d}w}{\mathrm{d}t} = \frac{\rho}{\bar{\rho}}g - g = g\left(\frac{\rho}{\bar{\rho}} - 1\right). \tag{4.3-18}$$

Substitution of the perfect gas law (Equation (2.4-7)) into Equation (4.3-18) furnishes:

$$\frac{\mathrm{d}w}{\mathrm{d}t} \equiv g\left(\frac{\bar{T}}{T} - 1\right), \tag{4.3-19}$$

where $\overline{T}$ is the absolute temperature of the rising air and T that of the surrounding still air.
For example, if $T = 15°C = 288.15$ K and $\overline{T} = 25°C = 298.15$ K, we obtain: $dw/dt = 0.34$ m/s^2. This result indicates that the initial velocities of the rising air may be very low. Moreover, the vertical motion quickly may further diminish through mixing of the ascending warm air with the cooler air surrounding it.
This behavior, typically, is found for the ascent of small-scale, irregular leaks of warm air over a large uniform plane. Higher velocities of ascent will occur over more separated surfaces, above which the rising air mass is too big and concentrated to be promptly mingled with the surrounding air.

The vertical speed of an element of air moving in upward direction, tends to increase (unstable atmosphere) if, starting from the initial level, the ratio $\overline{T}/T$ increases, i.e., when $dT/dh < d\overline{T}/dh$ (see Figure 4.3-3).
For unsaturated air this requires that the prevailing temperature gradient is smaller than the dry adiabatic rate, hence the temperature of the atmospheric air should decrease more than 1K/100 m with increasing altitude:

$$\left(\frac{dT}{dh}\right) < \left(\frac{d\overline{T}}{dh}\right)_d \approx -1\,\mathrm{K}/100\mathrm{m} .$$

Obviously, in an unstable atmosphere lifting as well as sinking of air occurs, which motions are known as convection. The altitude at which the rising air reaches its saturation temperature is called the *convective condensation level*. If above this level conditional stability exists, the air will continue its upward motion thereby cooling at the saturation adiabatic rate. This process may lead to the formation of clouds above the cloud ceiling (see Figure 4.3-2).

It should be appreciated that when a moist, unsaturated air mass ascends, its temperature decreases with the altitude according to the dry adiabatic rate, but that the ratio between vapor pressure and air pressure (e/p) remains constant. This means, according to Equation (4.2-7), that specific humidity q is conserved during ascent. This condition implies that when determining the convective condensation level (*the cloud ceiling*), we must recognize that the dew point also decreases with increasing altitude. In Figure 4.3-5 the adiabatic expansion process is indicated in the *e-T* diagram as well as in the *H-T* diagram. The subscript 'o' denotes the initial state at ground level and '*s*' the saturation condition at the convective condensation level.
To derive an expression for the dew point lapse rate we consider again Equation (4.1-4), repeated below:

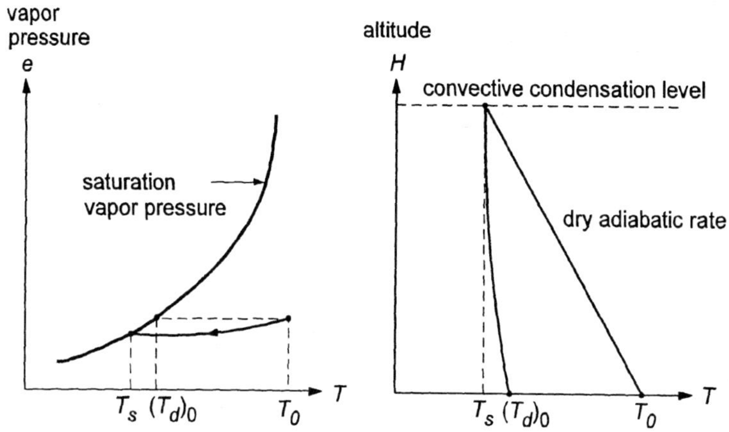

Figure 4.3-5. Determination of convective condensation level.

$$\frac{\mathrm{d}e_{max}}{\mathrm{d}T} = \frac{r_v}{R_w T^2} e_{max} \;. \tag{4.1-4}$$

Since vapor pressure e of an unsaturated air mass is equal to the saturated vapor pressure at the dew point T_d of an air mass (see Figure 4.2-1), we also may use Equation (4.1-4) to describe the relationship between vapor pressure and dew point:

$$\frac{\mathrm{d}e}{\mathrm{d}T_d} = \frac{r_v}{R_w T_d^2} e \;. \tag{4.3-20}$$

From Equations (4.3-20), (4.2-7) and the perfect gas law, $p/\rho = RT$, and assuming aerostatic equilibrium ($\mathrm{d}p = -\rho g \mathrm{d}h$), the dew point lapse rate is found to be:

$$\frac{\mathrm{d}T_d}{\mathrm{d}h} = -\frac{g}{r_v}\frac{R_w}{R}\frac{T_d^2}{T} \;. \tag{4.3-21}$$

Since dew point and air temperature differ only slightly ($T \approx T_d$), Equation (4.3-21) reduces to:

$$\frac{\mathrm{d}T_d}{\mathrm{d}h} = -\frac{g}{r_v}\frac{R_w}{R} T_d \;. \tag{4.3-22}$$

From this equation we find, for example, that at a temperature of $T_d = 280$ K, the dew point drops at a rate of about 0.002 K/m.

To explain the statement that the ratio e/p is constant, we next consider a mixture of ideal gases. Then, for species 'i', the equation of state relates partial pressure p_i, volume V, and temperature T (in kelvin), of n_i kmol by the equation:

$$p_i V = n_i R_a T \, , \tag{4.3-23}$$

where R_a is the universal gas constant (R_a = 8314.32 J/K kmol).
According to the experimental law stated by the English scientist John Dalton (1766 - 1844), the total pressure of a mixture of gases is equal to the sum of the partial pressures:

$$p = \Sigma p_i \, , \tag{4.3-24}$$

and the equation of state for the whole mixture:

$$p V = \Sigma n_i R_a T \, . \tag{4.3-25}$$

Equation (4.3-25) can be written as:

$$p V = n R_a T \, , \tag{4.3-26}$$

where $n = \sum n_i$ is the total number of kilomole of the mixture.
So, we obtain from Equations (4.3-23) and (4.3-26):

$$\frac{p_i}{p} = \frac{n_i}{n} \, , \tag{4.3-27}$$

where n_i / n is the kmol fraction of species *'i'*.
From physics we know that the number of kilomole, *n*, is equal to the ratio of the total number of molecules in a gas, *N*, and the number of molecules per kmol, N_A:

$$n = \frac{N}{N_A} \, , \tag{4.3-28}$$

where N_A is *Avogadro's number*; $N_A = 6.02 \times 10^{26}$ molecules per kmol, which number is based on the statement of the Italian scientist Amedeo Avogadro (1776 - 1856) that equal volumes of gas at the same pressure and temperature, contain equal numbers of molecules.
From Equations (4.3-27) and (4.3-28) it easily follows that:

$$\frac{e}{p} = \frac{N_v}{N} \, . \tag{4.3-29}$$

The latter equation makes clear that *e/p* remains constant in a rising air mass because it is related to the ratio of the number of molecules per kmol in the water vapor, N_v, and those in the moist air, *N*.

4.4 CLOUD FORMATION

When moist air is cooled below its dew point, condensation may take place so that clouds or fog can be formed. Clouds and fog consist of droplets of water or ice crystals.
When moist air reaches its saturation vapor pressure, the water vapor will not immediately form water or ice crystals, unlike suggested by Figure 4.1-1. The reason for this delay is that the water vapor only condenses if a body is present onto which it can condense. At ground level it will condense onto a body or surface, but in the air it is necessary to have available so-called *condensation nuclei*, which are microscopic small, solid particles. These particles include dust, i.e., soil particles blown into the air by the wind, smoke and soot from combustion products, and sulfur dioxide and salt crystals from evaporating sea spray. Not all particles are suitable as cloud condensation nuclei. The larger ones fall due to their own weight, and so they are unlikely to remain airborne long enough for water to condense onto them.
Water vapor in purely clean air, i.e., air entirely free from pollution by solid or fluid particulate matter, may become supersaturated or *supercooled*. The latter condition implies that the water vapor pressure exceeds the saturation vapor pressure at the actual temperature ($e > e_{max}$).
When clouds, formed at temperatures above freezing points are cooled slowly, they usually become supercooled. If the supercooled water droplets in a cloud encounter solid particles, partial or complete freezing may occur, depending onto to what extent supercooling has taken place. Even at temperatures below -10 to $-15°C$ there still may be water droplets present in so-called *mixed clouds*.

The problem of ice formation on outer surfaces of flying aircraft is also caused by the phenomenon of supercooling and occurs when water droplets freeze through the interception by aircraft parts.

Clouds in which temperatures below freezing points occur, are denoted as *cold clouds*. If the temperature of the cloud is entirely above freezing points, it is called a *warm cloud*.

The higher one ascends in the atmosphere, the less water vapor the air contains (see Figure 4.2-2), and the cleaner the air becomes. About half of all water vapor is contained within the layer up to an altitude of 2000 m. The latter altitude is also about the upper limit for industrial haze. Above an altitude of about 10 km, the air is practically dry and often without pollutant substances. Hence, it may be obvious that cloud formation virtually only takes place below the tropopause.

Nevertheless, in the southern polar region the temperature may become so much low that at an altitude of about 20 km *polar stratospheric clouds* are formed. Also, in the middle latitudes at an altitude of about 25 km, the combination of a very low temperature and water vapor may lead to the formation of the so-called *mother-of-perl clouds*. Other very high clouds, but fairly rare, are the *noctilucent ("night-shining") clouds*, occurring at altitudes of about 80 km. The latter trail-shaped clouds are presumed to be composed of small ice-coated particles, and produce silvery luminosity. They are occasional visible on the northern hemisphere from mid-May to mid-August, during the deep twilight's when the Sun lies just below the horizon.

The water vapor required for the formation of these tenuous clouds may be the product of the breakdown of atmospheric methane (Ref. 18):

$$CH_4 + 2\,O_2 \rightarrow CO_2 + 2\,H_2O \,. \tag{4.4-1}$$

The nuclei onto which this water condenses may be dust particles of interplanetary origin.
These special cloud forms may be of importance because of the effect that they can have on the radiation budget of Earth and atmosphere (see Chapter 5).

According to the altitude regions where they are found, tropospheric clouds are divided into four general families: I. *high clouds*, II. *middle clouds*, III. *low clouds*, and IV. *convection clouds* (clouds of vertical development). Four basic types of tropospheric clouds are distinguished: *cirrus* or curl cloud, *cumulus* or wool-pack cloud, *stratus* or layer cloud, and *nimbus* or rain cloud.
Depending on their particular forms, they are subdivided into ten subtypes, recognized internationally (Figure 4.4-1). Some of the subtypes shown in Figure 4.4-1 are subdivided again, while various species exist within these divisions and subdivisions.
The high, middle and low clouds are characterized by a layered structure, while cumulus is a convective cloud, having a substantial vertical development over a restricted area.
Only the high clouds consist exclusively of ice crystals. The middle, low and cumulus clouds are composed of water droplets.
When the top of a cumulus cloud begins to flatten, it has become a

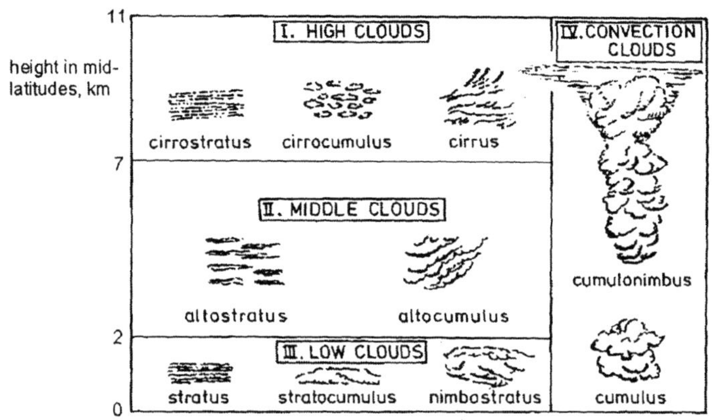

Figure 4.4-1. Cloud classification.

cumulonimbus cloud, containing water in the main body and ice in the top part. At the moment that the top of the cloud has spread (anvil-top), we recognize a *thunderstorm* with updrafts and downdrafts, and producing convectional showers, lightning and thunder (Figure 4.4-2). Lightning is the visible discharge of electricity associated with a thunderstorm. Thunder is the noise that usually follows lightning. It is produced by a violent expansion of air. These phenomena occur due to the currents of rising and falling air in cumulonimbus clouds.

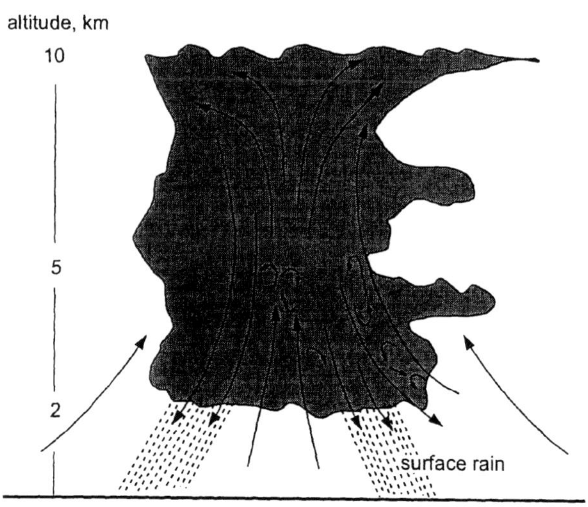

Figure 4.4-2. Vertical cross-section of a thunderstorm.

Fog, finally, is low stratus cloud formed on or near the ground. It arises from cooling of the air below its dew point. Well-known types are *radiation fog* and *advection fog*.
Radiation fog may occur near the centers of high pressure areas during clear nights, when the surface of the Earth cools down extensively through radiation of heat to the upper atmosphere. Advection fog results from warm moist air blowing over a colder surface. The cooling causes stability that helps hold the fog at low levels. A related form is *steam fog*, which is formed through the evaporation of the water when very cold air moves over a warmer water surface.

4.5 PRECIPITATION AND WIND

The condensation of water vapor caused by the cooling of moist air not only produces clouds but also the various forms of liquid or solid water falling from these clouds.
Forms of precipitation are:

- Rain, which is the falling of drops of water from clouds.
- Freezing rain, which is formed by supercooled water droplets freezing to objects in the open and forming glaze.
- Drizzle, which is the uniform falling of very small and very numerous droplets.
- Freezing drizzle, which differs from drizzle in that the droplets freeze to objects in the open and form glaze.
- Snow, which is the falling of white or translucent ice crystals, grown by sublimation, which can possibly become aggregated into snowflakes.
- Sleet, which is the falling of globular hard grains of ice, formed by the freezing of raindrops.
- Hail, which is the falling from clouds of ice balls or stones with relatively large diameters, and occurs almost exclusively in thunderstorms and at temperatures below freezing at ground level.
- Snow pellets, which is the falling of grains of snowlike structure, occurring almost exclusively in showers.

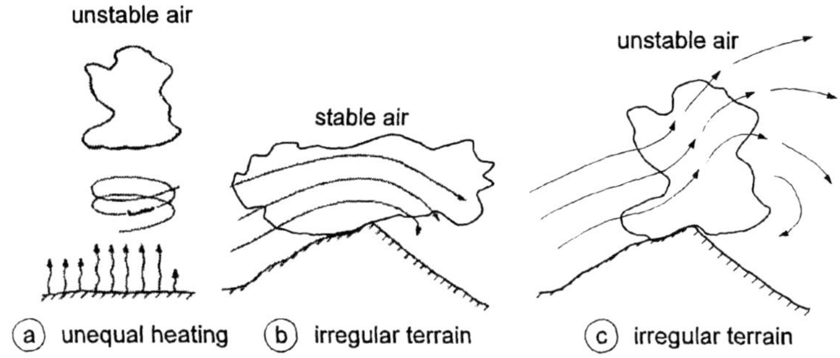

Figure 4.5-1. Effect of ground surface features on cloud formation.

It is interesting to note that if a raindrop starts to descend through the atmosphere it will grow and speed up under influence of the gravity force. At the same time, a drag force from the air will act in opposite direction. Since the latter force, however, increases with increasing velocity, after some time the two forces become equal, yielding from thereon a constant *terminal velocity*. Evidently, the larger the diameter, the higher the terminal velocity (Ref. 10).

Figure 4.5-1a pictures that the rapid updrafts due to unequal heating of the ground or water surface cause the development of convection clouds (cumulus and cumulonimbus). Figures 4.5-1b and 4.5-1c illustrate that the cooling and the accompanying condensation leading to the formation of low clouds may be the result of a slow rising of a whole layer of air, where the moist air is successively stable and unstable.

From Chapter 2 we know that a region of low atmospheric pressure at sea level with its attendant system of winds is called a *cyclone*. When occurring near the Equator, the rising air may develop into a tropical storm and even into a *hurricane*, which is very intense depression that forms over tropical oceans and brings extremely strong wind and heavy rain. As the lower air rises, its moisture condenses as clouds and releases latent heat, conditions that stimulate strong rising air currents to develop. In the center of a hurricane there is a circular area of very low pressure called the *eye* of the storm. There, light winds and clear skies occur. Dependent on their birthplaces, hurricanes are also designated as *tropical cyclones* or as *typhoons* (Ref. 9).

A related form of a hurricane is a *tornado*, which is an intense, cloud-cored vortex extending from the base of a severe local storm to the ground surface. A tornado is caused by a very strong uplift of air, which sucks in the lower air and makes it whirls faster and faster. A tornado has a smaller size than a hurricane, but can be even more violent.

In the temperate latitudes, where the masses of polar and tropical air meet, cloud formation, and the associated wind and precipitation, are often a consequence of the occurrence of atmospheric fronts. Mid-latitude storms, normally, are much less intense than tropical storms, especially in their wind speed and precipitation. Nevertheless, in spring and summer violent weather may occur in the form of thunderstorms where tornadoes are present.
As portrayed in the previous Figures 2.5-2 and 2.5-3, fronts are discontinuities at which a colder, heavier air mass underlies a warmer, lighter air mass.

The most pronounced weather changes take place along cold fronts. Two types of cold fronts may be discerned, *fast-moving* cold fronts and *slow-moving* cold fronts.

Figure 4.5-2a indicates that the motions in a fast-moving cold front consist of declining movements at both sides of the frontal surface, causing variable surface winds behind the front. In the warm air ahead of the front upward movements occur.
If the warm air mass is stable, the sky becomes overcast with altostratus, nimbostratus and stratus clouds, from which steady rain falls.

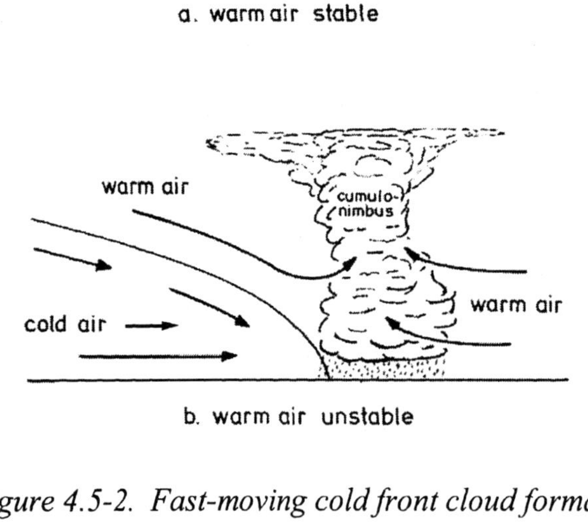

Figure 4.5-2. Fast-moving cold front cloud formation.

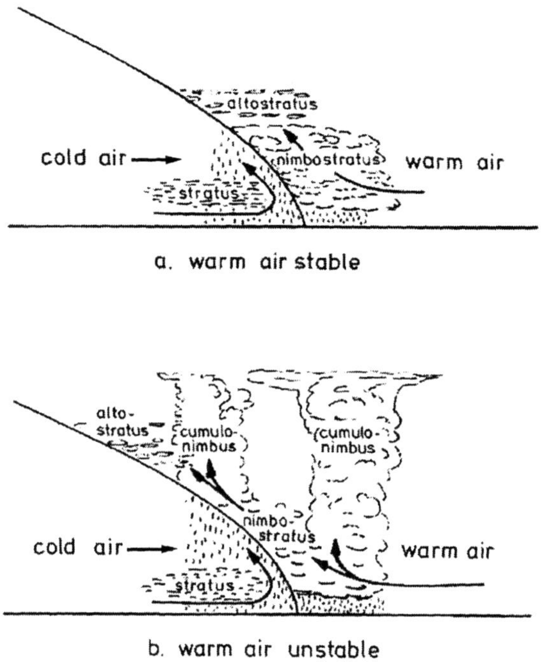

Figure 4.5-3. Slow-moving cold front cloud formation.

An unstable condition of the warm air mass leads to the development of cumulonimbus clouds, thunderstorm activity and showers along and ahead of the front (Figure 4.5-2b).

The cloud formation in a slow-moving cold front is pictured in Figures 4.5-3a and b. Now, warm air sloping upwards deliver nimbostratus and altostratus clouds. Also, there may be sufficient upward movements in the cold air mass to produce low stratus clouds behind the front. When the warm air mass is conditionally unstable, cumulonimbus clouds with thunderstorm activity may arise.

The cloud formation in a warm front is comparable with that in a slow-moving front. Figure 4.5-4a gives a sketch of the cloud system in a warm front with the warm air stable. In this case the warm air slowly advances over the sloping frontal surface and extensive cloud formation occurs. First comes into sight nimbostratus, and then altostratus, cirrostratus, and, finally, cirrus clouds. Precipitation may fall down from the nimbostratus clouds, through which the humidity of the cold air is increased to near-saturation and further cooling during the night may produce fog over large areas.

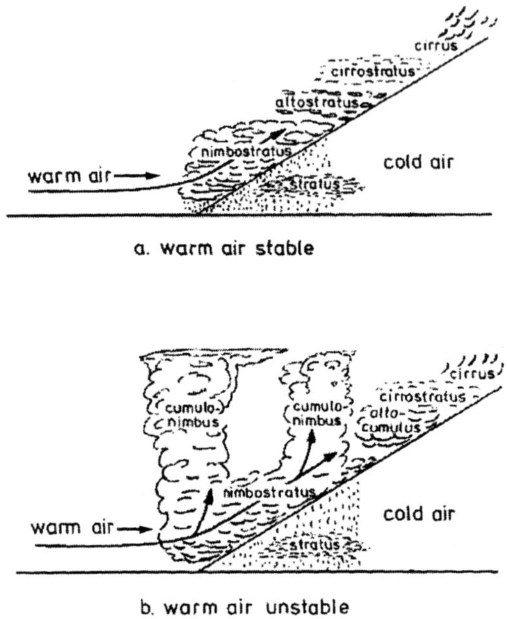

Figure 4.5-4. Warm front cloud formation.

As portrayed in Figure 4.5-4b, if the warm air is unstable, altocumulus and cumulonimbus clouds come after the nimbostratus clouds and thunderstorm activeness may be encountered ahead of the front, together with rain showers. Wind velocities and wind directions above and below the frontal surface commonly differ significantly.

In the case that a cold front meets a warm front, we may have the situation of an occluded front. The manner in which the occlusion works out depends on the relative temperature of the cold air masses underlying the two frontal surfaces.
In Figure 4.5-5a the weather conditions in a cold front occlusion is depicted. The air mass behind the cold front is the colder and therefore displaces the cool air behind the warm front. On the other hand, as shown in Figure 4.5-5b, in a warm front occlusion the air behind the cold front is less cold than the air underneath the warm front, so that now the cold front climbs over the wedge of the warm front.
Note that the direction of motion of the fronts in Figures 4.5-2 to 4.5-5 is indicated by means of arrows. On weather maps, fronts are marked by symbols or colored lines; see the legend in Figure 4.5-6. Front movement is in the direction of the triangular points.

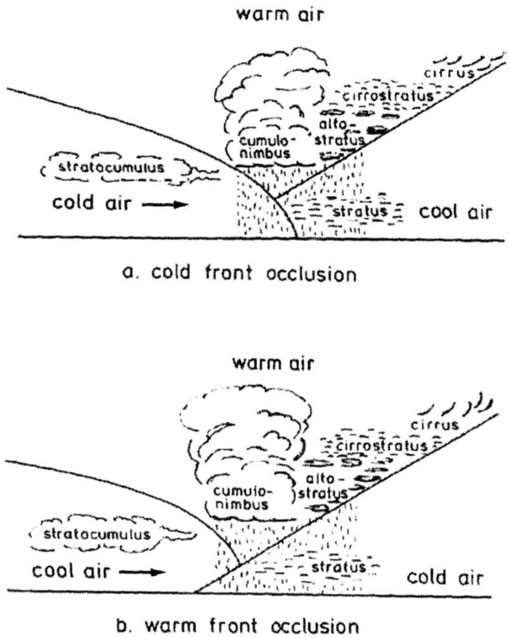

Figure 4.5-5. Occluded front cloud formation.

The study of fronts and their effects on climates of the world's regions is a subject of the scientific discipline 'climatology'. In this respect, it is worthy to remember that not only fronts but also ocean currents influence the weather conditions.

The high specific heat of water ($c_p = 4186\ m^2/s^2\ K$) and the gigantic volumes of the oceans make them an efficient heat store. In summer, the sea-surface temperature is lower than land-surface temperature, and air blowing from land to sea is cooled. In winter, it is warmed, because the sea-surface temperature is higher than the temperature on land.

It is only in recent years that the warm ocean currents have been discovered as the immediate cause of these climate extremes. For instance, these currents are

type of front	symbol		color
	ground level	higher altitudes	
warm front	⯊⯊⯊	◠◠◠	red
cold front	▲▲▲	△△△	blue
occluded front	▲⯊▲	◠△◠	purple

Figure 4.5-6. Key to reading of fronts on weather maps.

responsible for a meteorological situation termed 'El Niño-Southern Oscillation (ENSO). Usually, it is termed 'El-Niño, as abbreviation of El-Niño de Navidad, Spanish for 'Christ-child'. This name refers to the phenomenon that about every four years around Christmas the water temperature near the coast of South America, especially Peru, rises by several degrees. The weather associated with an El-Niño, may show a strong change for the worse over the entire globe. Especially 1982 and 1998 were years in which many parts of the world experienced extreme bad weather in the form of storms, heavy rainfall and serious flooding.

4.6 SELF-CLEANING CAPABILITY OF THE ATMOSPHERE

Obviously, the composition of the atmosphere is continuously changed by both natural and man-made gases and solid particles, which enter the atmosphere and may leave it again after some time. The amount of a particular substance in the atmosphere, of course, is determined by the difference between its rate of input and of removal.

Removal may occur by means of physical processes such as diffusion, dispersion, precipitation, surface deposition and chemical transformations.

In summary, some important processes removing pollutant substances from the troposphere are summarized below.

- Precipitation of solid particulate matter and droplets.
- Dissolutions of gases into rain droplets (wet deposition).
- Absorption of gases by surface water.
- Absorption of gases by soil and vegetation (dry deposition).
- Dilution by diffusion.
- Dispersion by convection and advection (wind).
- Chemical and biochemical reactions.

It is difficult to say precisely how rapidly gases and particles move between the air and ground surface. Only rough estimates are known of the times required for the cycling of atmospheric substances. The average length of time a molecule or particle is likely to remain in the air, is termed *residence time*. Residence times for the different substances may vary from one minute to years, and even to hundreds of years.

Important is that the hydrological cycle, portrayed earlier in Figure 1.7-1, is relatively short. The mean time span between evaporation of water and its precipitation is only eight to ten days. Therefore, owing to the various weather conditions the life span of most pollutants in the troposphere ranges from several hours to approximately 10 days.

Solid particles seldom remain airborne in the troposphere for more than a couple of hours or, at most, days, before they are returned to the ground. If the particles are relatively large, they fall gravitationally and settle and adhere to surfaces with which they come into contact. This behavior is called *dry deposition*.

Smaller particles may act as condensation nuclei for the water vapor in the air. The water droplets then may dissolve soluble gases, and the rain washes them, together with the particles, to the ground surface, producing *wet deposition*.

It should be noted that a particle that is carried into the layers above the tropopause may remain there for more than years. These long residence times imply that any effect that pollutants may have on the climate or on the ozone layer, is greatly prolonged, and this is why man especially is concerned about air pollutants that are emitted by airplanes flying just below or above the tropopause.

In spite of their proximity, the troposphere and the stratosphere are very different layers. To illustrate this, we may consider the convective flows as arising from the prevailing temperature gradients in the atmosphere (see Section 4.3). Therefore, in Figure 4.6-1 the typical variation of air temperature with the altitude is sketched again, emphasizing that the stratospheric layer between 11-20 km may be regarded as a boundary between the troposphere with a negative temperature gradient and the upper part of the stratosphere, where the temperature gradient is positive.

The vertical temperature profile in the atmosphere is determined by thermal processes, activated by solar radiation. The layer in which we live, the troposphere, is characterized by a temperature decrease with increasing height, from about 15°C near the ground, to approximately −60°C at the tropopause (see also Section 1.3).

In the troposphere the phenomena which we call *weather* (wind, cloudiness, precipitation) occur, and both relatively fast circulation and vertical mixing of air take place. Owing to these convectional flows the atmosphere is disposed of pollutant substances by returning them to Earth. Moreover, all forms of precipitation occur in the troposphere, also contributing to their disposal.

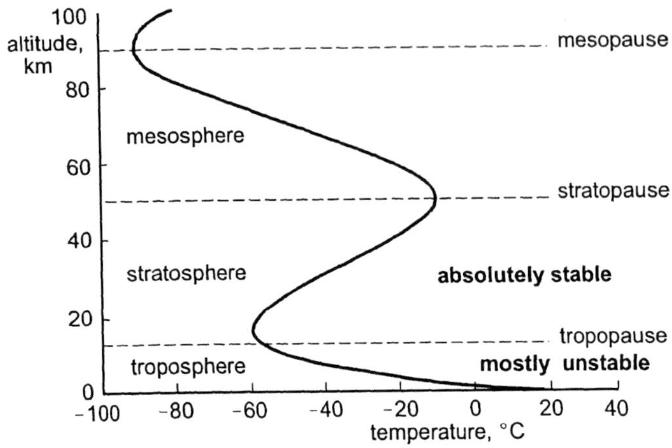

Figure 4.6-1. Typical variation of temperature with altitude.

In the stratosphere, the temperature increases to a maximum of approximately -10°C at an altitude between 30 and 50 km. Due to the positive temperature gradient, the stratosphere is vertically stable, so that vertical mixing is largely suppressed. No forms of purifying precipitation occur, so that pollutants may migrate here for very long times.
In conclusion, the layer between 11-20 km is an important boundary, through the fact that the temperature gradient changes from negative to positive.
The main cause of the temperature inversion above the troposphere is the ozone layer, which has its maximum concentration in the upper part of the stratosphere at an altitude of about 30 km.
Considering this, it might be desirable to restrict air traffic to the troposphere, where vertical mixing and precipitation frequently take place.

A first contribution to the dilution of pollutants comes from the mixing that occurs because of the random movement of the molecules, which is called *diffusion*. The effect is that the diffusing substance moves from a region where its concentration is high to one where its concentration is low.
The German physiologist Adolf Fick (1829 - 1901) determined experimentally that the rate of mass flow across a unit area, J, is directly proportional to the *concentration gradient*, i.e., the change in concentration in kg/m^3 per unit distance, dc/dx:

$$J = -D\frac{dc}{dx}\ ,\ \mathrm{kg/s\,m^2}\ , \tag{4.6-1}$$

where D (in m^2/s) is a constant of proportionality.

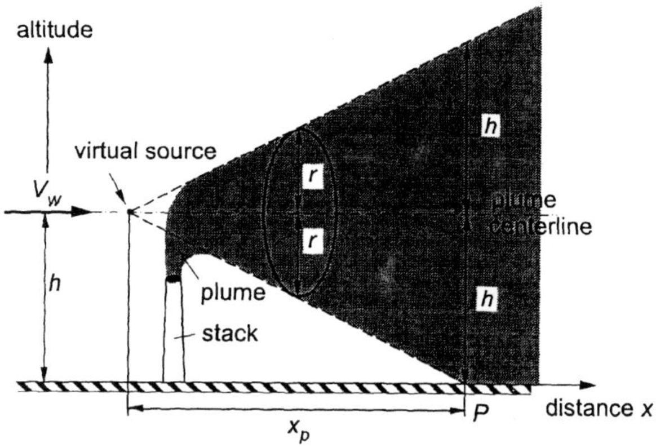

Figure 4.6-2. Dispersion of atmospheric pollutants by wind.

Numerical values for the diffusion constant D can be found in standard textbooks on physics (see, e.g., Ref. 19).

Generally, air currents play a much greater role in diluting pollutants than does diffusion. As an example, the dispersion of pollutants from a point source by wind will be considered. The source, in general terms, is the mechanism expelling the pollutant as a gas or cloud. In the following we assume that the polluted substance consists of finely divided particles, which are sufficiently small as to be capable of suspension in the atmosphere without appreciable fallout.

The emission is represented in Figure 4.6-2 by the exhaust plume of pollutants discharged into the atmosphere through a chimney stack placed on a plane ground surface.

Initially, the plume rises and then the wind carries the pollutants away from the source in downwind direction. As the plume travels, a cone of pollutant substances arises, with its axis coinciding with the direction of the wind. Plume spread occurs, primarily resulting from atmospheric turbulence.

For the sake of simplicity, steady flow through a clear-cut, circular cone is considered, so that through each cross-section of the cone perpendicular to its horizontal axis, just as much pollutant passes as is discharged by the source per unit time.

When uniform flow is assumed, we have a constant concentration of pollutants throughout each cross-section of the cone. Hence, the mass flow of the pollutant substances per unit time is given by:

$$Q = \pi r^2 c_v V_w , \qquad (4.6\text{-}2)$$

where:

- Q is the source strength, i.e., the mass flow rate emitted by the source in kg/s.
- V_w is the (horizontal) wind velocity through the plume in m/s.
- c_v is the local concentration of suspended matter in kg/m^3.
- πr^2 is the area of cross-section of the cone, measured perpendicular to the direction of flow in m^2.

For the concentration c_v follows from Equation (4.6-2):

$$c_v = \frac{Q}{\pi r^2 V_w} . \qquad (4.6\text{-}3)$$

Obviously, wind dilutes the pollutants as their concentration decreases with increasing wind velocity and distance from the source.
On the ground surface near the chimney the concentration equals zero. Only at distances greater than x_p in the direction of the wind (point P in Figure 4.6-2), the air will contain pollutant substances. Ignoring contributions from ground reflected pollutants, at this location also the maximum concentration of the pollutant substances at ground level is found. Furthermore, the maximum concentration on ground level is inversely proportional to the square of the chimney height. Consequently, the higher the chimney the lower the pollution concentration on the ground.
To determine distance x_p in Figure 4.6-2, it will be necessary to know how the shape of the cone is affected by wind velocity, vertical wind gradient, vertical air motions, and atmospheric turbulence. The exhaust plume not only will be dragged along by the wind, but will also be pulled down by the gravitational force. The effects of convective flow and the temperature differences between the hot gases and the surrounding air on the shape on the exhaust plume flow make this a difficult question (see, e.g., Figure 4.6-5).
Obviously, a rising plume at the source is desirable, which will be promoted by the application of a divergent exhaust channel (see Chapter 7, Section 7.7).

The coning plume depicted in Figure 4.6-2 occurs when the atmosphere has neutral stability. Plume behavior, however, is strongly affected by the prevailing lapse rates in the atmosphere. From Section 4.3 we know that if the prevalent lapse rate is smaller than the dry adiabatic rate, the air is absolutely unstable for vertical disturbances. Vertical motions, initiated by atmospheric turbulence will then escalate. Under these circumstances the smoke plume will tend to meander considerably, as outlined in Figure 4.6-3a.

If the prevalent lapse rate is larger than the saturation adiabatic rate, the atmosphere is absolutely stable. Only minor dispersion in vertical direction occurs (Figure 4.6-3b). Pollutant substances reach the ground only after a long time, so that the maximum concentration occurs at a long distance from the chimney. Since the pollutants may already have dispersed in horizontal direction, the concentrations at ground level will be much lower than in case of

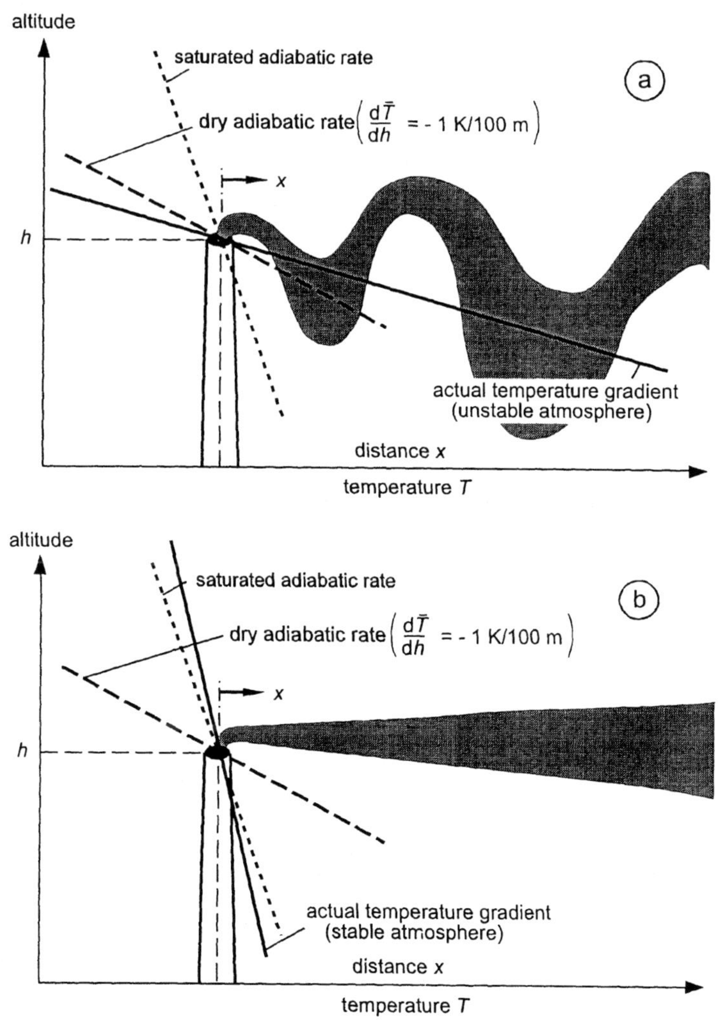

Figure 4.6-3. Effect of vertical stability on plume behavior.

an unstable atmosphere.

When a temperature inversion layer prevails, the atmosphere will be extremely stable. All vertical motion of air will be suppressed, leading to even less dispersion of the plume than portrayed in Figure 4.6-3b.

In the troposphere, temperature inversions usually are present in several layers. When, as shown in Figure 4.6-4a, the outlet of the chimney is in an unstable layer, just between two inversion layers, the pollutants will be trapped in this layer.

A favorable situation occurs when the inversion layer is just below the chimney

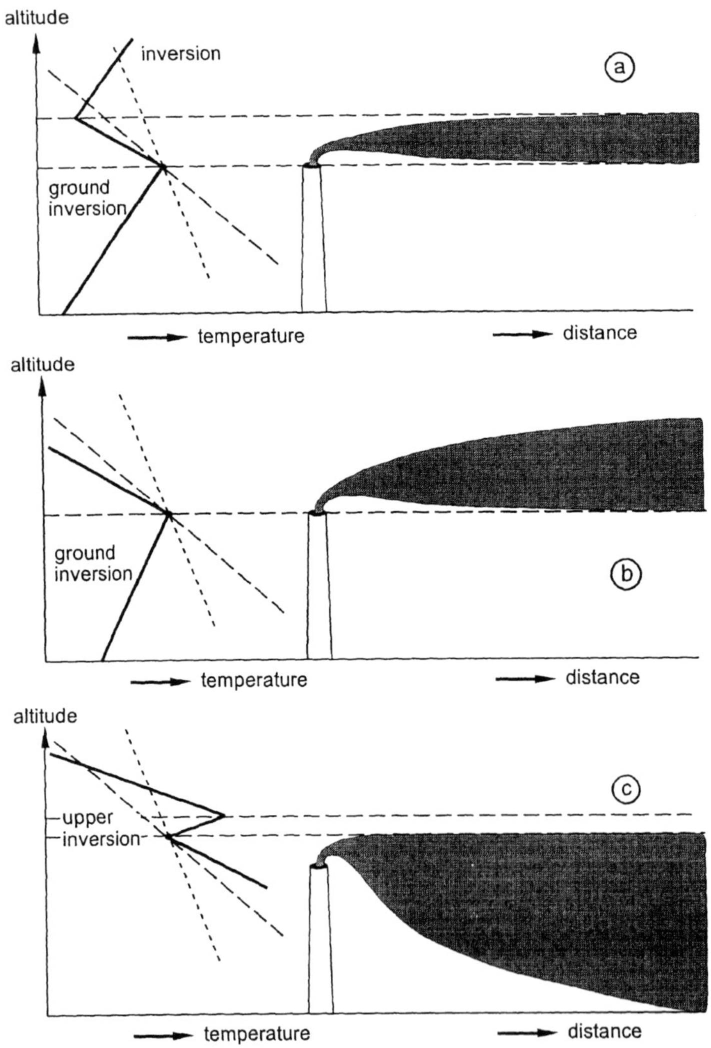

Figure 4.6-4. Effect of temperature inversion on plume behavior.

outlet (Figure 4.6-4b). Under these circumstances the pollutants will hardly reach the ground surface.
When the inversion layer is just above the chimney outlet, dispersion of pollutants in upward direction will be obstructed, and all pollutants will not go beyond the inversion layer (Figure 4.6-4c). This condition causes much higher ground level concentrations than occur in an atmosphere with a negative temperature gradient.
While in case of absolute stability the concentration of pollutants discharged by a chimney stack remain small near the ground surface (Figure 4.6-4b), in urban areas it may become unfavorably high. Here the pollutants are emitted near the ground, and cannot go beyond the inversion layer. This explains the strongly polluted air in cities situated near the sea, as the layer below the cold coastal air flow is mostly very stable.

Naturally, plume behavior not only depends on wind velocity, but also is affected by atmospheric turbulence, which type of air motion has two contributions (see also Chapter 2, Section 2.9). The first contribution is the turbulence due to convection or *thermal turbulence*, which is of considerable significance for vertical dispersion. The second contribution is the so-called *friction turbulence*. This type prevails near ground level and may be of importance to the occurrence of horizontal turbulence disturbances.
Therefore, more sophisticated models for the dispersion of emissions are developed. A well-known example is the *Gaussian plume formula* for the prediction of the pollutant concentrations in a uniform flow with homogeneous turbulence (see, e.g., References 5 and 46):

$$c_v(x,y,z) = \frac{Q}{2\pi V_{w,H}\sigma_y\sigma_z}\exp\left[-\frac{y^2}{2\sigma_y^2}-\frac{z^2}{2\sigma_z^2}\right], \tag{4.6-4}$$

where the x-axis points in the direction of the wind velocity (Figure 4.6-5). Q is the emission rate, $V_{w,H}$ is the wind velocity at a height H, and σ_y and σ_z are the lateral and vertical Gaussian plume dispersion parameters. Apparently, in deriving Equation (4.6-4) the assumption is made that the effects of reflection of pollutants impacting the ground surface are ignored.

From the preceding discussion it may be concluded that the dispersion of pollutants depends on the following factors:

- Meteorological and lapse rate conditions.
- Mass flow rate of the emitted pollutants.

- Height at which the emission takes place.
- Ground surface features.
- Type of source (point, line, or diffuse source).

As defined before, the length of time a particle remains in the air, is its residence time. In this respect use is made of the so-called *dissipation speed*, which is defined as the decline of the mass of pollutant substance per unit time, e.g., expressed in kilogram per second.
For a stationary concentration the dissipation speed is equal to the emission speed, i.e., the emitted mass per unit time. Related to the total air mass in the troposphere, the mean residence time of a pollutant follows from the equality:

$$Q\,t = c_m M_t \quad \text{or} \quad t = \frac{c_m M_t}{Q}, \tag{4.6-5}$$

where t is the residence time in seconds, c_m is the concentration in kg/kg, M_t is the total air mass in the troposphere in kg ($M_t = 4\times10^{18}$ kg) and Q is the emission rate in kg/s. The product $c_m\ M_t$ is the total mass of the pollutant substance in the air.
Also, we have the *chemical lifetime* of an atmospheric substance, which is the time interval by the end of which it has been transformed into a different substance. Especially, oxygen is an effective cleanser because it oxidizes many different gas molecules.
Certainly, the chemical lifetime is also a key parameter for the evaluation of the impact of natural and anthropogenic emissions. It determines the length of time

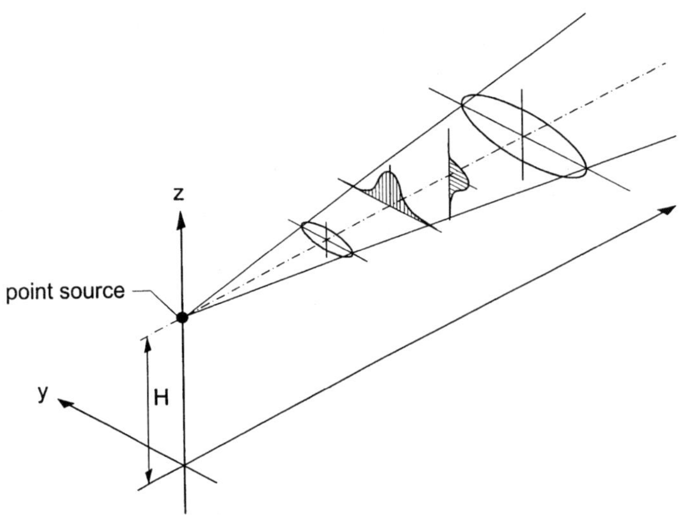

Figure 4.6-5. Gaussian plume model.

available for transport before chemical removal.
The chemical lifetimes of trace gases in the atmosphere may vary from one second to more than hundred years (Ref. 20). Of particular interest is the presence of substances like chlorofluorocarbons (CFCs) and hydrofluorocarbons (HFC's) in the upper part of the stratosphere, which may show extreme long chemical lifetimes of even more than 500 years (Ref. 21).

5 SOLAR ENERGY

5.1 ELECTROMAGNETIC RADIATION

A physical body with a finite temperature constantly emits energy, transmitted by means of electromagnetic waves. Due to the emitted radiant energy the atomic and molecular motions within the body diminish, decreasing its temperature.
However, radiation not necessarily leads to temperature decrease, as in general a body also experiences radiation from its environment. When a body receives as much energy as it emits, it is in thermal equilibrium with its surroundings.
In theoretical physics the term *blackbody* is used to describe the radiation phenomena. A blackbody behaves as a so-called *adiabatic wall*, which is a perfect thermal insulator that allows no flow of heat through the body. In other words, it reflects and/or transmits no electromagnetic radiation, all the radiation energy received by the body is absorbed entirely.
In 1901 the German scientist Max Planck (1858 - 1947) showed that for a perfect black-body with absolute temperature T, the radiation intensity at wavelength λ is given by:

$$I_\lambda = \frac{2\pi h c^2}{\lambda^5 \left[e^{\frac{hc}{\lambda k T}} - 1 \right]}, \tag{5.1-1}$$

where h is *Planck's constant*, c is the speed of light (in vacuum), and k is the *Boltzmann constant*, the ratio of the universal gas constant to Avogadro's number and named after the Austrian physicist Ludwig Boltzmann (1844 - 1906).

The radiation intensity is the energy per unit time passing through a unit area transmitted by electromagnetic waves, i.e., the electromagnetic power per unit area in watt per square meter (W/m^2). The value of h was estimated by Planck by fitting his formula for the blackbody radiation curve to experiment.
The numerical values adopted today are as follows:

- Planck's constant: $h = 6.626196 \times 10^{-34}$ J s,
- Boltzmann's constant: $k = 1.380622 \times 10^{-23}$ J/K,
- speed of light (in vacuum): $c = 2.99792458 \times 10^{8}$ m/s.

Integration of Equation (5.1-1) between $\lambda = 0$ and $\lambda = \infty$ furnishes the *Stefan-Boltzmann law* (found experimentally by the Austrian physicist Joseph Stefan

(1853 - 1893) in 1879 and proved theoretically by Boltzmann in 1884 for the emitted intensity:

$$I = \sigma T^4 \ . \tag{5.1-2}$$

Here σ is a universal constant called the *Stefan-Boltzmann constant*, which is given by:

$$\sigma = \frac{2\pi^5 k^4}{15 h^3 c^2} \ . \tag{5.1-3}$$

Substitution of the constants yields: $\sigma = 5.66961 \times 10^{-8}$ W/m^2 K^4.

A blackbody radiates heat in a wide range of wavelengths. The maximum radiation intensity per unit wavelength for a blackbody with temperature T is found at a particular wavelength λ_m, which results from the condition:

$$\frac{\mathrm{d}I_\lambda}{\mathrm{d}\lambda} = 0 \ . \tag{5.1-4}$$

Applying this operation to Equation (5.1-1) furnishes *Wien's law*, named after Wilhelm Wien, German physicist (1864 - 1928):

$$\lambda_m T = \text{constant}, \tag{5.1-5}$$

where the constant amounts to 2.8978×10^{-3} m K.

The magnitude of the radiation intensity according to Equations (5.1-1) and (5.1-5) is shown diagrammatically in Figure 5.1-1 in terms of wavelength and

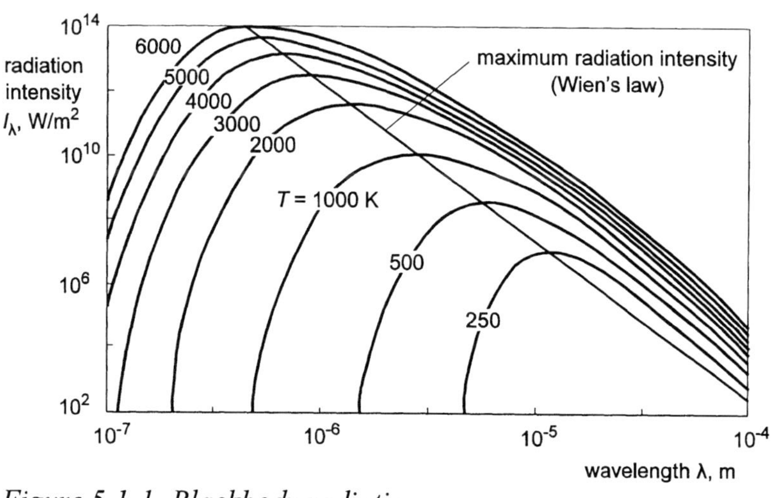

Figure 5.1-1. Blackbody radiation.

temperature. The emitted energy has the following properties:

- The radiation intensity is spread over the region of wavelengths in a specific way, depending on the body temperature.
- The maximum value for the radiation intensity moves with increasing temperature to shorter wavelengths.
- For a particular wavelength, the emitted energy increases with increasing temperature of the body.
- For a specific body temperature a maximum in radiation intensity occurs within the electromagnetic spectrum.

If a body radiates according to the Stefan-Boltzmann law, Equation (5.1-2), it is called a "perfect radiator". In real life, however, such an idealized body does not exist. Then, the emitted energy may be written, using the Stefan-Boltzmann law, as:

$$I = \varepsilon\sigma T^4 \, (0 \le \varepsilon \le 1) \; , \tag{5.1-6}$$

where ε is the *emittance*, the ratio of the actual radiation by the body, to the radiative emission of a perfect blackbody having the same temperature.
Obviously, a perfect blackbody with $\varepsilon = 1$ is an ideal radiator, as for a constant temperature the radiation intensity is always larger than for any arbitrary body. Very black surfaces have an emittance close to one, whereas polished surfaces exhibit a radiation with ε close to zero and hence emit correspondingly less radiant energy.
In general a body will partly reflect, and/or let through the radiation. For an arbitrary body the term *absorptance*, α, is used, indicating its actual absorption in relation to the absorption of a perfect blackbody with the same temperature.
According to *Kirchhoff's law*, established by the German physicist Gustav Kirchhoff (1824 - 1887), at a given temperature and wavelength all substances are able to absorb the same radiation as they emit, i.e.,

$$\alpha_\lambda = \varepsilon_\lambda \; , \tag{5.1-7}$$

where α_λ and ε_λ are the monochromatic absorptance and the monochromatic emittance, respectively.
As, in general, the spectrum of absorbed and emitted radiation substantially may vary due to differences in temperature, α and ε (obtained from integration of α_λ and ε_λ over the entire spectrum of wavelengths) will be different. Moreover, the coefficients α and ε also depend on surface conditions of the body. For a blackbody, by definition, $\alpha = \varepsilon = 1$.

The behavior of gases differs from solid and fluid substances, regarding absorption and radiation. They only absorb particular wavelengths, which change as the temperature changes.

5.2 SOLAR RADIATION

The Earth's climate system, consisting of atmosphere, hydrosphere, biosphere, land and water surfaces, and lithosphere, is powered by the radiation emitted by the Sun. *Solar radiation* is the spatial propagation of electromagnetic waves, carrying energy from one region of space to another. Electromagnetic waves are transverse waves produced by electric charges that are oscillating. They propagate with an electric field strength, E, and a magnetic field strength, B, which oscillate in planes perpendicular to each other and to the direction of motion (Figure 5.2-1).
The features of electromagnetic radiation are analogous to a sound field, brought about by a sound source (Ref. 22). The radiant energy transmitted by the electromagnetic waves per unit time per unit area normal to the solar beam, i.e., the solar power per unit area at a given distance from the Sun, is the solar intensity or *irradiance* in W/m^2. From its definition it readily follows that the irradiance diminishes as the square of the distance from the Sun.
Most of the solar radiation is emitted from the photosphere, which is the visible outer shell of the Sun's atmosphere where the temperature amounts to about 5800 K (see the previous Figure 1.3-6). This temperature is a so-called *effective temperature*, as its value is derived from the Sun's observed radiation, assuming that the Sun radiates as a perfect blackbody for which the radiation laws treated in Section 5.1 are valid.
The oscillations associated with electromagnetic radiation are characterized by wavelength λ and frequency f. The velocity of propagation is equal to the speed of light, c, in a vacuum. As stated earlier in Section 5.1, the accepted value today for this speed is 2.99792458×10^8 m/s. Usually, one rounds this off to:
$c = 3.0 \times 10^8$ m/s.

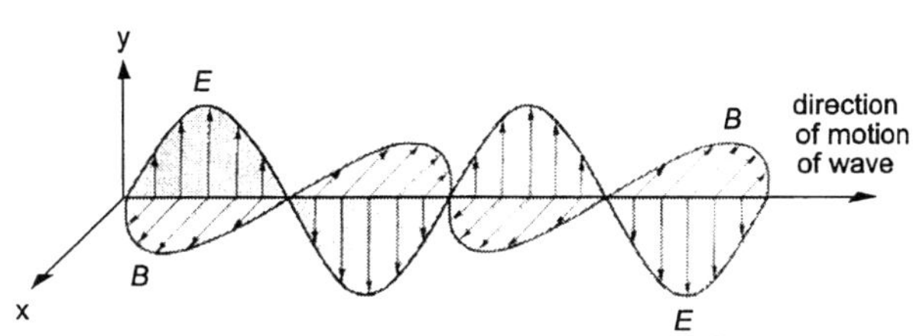

FigureFigure 5.2-1. Representation of an electromagnetic wave.

Wavelength and frequency are related by:

$$\lambda f = c \, . \tag{5.2-1}$$

Evidently, as the travel speed c is a constant, the higher the wavelength, the lower the frequency of vibration.

An explanation of blackbody radiation is given by Planck's quantum hypothesis, in which electromagnetic radiation is represented as a stream of particles called *quanta*, traveling at the speed of light and having only energy of their motion and no mass. The energy E_p of a quantum is proportional to the frequency of the radiation:

$$E_p = hf = h\frac{c}{\lambda} \, , \tag{5.2-2}$$

where h is Planck's constant ($h = 6.626196 \times 10^{-34}$ J s).
The quantum of energy given by Equation (5.2-2) is the smallest amount possible. Clearly, the shorter the wavelength of the radiation, the greater the energy a quantum has.

According to Planck's theory, the energy of any molecular vibration is equal to some whole number multiple of the quantum of energy. Continuing with this suspension of common sense, a quantum must also be a wave. The wave extends over its whole travel distance, but most of the wave characteristics are

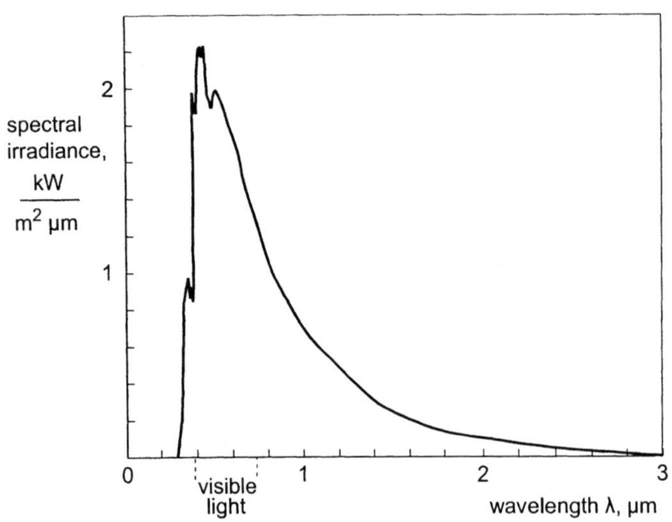

Figure 5.2-2. Spectral distribution of solar irradiance just outside the Earth's atmosphere.

confined to a small region, and it is this confinement that gives quanta some of the properties of radiation energy. Note that in Equation (5.2-2) the subscript 'p' is used, denoting that the quantum of energy may also be termed a *photon*.
Figure 5.2-2 shows the distribution of the solar radiation (in W/cm^2 μm) over the different wavelengths λ just outside the Earth's atmosphere. The range of electromagnetic wavelengths is called the *solar spectrum*.
Since the Earth's orbit around the Sun is slightly elliptical, a variation in the solar irradiance occurs. In addition, the Sun exhibits periods of increased solar activity observed in numbers of sunspots, with greater radiative output, magnetic activity, and emission of high energy particles.
The annual average solar irradiance just outside the Earth's atmosphere is called the *solar constant, S*. This amount of intensity is given by the total area under the spectrum of solar irradiance in Figure 5.2-2. From measurements with artificial satellites it is known that the average value of the actual solar constant may be taken as 1360 W/m^2.

If, as depicted in Figure 5.2-3a, the approximation is used that the Earth intercepts solar radiation almost exactly as a disc with radius equal to the Earth's mean radius R_e, the power of the energy flux reaching the outer surface of the Earth's atmosphere is given by:

$$W = \pi R_e^2 S \ . \tag{5.2-3}$$

Clearly, this expression is based on the assumption of nearly parallel solar radiation.
Averaged over the Earth's surface, the mean solar intensity impinging on the

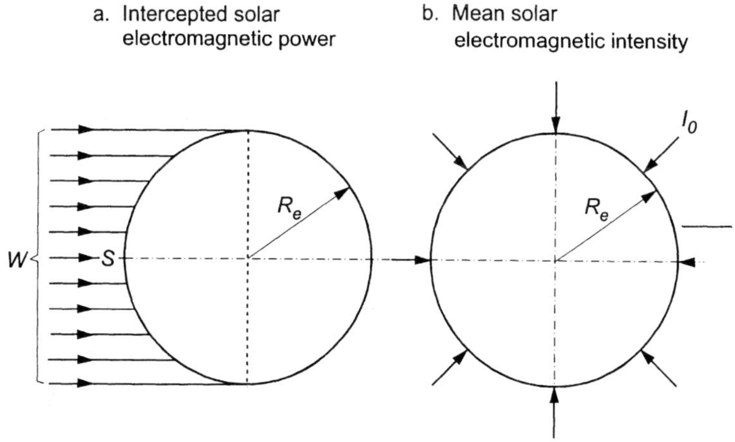

Figure 5.2-3. Solar radiation transmitted to the Earth.

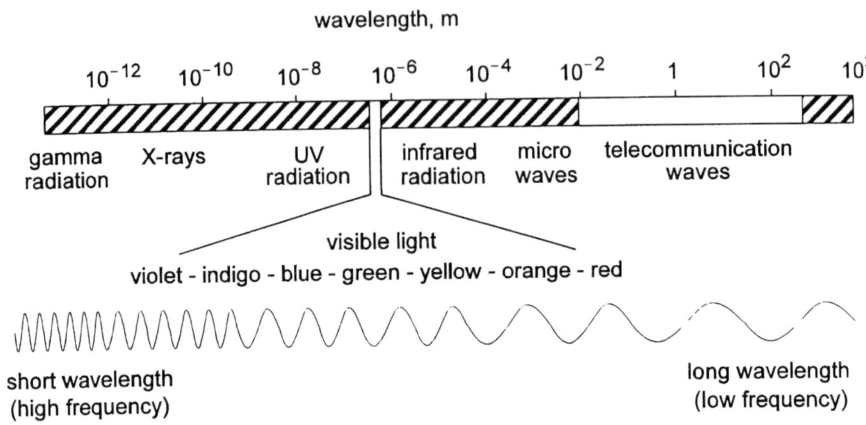

Figure 5.2-4. The electromagnetic spectrum.

Earth, I_0, amounts to (Figure 5.2-3b):

$$I_0 = \frac{W}{4\pi R_e^2} = \frac{S}{4} . \qquad (5.2\text{-}4)$$

Inserting $S = 1360$ W/m^2 into Equation (5.2-4), we find $I_0 = 340$ W/m^2.

The amount of radiation given by Equation (5.2-3) is the power that would reach the surface of the Earth when there were no reflections, deflections, and absorptions on its way through the atmosphere. However, about 30% of the intercepted power W is reflected from the atmosphere to outer space, and the remainder 70% reaches the Earth's surface; see the discussion on the radiation balance in Section 5.4.

The categories into which the electromagnetic spectrum is divided, correspond to different ranges of wavelengths, and are named after their properties or sometimes after their application. Figure 5.2-4 indicates that the wavelengths of the solar radiation vary from 10^{-14} to 10^3 m. With increasing wavelengths we successively find: gamma rays, *X*-rays, ultraviolet radiation (UV), visible light, infrared or heat radiation, microwaves, and telecommunication or radio waves. Ultraviolet radiation is commonly split into three bands: UVA (320-400 nm), UVB (280-320 nm), and UVC (100-280 nm).

phenomenon/application	name	wavelength, m
ionizing radiation	gamma	10^{-14} - 10^{-10}
	X-rays	10^{-11} - 10^{-7}
ultraviolet radiation	UVC	< 2.8×10^{-7}
	UVB	2.8×10^{-7} - 3.2×10^{-7}
	UVA	3.2×10^{-7} - 4.0×10^{-7}
visible light	violet	3.8×10^{-7} - 4.5×10^{-7}
	indigo	4.5×10^{-7} - 5.0×10^{-7}
	blue	5.0×10^{-7} - 5.2×10^{-7}
	green	5.2×10^{-7} - 5.5×10^{-7}
	yellow green	5.5×10^{-7} - 5.7×10^{-7}
	yellow	5.7×10^{-7} - 6.0×10^{-7}
	orange	6.0×10^{-7} - 6.3×10^{-7}
	red	6.3×10^{-7} - 7.8×10^{-7}
heat	infrared	7.8×10^{-7} - 10^{-3}
communication	microwaves	10^{-3} - 2×10^{-2}
	radio waves	10^{-2} - 10^{2}
	low frequency	10^{3} - 10^{5}

Table 5.2-1. The solar spectrum.

As can be seen from the Figures 5.2-2 and 5.2-4, the part of the solar spectrum of which we are aware - visible light, centered on a wavelength of about 5×10^{-7} m - is quite narrow, but it is also the region in which the Sun radiates most intensely. To explain this, we may refer to Wien's law, Equation (5.1-5), which indicates that the wavelength for maximum radiation intensity is inversely proportional to the temperature of the body. In other words, very hot bodies emit their radiant energy at short wavelengths. Using the datum that the effective temperature in the Sun's photosphere amounts to about 5800 K, a comparison of the solar spectrum in Figure 5.2-2 with the emission spectra of a blackbody in the previous Figure 5.1-1, reveals that the Sun may be regarded as an almost perfect blackbody at the temperature of the photosphere.

A detailed division of the complete spectrum into wavelengths is listed in Table 5.2-1.

The most energetic end of the solar spectrum consists of *gamma radiation* at wavelengths between about 10^{-14} and 10^{-10} m. This is the radiation emitted by events that occur inside atomic nuclei. When any atom is bombarded by

electrons, it emits *X*-rays (10^{-11} m < λ < 10^{-7} m). Wavelengths of about 10^{-7} to 4.0×10^{-7} m comprise the ultraviolet (UV) part of the spectrum.
The wavelength of the visible part of solar radiation ranges from about 4.0×10^{-7} to 8.0×10^{-7} m. Within this range the human eyesight perceives wavelengths in the form of colors.

Besides UV and visible light, the Sun emits a substantial amount of infrared radiation as well. Infrared radiation waves (7.8×10^{-7} m < λ < 10^{-3} m), whose shortest wavelength just exceeds that of visible light, is mainly responsible for the heating effect by the Sun. As a reaction on infrared frequencies the molecules of our skin become into motion. Apparently, these frequencies are preferentially absorbed and thus warm us up.
Contrary to higher frequency waves, lower frequency waves such as radio waves, are not very difficult to produce electronically. According to the application of radio waves, these are categorized as follows:

- Radar ('Radio Detection and Ranging') waves in the centimeter-range.
- Television and FM (Frequency Modulation) radio waves in the meter-range.
- World radio broadcasting waves in the decameter-range.
- AM (Amplitude Modulation) radio waves in the hectometer-range.
- Long range navigation aids in the kilometer-range.
- Global navigation systems in the range 10 km - 100 km.

5.3 REFLECTION AND REFRACTION OF SUNLIGHT

Light advances straight-lined in a vacuum and in an isotropic medium. This observation has led to the *ray model* of light. The straight-line paths that light follows are called *light rays*, which in fact represent an infinitely narrow beam (wave front) of electromagnetic waves.

Another feature of light is that it is reflected and refracted when it passes from one medium to another. These features are described by the laws of reflection and refraction, which are based on the *Principle of Huygens*. This Principle, which was developed by the Dutch scientist Cristian Huygens (1629-1695),

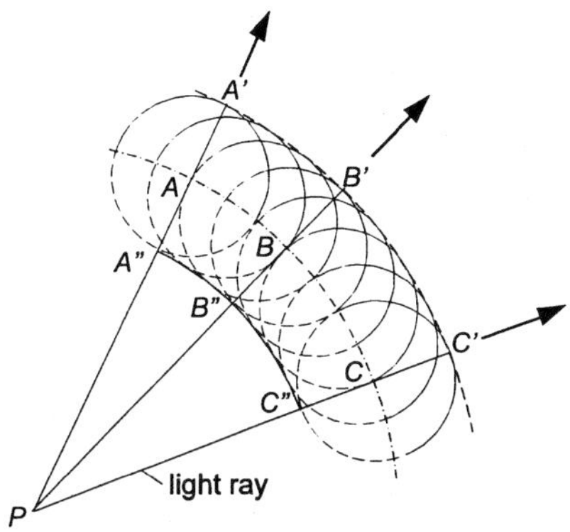

Figure 5.3-1. Huygens' principle.

concerns a technique for predicting the future position of a wave front when an earlier position is known. It can be stated as follows:

Every vibrating point on a wavefront can be considered as a point source of secondary wavelets that spread out at the speed of the original wave; the new wavefront at any time is the envelope of these wavelets.

In Figure 5.3-1, *ABC* is the position of a wavefront at time t and *A'B'C'* is the position at time $t + \Delta t$. The normals to the wavefronts (*AA', BB',CC'*, etc.) may be interpreted as light rays emerging from the point *P*. In this model for the propagation of light there is no wavefront moving backward (position *A"B"C"*), but only an envelope of wavelets in direction of propagation of the light rays is possible.

To derive the *law of reflection*, in Figure 5.3-2 is considered a small (parallel)

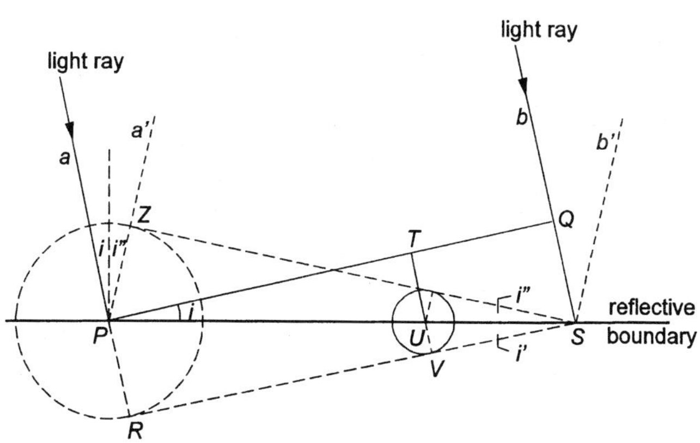

Figure 5.3-2. Law of reflection.

beam of light, confined by the rays a and b, incidenting at an angle i to the surface of a reflecting plane. A cross-section perpendicular to the secant of wavefront PQ and the surface, just when it impinges is contemplated.
When a light ray a arrives in the point P at the surface, light ray b still has to travel distance QS to reach the surface (interface), which takes time $\Delta t = QS/c$. After this interval Δt, the light ray a has reached the point R ($PR = QS$).
However, point P may be considered a point source from which secondary wavelets start to spread, forming a new wavefront with radius PR at time Δt.
Another point T on the wavefront PQ will first travel distance TU, and subsequently will form a secondary wavelet with radius UV, during the time interval which would be needed to cover distance UV. The envelope of all these wavelets forms a new wavefront ZS, which is the opposite image of RS with respect to the interface. The outer (specular reflected) light rays a' and b' are formed by lines perpendicular to the new wavefront ZS. From geometry it is clear that $i = i' = i''$. Hence, the angle of incidence equals the angle of reflection. Moreover, the incident rays and the reflected rays (e.g., a and a') lie in the same plane with the normal to the surface.
A part of the solar radiation is reflected by clouds and dust particles in the atmosphere. Surfaces of water and snow may reflect light thereby increasing its local intensity considerably.

On a beam of light passing from one medium into another, the same principle and outlining may be applied. Suppose the speed of light in medium 1 and 2 is respectively c_1 and c_2 ($c_1 > c_2$), and both media are isotropic (Figure 5.3-3). During time interval Δt required to cover distance QS in medium 1, the wavelet arising from P will expand to form a wavelet with radius $PK = PZ = QS\, c_2 / c_1$

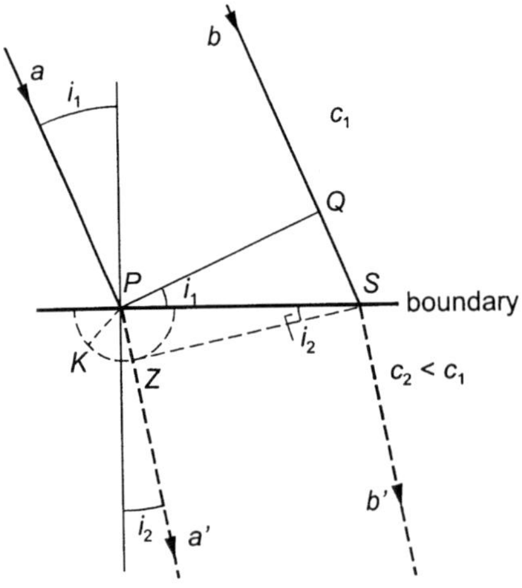

Figure 5.3-3. Law of refraction.

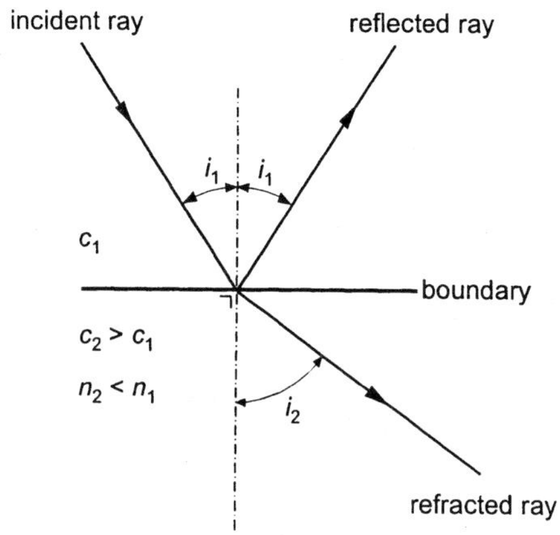

Figure 5.3-4. Reflection and refraction.

in medium 2. The envelope of the secondary wavelets forms the new wavefront *ZS*. Light rays *a'* and *b'* are perpendicular to *ZS*. From the geometry the following relation between i_1 and i_2 follows:

$$\sin i_2 = \frac{c_2}{c_1} \sin i_1 \ . \tag{5.3-1}$$

Apparently, the rays are bent as they enter the new medium. The bending of light in passing through optical media and through the atmosphere is called *refraction.*

The angle of incidence i_1 and the angle of refraction i_2 are found in Figure 5.3-3 between the normal to the surface and the light ray. The incident ray and the reflected ray are on each others opposite side of the normal, all three lie in the same plane. Notice that the rays bend toward the normal when entering a medium with $c_2 < c_1$.

The ratio of the speed of light in vacuum, $c = 3.0\times10^8$ m/s, and the speed of light in a particular medium, c^* $(c^* \le c)$, is called the *index of refraction*, *n*, of the medium:

$$n = \frac{c}{c^*} \ . \tag{5.3-2}$$

In air the speed of light is only slightly less than in vacuum. In other transparent media the speed of light is always considerably less.

The refraction may be expressed through the following relationship:

$$n_1 \sin i_1 = n_2 \sin i_2 \ . \tag{5.3-3}$$

Equation (5.3-3) is known as *Snell's law of refraction*, named after its discoverer the Dutch mathematician Willebrord Snellius (1580-1626). In 1621 he established this relationship between angle of incidence, angle of refraction and the speed of light in the two media. For a particular material, light is refracted in a characteristic way, expressed by the index of refraction n.
A larger index of refraction means that light rays are refracted more strongly.
As depicted in Figure 5.3-4, when a light ray passes obliquely from a denser to a less dense medium ($n_2 < n_1$), it is bent from the normal as according to Equation (5.3-3): $i_2 > i_1$. When the angle of incidence i_1 is increased, a certain critical angle is reached for which the angle of refraction $i_2 = 90°$. For this angle $(i_1)_{max}$ follows:

$$(\sin i_1)_{max} = \frac{n_2}{n_1} \; . \qquad (5.3\text{-}4)$$

For angles larger than $(i_1)_{max}$ no refraction occurs, the light then is reflected completely.
As mentioned earlier, the color of the light is related to the wavelength of the radiation. Visible light lies in the wavelength range of about 4.0×10^{-7} m to 7.4×10^{-7} m, comprising the different colors from violet to red (see Table 5.2-1).
It is a well-known fact that when a beam of white light impinges on a prism, it is refracted and divided into a number of beams, having the colors of the visible spectrum (Figure 5.3-5). The breaking up of white light into the complete spectrum is termed *dispersion*, and is caused by the fact that the index of refraction of a material, $n = c/c^*$, is determined by the wavelength of the light. Since the index of refraction is greater for the shorter wavelengths, violet light is turned off the most and red light the least. For example, violet light is refracted more strongly than red light, as $n_{violet} > n_{red}$ ($c_{violet} < c_{red}$), as shown in Figure 5.3-5.
Rainbows are well-known examples of dispersion of light by water droplets.
Bending of light also occurs in the atmosphere when it passes through layers

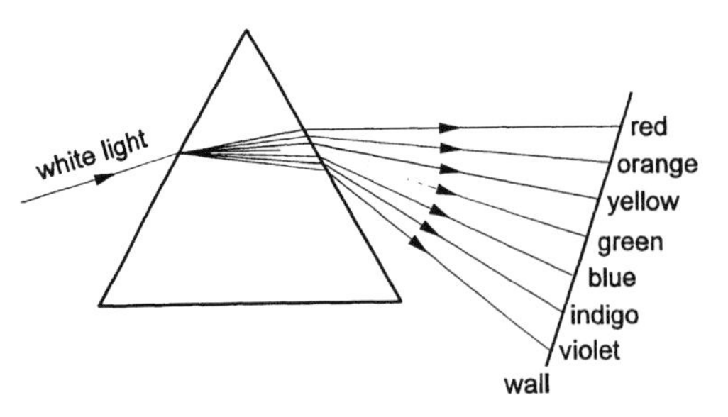

A prism is a wedge-shaped glass solid with three or more plane faces. Prisms are used in spectrometers for the analysis of light.

Figure 5.3-5. Dispersion of white light by a prism.

with varying density.
It should be noted that when a light ray is incident upon a rough surface, it is reflected in many directions (Figure 5.3-6a). This is called *diffuse reflection.* Equation (5.3-3) still holds, however, at each small section of the surface. Because of diffuse reflection in all directions, an object can be seen from many different angles, contrary to the situation that a narrow beam of light is shining on a mirror. Then the light will not reach your eye unless it is placed at just the right place where the law of reflection is satisfied (Figure 5.3-6b).
In this respect, the phenomenon *scattering of solar radiation* must be recognized. Scattering of light by the Earth's atmosphere occurs when light is diffused by molecules of the atmosphere or particulate matter in the atmosphere, in no particular direction. The shorter the wavelengths and the smaller the scattering objects are compared to the wavelengths of the light, scattering takes place to a larger extent. This implies that violet and blue light is affected more strongly than orange and red light. Therefore, solar radiation tends to become more red on its way to Earth.

During sunset, this phenomenon is highly perceptible. At first the Sun is yellow, but while approaching the horizon, it becomes more red, as the solar rays have to travel a longer distance through the atmosphere.
The blue color of the sky is due to the fact that the scattered blue light is diffused even further, so it reaches the Earth's surface partly from all directions. Clouds contain water droplets of which the dimensions are much larger than the wavelengths of the light, through which all frequencies of light are scattered nearly uniformly. For this reason, clouds are colored gray/white.

Finally, mention must be made that substances and materials may show *transparency*, which property allows light to pass through an object to a certain extent. The light is strongly scattered and therefore objects may not be very

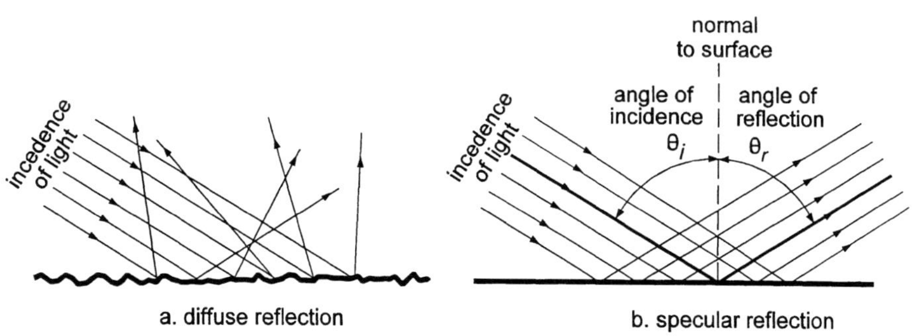

Figure 5.3-6. Types of reflection.

visible on their rear sides. Clearly, the transparence of a particular substance depends on its thickness and on its state (vapor, liquid or solid).
If white light illuminates an object that absorbs all wavelengths except a particular one, the object will become the color of that particular wavelength. Therefore, the color of an object depends on its absorption of the various wavelengths. For a particular substance an absorption spectrum may be determined. This is a graph of the absorbed wavelengths (colors) of light.
When an object is non transparent, for instance metal, it is called *opaque*. Such an object reflects all the wavelengths of the light impinging on it. Shiny metal objects virtually reflect all light, and are therefore bright. Conversely, Indian ink reflects almost no light, and therefore appears black.

5.4 RADIATION BALANCE OF EARTH AND ATMOSPHERE

The Earth receives its energy through solar radiation. Due to thermal equilibrium, i.e., the balance between the absorbed solar radiation and the emitted terrestrial radiation, there is a global equilibrium temperature on Earth.
As shown earlier in Figure 5.2-2, the solar energy supplied to the atmosphere consists mainly of short-wave radiation.
In principle, the solar constant may be determined by application of Stefan-Boltzmann's law (Equations (5.1-2) and (5.1-3)):

$$S = I_s \frac{4\pi R_s^2}{4\pi R_z^2} = \sigma T_s^4 \left[\frac{R_s}{R_z} \right]^2 , \tag{5.4-1}$$

where $I_s = \sigma\ T_s^{\ 4}$ is the *photospheric irradiance*, and the denotation of the constants is:

σ	is the Stefan-Boltzmann constant:	$\sigma = 5.67\times10^{-8}$ W/m^2 K^4,
T_s	is the effective temperature of the Sun:	$T_s = 5800$ K,
R_s	is the radius of the Sun:	$R_s = 6.96\times10^8$ m,
R_z	is the mean distance Earth to Sun:	$R_z = 1.496\times10^{11}$ m.

Substitution of all constants and radii yields S = 1390 W/m^2. As mentioned earlier in Section 5.2, the measured average value of solar constant may be

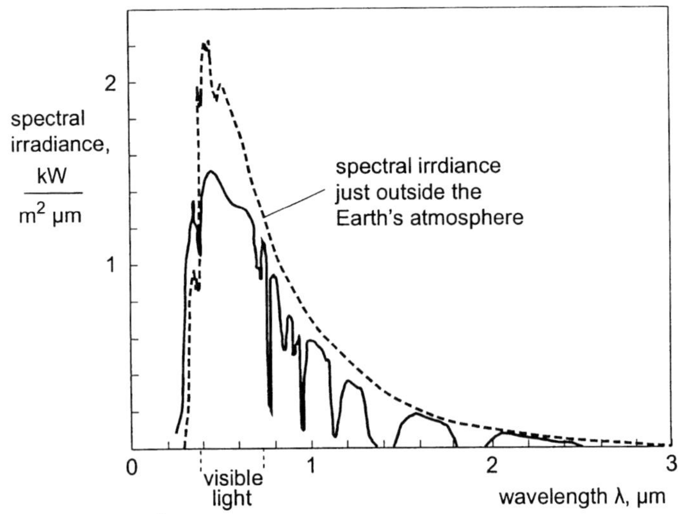

Figure 5.4-1. Spectral distribution of solar irradiance at the Earth's surface for clear atmosphere.

taken as 1360 W/m^2. Averaged over the surface of the Earth, the mean annual solar intensity amounts to $I_0 = S/4 = 340$ W/m^2 (Equation (5.2-4)).

Figure 5.4-1 shows the spectral distribution of the solar irradiance at the Earth's surface with the Sun in the zenith of a clear sky. For comparison, the incoming spectral radiation just outside the Earth's atmosphere given in the previous Figure 5.2-1, is also included in Figure 5.4-1.
The two spectra in Figure 5.4-1 show that a standard cloudless atmosphere affects the radiation transmitted to the Earth's surface quite significantly. Of course, the presence of clouds will further reduce the received solar energy.
Obviously, the power W of the energy flux reaching the outer surface of the Earth's atmosphere given by Equation (5.2-3), must be re-radiated by the Earth and atmosphere into outer space.
An outline of the radiation balance of Earth and atmosphere is given in Figure 5.4-2, where 100% represents the total energy received from the Sun. The figures used, must be considered as orientation estimates only.

Of the perceived solar radiation at the top of the atmosphere, about 25% is reflected directly back into space. Another 5% is reflected from the ground surface. These figures indicate that the Earth as a whole reflects about 30 percent of the short-wave solar radiation falling upon it back toward space. Thus, after traveling through the atmosphere, about 70% of the incoming solar radiation is absorbed by the Earth surface, of which 45% is received as short-wave radiation and 25% in the form of long-wave radiation.

For thermal equilibrium, the above implies that also 70% has to leave the atmosphere. This indeed occurs, since all objects with a finite temperature emit electromagnetic radiation. At the prevalent temperature of the Earth's surface this involves mainly long-wave (infrared) radiation, consistent with blackbody emission at a temperature of about 288 K (15°C). This value equals the global equilibrium temperature on Earth, averaged over its whole surface and a full year. An estimate of 115% of the incoming energy is emitted by the ground consisting of 5% short-wave and 110% long-wave radiation, The magnitude of the incoming and outgoing energy flux at the Earth's surface is governed by a diffuse infrared radiation, directed downwards, arising from atmospheric thermal processes. As described earlier in Chapter 1, this is known as the *greenhouse* or *blanketing effect*, through which an extra amount of about 40% long-wave infrared radiation energy is reflected backwards to the Earth.

Based on the figures used in Figure 5.4-2, the balance between incoming and outgoing energy requires that the radiation intensity emitted by the Earth, I_e, equals $1.15 \times 340 = 391\ \mathrm{W/m^2}$. This furnishes, applying Stefan-Boltzmann's law (Equations (5.1-2) and (5.1-3)), the following sea-level temperature:

$$T = \left(\frac{I_e}{\sigma}\right)^{\frac{1}{4}} = \left(\frac{391}{5.67 \times 10^{-8}}\right)^{\frac{1}{4}} = 288\ \mathrm{K}\ .$$

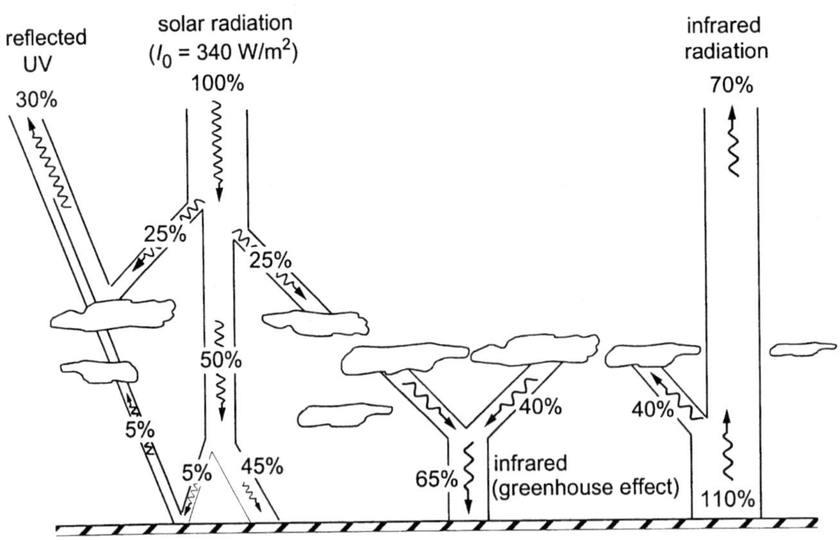

Figure 5.4-2. Radiation balance of Earth and atmosphere.

It may be noted that, technically, the reflectiveness of a body is known as its *albedo*, α, normally expressed as a decimal fraction ($0 \leq \alpha \leq 1$). Hence, the planetary albedo amounts to about $\alpha = 0.3$. For an entirely white body the albedo is equal to one. The albedo of snow is 0.52, for clouds 0.36 - 0.78, and for forests and oceans 0.04. Clearly, the presence or absence of clouds will have a profound effect on the temperature and visibility at the Earth's surface. In extreme cases, as we all know, big cumulonimbus clouds block almost all the solar radiation, turning daylight into darkness, and also a layer of snow in winter has a strong influence on the proportion of solar radiation that is reflected. Consequently, the composing albedos will change from place to place and from time to time during the year.

As stated above, the Earth emits radiation energy as if it were a blackbody with a surface temperature of about 288 K (15°C).
This temperature differs from the temperature that the Earth would have if the intensity absorbed by the atmosphere equals the intensity radiated by the Earth into outer space. Then we have:

$$I_e = I_0 (1 - \alpha) = 340 (1 - 0.3) = 238 \text{ W/m}^2 .$$

From this equality and Equations (5.1-2) and (5.1-3) the latter temperature is given by:

$$T = \left[\frac{238}{5.67 \times 10^{-8}} \right]^{0.25} = 255 \text{ K} .$$

This result predicts a temperature decrease of more than 30°C compared to the actual temperature. Obviously, without the occurrence of the natural greenhouse effect, the mean surface temperature of the Earth would be appreciable lower, making many life-forms impossible (see also Chapter 6, Section 6.4).

By the Intergovernmental Panel on Climate Change (IPCC), two global warming describing parameters have been developed, which are closely related to each another, namely:

- the Radiative Forcing of a pollutant,
- the Global Warming Potential of a pollutant.

These parameters will be explained briefly in the Sections 5.5 and 5.6.

5.5 THE CONCEPT OF RADIATIVE FORCING

As described in Section 5.4, on a global average, the Earth maintains a radiative balance between incoming solar radiation minus outgoing reflected solar radiation and outgoing infrared radiation emitted by the Earth-atmosphere system.
When a particular human activity increases the greenhouse gases in the atmosphere or affects the albedos for the various surfaces covering the Earth, the resulting radiative imbalance is termed *Radiative Forcing*, *RF*.
Such an imbalance will not sustain for long, and the climate system - primarily temperature and clouds of the lower atmosphere - regulates to restore the balance between incoming and outgoing energy.
To describe the climate impacts of human activities, the parameter Radiative Forcing is employed by the IPCC as a means of estimating the externally imposed perturbation in the radiative energy budget of the Earth.
The index *RF* is computed from the changes in greenhouse gases, albedos, and cloudiness, and is expressed in terms of radiation in W/m^2. A positive Radiative Forcing leads to global warming.
According to IPCC, Radiative Forcing can be defined as follows:

"*The global annual average of a radiative imbalance of the atmosphere-land-ocean system caused by anthropogenic perturbations*".

The concept of Radiative Forcing is based on climate model calculations that show that there is an approximately linear relationship between global-mean *RF* at the tropopause and the change in the equilibrium global mean sea-level temperature, $T = 288$ K, derived in Section 5.4.

Consequently, the *RF* providing the best metric of climate change is the radiative imbalance of the land-ocean-troposphere climate system, that is, the *RF* determined at the tropopause.

Increases in greenhouse gas concentrations during the last decades have led to a positive Radiative Forcing, tending to warm the lower atmosphere in order to increase the terrestrial radiation and restore radiative balance. As remarked earlier, an increased global mean surface temperature will lead to other changes of the climate, like altered cloud distributions, storms, extreme rainfall events, danger of flooding, and prolonged periods of dryness.

Obviously, an increase of the global mean sea-level temperature due to a positive *RF* will not be distributed uniformly across the world. Some regions warm or cool more than others, so that the global mean surface temperature does not provide information about local aspects of climate change.

For a detailed consideration of the calculation models for Radiative Forcing, the interested reader is referred to References 1 and 24 to 27.

In Ref. 1 the index *RF* is used to provide an indication of the contribution of each aviation-induced pollutant to the global warming process and to make a comparison between a number of future aviation-scenarios.
With respect to the contribution from aviation, certainly, the Radiative Forcing driven by emissions of CO_2, H_2O, tropospheric O_3, and contrails, occurs predominantly at mid-latitude regions on the northern hemisphere and is therefore not globally homogeneously distributed. Moreover, the aviation-induced atmospheric perturbations take place primarily in the upper troposphere and lower stratosphere, and thus may affect the vertical temperature profile of the troposphere. Nevertheless, IPCC takes the summed *RF* from aviation as a first-order measure of the global mean climate change.

It is important to recognize that Radiative Forcing does not include the time period that a pollutant is affecting the climate. To improve this problem, also the parameter Global Warming Potential is used. This index will be discussed in the next section.

5.6 THE CONCEPT OF GLOBAL WARMING POTENTIAL

Global Warming Potential, GWP, is an index that attempts to integrate the overall climate impacts of well-mixed greenhouse gases.

The concept of *GWP* has been defined as the ratio of the time-integrated radiative forcing from the instantaneous release of a unit mass of a given greenhouse gas in the today's atmosphere, relative to that of 1 kg of the reference gas carbon dioxide. The duration of the perturbation is included by integrating the effect of a global, annual, mean radiative imbalance to the Earth's climate.
Hence, *GWP* may be used to compare the ability of a given greenhouse gas to trap heat in the atmosphere, taking into account the lifetime of each pollutant,

i.e., the period of time that a pollutant is effective in the atmosphere. For example, a lifetime of 100 years means that pollutants emitted today will affect our climate for the next 100 years.

Global Warming Potential has been developed in order to provide the possibility to weigh the relative importance of the various contributions to the greenhouse effect (Ref. 24).
According to the above definition that the *GWP* of any substance is the integrated radiative forcing of 1 kg of that gas relative to the integrated forcing of 1 kg of carbon dioxide over the same time horizon, the following equation holds:

$$GWP_i = \frac{\int_0^T a_i c_i(t)\,\mathrm{d}t}{\int_0^T a_r c_r(t)\,\mathrm{d}t}, \qquad (5.6\text{-}1)$$

where T is the time horizon over which the integration is performed, a_i is specific radiative forcing due to the release of a unit mass of the gas in question in W/ m^2 kg, $c_i\,(t)$ is the concentration of gas i at time t. The numerator of the above equation is the absolute *GWP* of the substance considered, referred to as the *AGWP*. The corresponding quantities for the reference gas CO_2 are in the denominator. Some examples of *GWP* values will be given in Chapter 6.

To conclude this section, we may note that local climate variations can also occur in the absence of a change in radiative forcing as a result of natural interactions between components of the climate system such as between the oceans' surface and the atmosphere above it. A striking example of such a natural climatological situation is the 'El Niño-Southern Oscillation' (ENSO) phenomenon, an enormous amount of warm water flowing periodically from west to east along the coast of South America. This oceanic event is associated with a fluctuation of the inter-tropical surface pressure pattern and circulation in the Indian and Pacific oceans (see also Section 4.5).

5.7 GREENHOUSE GASES

A greenhouse gas is a gas that absorbs radiation energy of specific wavelengths within the infrared spectrum emitted by the Earth's surface and clouds. The gas

in turn emits infrared radiation in all directions, including downwards to the Earth's surface. The resulting effect is a warming of the planetary surface.
In order to explain the ability of a greenhouse gas to 'trap heat', we consider the internal energy, E, of an ideal gas (see Chapter 7 for some notes on thermodynamics).
For an ideal monatomic gas, this energy is equal to the number of molecules times the average translational kinetic energy per molecule:

$$E_{\text{monatomic}} = N_A \, \tfrac{1}{2} m \, \overline{v}^2 \quad \text{per kmol}, \tag{5.7-1}$$

where N_A is the Avogadro Number, the number of molecules in one mole of a substance, $N_A = 6.02257 \times 10^{26}$, and $\bar{v}$ the mean air-particle speed (see Section 3.3).

According to *kinetic theory* the energy $E_{\text{monatomic}}$ can be expressed as:

$$E_{\text{monatomic}} = \frac{3}{2} R_a T = \frac{3}{2} N_A k T \,, \quad \text{J/kmol}, \tag{5.7-2}$$

where $R_a = 8314.32$ J/K kmol, is the universal gas constant and $k = R_a / N_A = 1.380622 \times 10^{-23}$ J/K is the Boltzmann constant.
Equation (5.7-2) indicates that the average translational kinetic energy of molecules is directly proportional to the absolute temperature of the gas.

Water vapor H_2O, carbon dioxide CO_2, nitrous oxide N_2O, methane CH_4, and ozone O_3, the primary greenhouse gases in the Earth's atmosphere contain three or more atoms per molecule. The internal energy of these more complex molecules is composed of the *translational kinetic energy*, the energy associated with particular internal modes such as molecular *rotation* and *vibration*, and *electronic excitation* (Ref. 19). This means that at a given temperature the internal energy will be greater than for a monatomic gas. Depending on the temperature of the gas, their molecules can absorb and emit the following additional forms of energy:

a. Rotational energy (energy from rotational motion).
b. Vibrational energy (energy from vibrational motions).
c. Electronic energy (energy from the excitation of electrons).

The rotational degrees of freedom of the atoms are already fully excited at temperatures above a few degrees kelvin. At temperatures of the order of 1000 K, the vibrational contribution to the internal energy may become important.

The vibrations are oscillations of the atoms about the lines joining their centers. Electronic energy is emitted when electrons jump from one stationary state in an orbit around the nucleus of the atom to another orbit of lower energy. When the temperature approaches zero kelvin ($T \rightarrow 0$), the electronic energy is in the lowest energy state or *ground state*.
For many applications, however, the electronic energy above the ground state can be neglected because its contribution is fairly small.

In any event, each of these energies can be associated with different wavelengths of the electromagnetic spectrum, according to the formula for the radiation energy from an object (Equation 5.2-2):

$$E_p = hf = h\frac{c}{\lambda}, \qquad (5.2\text{-}2)$$

where E_p is quantum of energy or photon in joule, h is Planck's constant ($h = 6.626196 \times 10^{-34}$ J s), f is frequency in Hertz, λ is wavelength in meter, and c is speed of light in vacuum ($c \approx 3.0 \times 10^8$ m/s).

The specific frequencies or wavelengths for the different energy forms are given below.
Rotational energy in the far-infrared and microwave region:

$$10^{11}\ \text{Hz} < f < 10^{12}\ \text{Hz} \ \text{ or } \ 3 \times 10^{-3}\ \text{m} > \lambda > 3 \times 10^{-4}\ \text{m}.$$

Vibrational energy in the near-infrared and infrared region:

$$10^{13}\ \text{Hz} < f < 10^{14}\ \text{Hz} \ \text{ or } \ 3 \times 10^{-5}\ \text{m} > \lambda > 3 \times 10^{-6}\ \text{m}.$$

Electronic energy in the visible and ultraviolet part:

$$10^{15}\ \text{Hz} < f < 10^{16}\ \text{Hz} \ \text{ or } \ 3 \times 10^{-7}\ \text{m} > \lambda > 3 \times 10^{-8}\ \text{m}.$$

These frequencies or wavelengths depend on the chemical background of the gas, i.e., on the type and bond of the atoms in the molecules. What makes a gas a greenhouse gas, is the fact that greenhouse gases have internal modes that absorb energy in the same infrared wavelengths as emitted by the surface of the Earth and by doing so, reflect heat. Only the gases that have this property are greenhouse gases.

5.8 THE OZONE LAYER

Ozone, O_3, is an unstable gas, composed of three oxygen atoms. It is particularly formed photochemically in the stratosphere and mesosphere, under the influence of solar ultraviolet radiation. Through this short-wave radiation, the oxygen molecules disintegrate into oxygen atoms, which also are called *oxygen radicals*. Subsequently, these radicals react with oxygen molecules, to form gaseous ozone. At the same time, a reversed process takes place, disintegrating ozone molecules into oxygen molecules and oxygen radicals. The formation and destruction of ozone are continuous processes, resulting in more or less a constant amount of ozone in the atmosphere.
Ozone is particularly formed at high altitudes between 10 and 75 km, which region contains as much as about 90 percent of the total amount of the ozone in the atmosphere. At mid-latitudes, the maximum ozone concentration is found at an altitude of approximately 30 km. This altitude decreases to about 25 km at higher latitudes. Concerning the global ozone distribution, on the whole, there is a clear minimum above the equatorial region. Toward the Poles, ozone concentrations increase to reach maxima between latitudes of 60° and the Poles. For this natural distribution the Antarctic ozone abundance is lower than the Arctic ozone abundance. Furthermore, the values of the ozone concentration depend strongly on the time of the year (Ref. 23).
The ozone layer is responsible for the creation of the temperature inversion in the stratosphere due to the absorption of ultraviolet radiation, and completely filters out the extremely dangerous UVC (100-280 nm). Also excessive levels of the Sun's UVB radiation (280-320 nm) are absorbed by ozone and molecular oxygen. Due to the absorption of energy in the stratosphere by ozone and normal oxygen, only UVA (320-400 nm) and a minor part of UVB radiation reaches the Earth surface.
Consequently, because of its absorption of harmful short-wave radiation, the ozone layer acts as a protective shield around the Earth's surface.

The fundamental theories for the prediction of ozone concentrations are the undermentioned *Chapman reactions*, as suggested already in 1930 by the British geophysicist Sydney Chapman (1888 - 1970).
From photo dissociation of molecular oxygen into atomic oxygen required for the production of the ozone, the following reaction takes place:

$$O_2 + hf(\lambda < 250 \text{ nm}) \rightarrow O + O\,, \qquad (5.8\text{-}1)$$

where $hf\,(\lambda < 250\ \text{nm})$ indicates that the wavelength of the radiation energy is shorter than 2.5×10^{-7} m.
Subsequently, the oxygen radicals react with oxygen molecules, to form gaseous ozone, via the following three-body collision:

$$O_2 + O + M \rightarrow O_3 + M\,. \tag{5.8-2}$$

Here M is a non reacting atom or molecule, (mostly N_2 or O_2), which acts as a third body carrying down the chemical excess energy of recombination. Therefore, in spite of the fact that M remains chemically unchanged, its kinetic energy and momentum will change.

At the same time a reversed process takes place, destroying ozone molecules into oxygen molecules and radicals and next ozone into molecular oxygen. Two types of reactions occur, either by photo dissociation:

$$O_3 + hf\,(\lambda < 320\ \text{nm}) \rightarrow O_2 + O\,, \tag{5.8-3}$$

or by collisions with atomic oxygen:

$$O_3 + O \rightarrow 2O_2\,. \tag{5.8-4}$$

At altitudes above 100 km, the intensity of ultraviolet solar radiation is tremendous, dissociating almost all oxygen molecules into radicals. Moreover, the air density is extremely low. Hence the probability of particle collisions, which is indispensable for oxygen formation, has become very small. Consequently, at these high altitudes virtually no molecular oxygen may be found, but only oxygen radicals. Just below an altitude of 100 km there is a transition layer. Here the air is more dense, which results in more particle collisions, forming molecular oxygen. At 95 km, practically all oxygen radicals are reunited into molecules again.
Below 75 km the oxygen molecules react with free oxygen radicals, leading to ozone formation by the reactions described above. With decreasing altitudes, more reactions will take place, resulting in an enhanced ozone formation. Because of the absorption of ultraviolet radiation in the ozone formation process, the number of oxygen molecules dissociating into free radicals will also be cut down with decreasing altitudes. As a consequence, the formation of ozone reduces again. This implies the attendance of a maximum in the ozone concentration, which occurs at altitudes between about 20 and 25 km. At low altitudes almost no ozone formation takes place.
The amount of ozone in the atmosphere varies considerably with time of the day and season, as well as geographically.

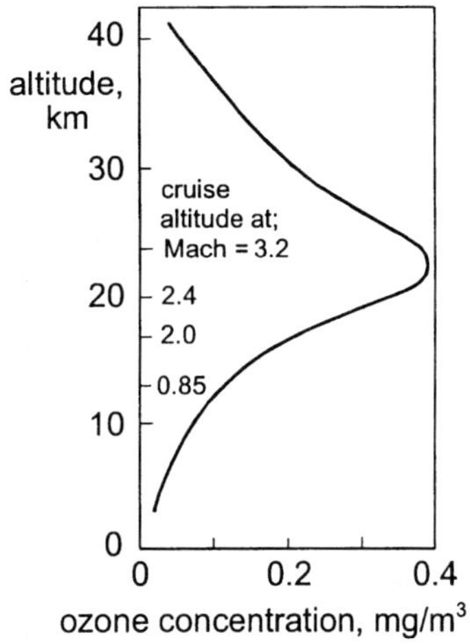

Figure 5.8-1. Atmospheric ozone profile (mid-latitude model).

In Figure 5.8-1 a model of the ozone profile is presented, as is typical for mid-latitude regions (Ref. 28). It reflects the ozone concentration as a function of the altitude. They are determined by measurements performed with an ozone probe, carried upwards by a balloon. Apparently, at altitudes above 10 km the ozone concentration increases enormously, to reach a maximum at about 25 km. Above this altitude the ozone concentration decreases rapidly.

If the total amount of ozone would be converted to standard sea-level pressure (p_0 = 101325 N/m^2) and a temperature of 0°C, a layer of only 3 mm would remain. Associated with this conception is the so-called *ozone column*, i.e., the total amount of ozone in a vertical column in the atmosphere.
The absorbing effect of the ozone layer eventuates in the absence of some wavelengths in the electromagnetic spectrum. Spectroscopy, i.e., spreading out the wavelengths contained in the solar radiation, can be exploited to measure the (local) thickness of the ozone layer, by determining which part of the ultraviolet radiation fails. Then, for any given change in the stratospheric ozone concentration, it is a straightforward matter to calculate the relative variation of ultraviolet radiation at the Earth's surface.
The required spectral analysis can be accomplished by means of a *spectrometer*, which can be used to disperse the radiant energy into a spectrum and measure the relative intensities of the wavelengths in the spectrum.
As shown in Figure 5.8-2a, the instrument contains a narrow slit through which the radiation enters. The slit is placed at the focus of the collimator lens to produce a parallel beam of rays. The rays fall on a prism separating the radiation into its frequency spectrum (see Figure 5.3-5). A series of prisms may be used to increase the amount of dispersion. The resulting spectrum can be observed by

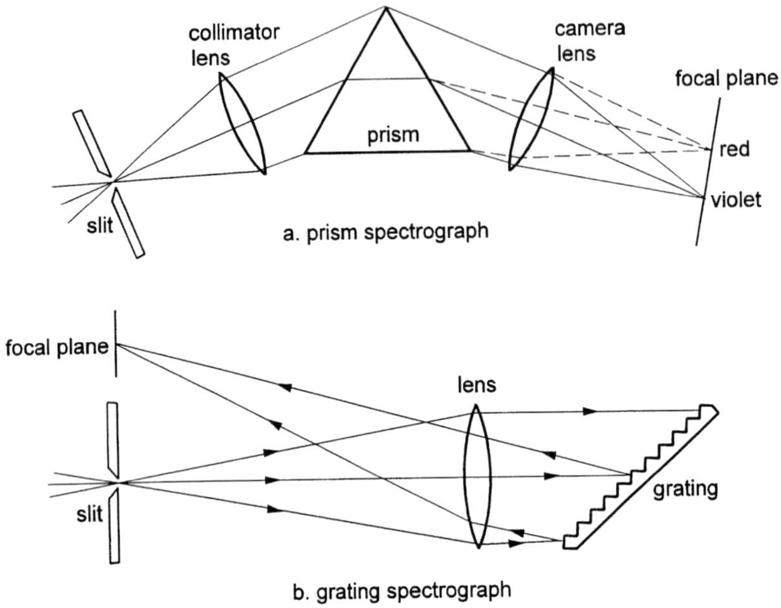

Figure 5.8-2. Spectrometers.

means of an ocular or can be recorded on a (photographic) plate. Some spectrometers use a *grating*, i.e., an aluminum-coated glass plate on which a number of fine grooves have been ruled, instead of a prism to disperse the radiation (Figure 5.8-2b).

The thickness of the ozone layer may be expressed in *Dobson units*, *DU*, named after the British physicist G. M. Dobson, who invented the first spectrometer suitable for ozone concentration measurements.
The number of Dobson units is the thickness in units of 10^{-5} m, which the ozone column would occupy if compressed into a layer of uniform density at a pressure of p_0 = 101325 N/m^2, and a temperature of 0°C. Accordingly, a layer thickness of 3 mm is equal to 300 *DU*. As stated above, the latter thickness, although very variable, is typical for the amount of ozone in a column of the Earth's atmosphere.

As remarked before, due to the absorption of ultraviolet radiation by ozone, only a minor part of it reaches the Earth's surface. In this way the ozone layer protects life in the biosphere against harmful ultraviolet radiation.

6 ENVIRONMENTAL MAJOR CONCERNS

6.1 AIR POLLUTANTS

Various pollutant substances are emitted into the atmosphere, both from natural and anthropogenic sources.
Emissions from natural sources include those from living and nonliving sources, such as vegetation, radiological decomposition, forest fires, volcanic eruptions, and emissions from land and water. The primary anthropogenic sources of the pollutants are agriculture, stock-breeding, power and heat generation, industry, and transport.
The most important air pollutants may be put together as follows:

- Carbon compounds.
- Nitrogen compounds.
- Sulfur compounds.
- Particulate matters.
- Water vapor.
- Compounds containing carbon and one or more of the elements: chlorine, fluorine, bromine, and hydrogen.

Carbon compounds can be divided into organic and inorganic compounds. A well-known organic compound is *natural gas*, which is a mixture of hydrocarbon gases consisting largely of methane (CH_4) and ethane (C_2H_6). From an environmental viewpoint, a major organic compound is *atmospheric methane*, which is a predominantly biogenic substance. Natural sources of atmospheric methane are wetlands and oceans. Man-made emission is almost twice that of the natural emission. Important anthropogenic sources of organic methane are fields of wetland in which rice is grown (paddy fields), intensive livestock keeping, refuse dumps, gas drilling, losses through pipeline leakage, coal mining and biomass burning on a global scale (mainly from fires in the tropics caused by man for land clearing purposes).
There are many other organic carbon compounds, which usually are denoted by *volatile organic compounds*, VOC's. Emissions of these compounds have increased strongly during the past decades. For instance, the emission of benzene C_6H_6 by road traffic in urban areas has become significant.

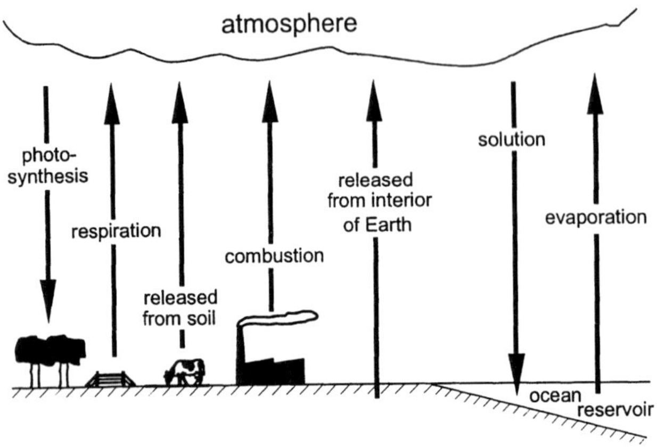

Figure 6.1-1. Illustration of the global carbon cycle.

The two dominant inorganic carbon compounds are carbon dioxide CO_2 and carbon monoxide CO, of which CO_2 is the most important. The behavior of these gases is embedded in the *global carbon cycle*. The cycle encompasses the carbon streams that are moving between the different parts of the *biosphere* (the entirety of those areas of the Earth's surface and atmosphere in which life exists), the *litosphere* (the solid rocky crust of the Earth), and the oceans. Figure 6.1-1 attempts to illustrate the complexity of the problem. The primary mechanisms involved in the exchange of carbon oxides between the atmosphere and vegetation and between the atmosphere and living organisms are respectively *photosynthesis* (the process whereby plants take CO_2 from the air and release O_2), and *respiration* (the process whereby living organisms convert O_2 to CO_2). Naturally, the exchange of CO_2 between atmosphere and oceans is primarily based on solution and evaporation.

It should be noticed that due to the accelerated increase in the concentration of CO_2 in the atmosphere after about 1950, each year also enormous amounts of carbon in the form of CO_2 are absorbed by trees, soils, and the oceans. These reservoirs, which absorb chemical elements or compounds from other parts of their cycles, are called *sinks*.

The anthropogenic contribution to the total global emission of carbon oxides differs strongly for CO and CO_2. For carbon monoxide, man-made emissions are estimated to be equal to the level of naturally produced emissions. The yearly anthropogenic carbon dioxide emissions, however, are estimated to amount to only 4% of the natural emissions (Ref. 21).

Natural sources of CO are methane oxidation, plant emissions, the oceans, forest fires and oxidation of natural hydrocarbons. Examples of CO-sources of

man-made emissions are forest clearing, agriculture and oxidation of anthropogenic hydrocarbons, mainly from motorized road traffic.
Man-made carbon dioxide is released into the atmosphere when solid waste, fossil fuels (oil, natural gas, and coal), and woods are burned.

The *nitrogen compounds* can be divided into nitrogen oxides NO_x (NO and NO_2), ammonia NH_3, and nitrous oxide (laughing gas) N_2O.
Although there is a considerable natural supply of NO_x into the atmosphere, it is known that the largest part of global NO_x-emission is caused by man-made activities, especially the combustion of fossil fuels.
An important feature of the anthropogenic NO_x-emission into the atmosphere is that the locations of the sources are strongly concentrated in the USA and in Europe, together responsible for more than 50% of the global NO_x emission from fossil fuel combustion. Emissions from airplanes are a major source of NO_x in the highest region of the troposphere, particularly in the northern hemisphere.
Important natural sources of NO_x-emission are microbiological activity in the soil, emissions from the oceans, oxidation of N_2O in the atmosphere, and lightning within thunderstorms. The primary anthropogenic sources of NO_x-emission are, besides the aforementioned fossil fuel combustion, also biomass burning and soil cultivation (Ref. 30).
The emissions of ammonia NH_3 are predominantly caused by man-made activities. Animal waste is the most important contributor. The chemical process where ammonia NH_3 is oxidized can also contribute to the formation of NO_x.
Emissions of nitrous oxide N_2O are relatively low compared to the emissions of NO_2. Relevant sources of N_2O are natural processes in the oceans and in the soil, from which it is released into the atmosphere, as well as human activities such as combustion of fossil fuels and solid waste, motorized traffic, power plants, industrial and agricultural activities, and livestock breeding.

The man-made emission of *sulfur compounds* mainly concerns sulfur dioxide SO_2 due to burning of coal and other fossil fuels, and is estimated to be about half the amount from natural sources.

Anthropogenic production of *particulate matter*, principally in the form of extremely fine dust or droplets (aerosols), is about one quarter that of natural emissions.

Man-made *water vapor* emissions into the lower atmosphere are virtually negligible when compared to natural emissions. In the stratosphere, which is practically dry, the exhaust of water vapor by aircraft engines results in the formation of *contrails*, contraction of *condensation trails*. These white line-clouds behind aircraft consist of ice particles which are formed as the water vapor condenses after leaving the engine exhaust nozzle. Aircraft contrails may have the effect of increasing the cloud cover at higher altitudes, which in turn will reduce the transmission of infrared radiation from the Earth and so upset the radiation balance of Earth and atmosphere (see Section 5.4). A discussion of the formation of aircraft contrails will be deferred until Section 10.8.

Compounds containing *carbon* (C), *chlorine* (Cl), and either *fluorine* (F) or *bromine* (B), can act as powerful greenhouse gases in the atmosphere. The carbon, chlorine, and bromine containing compounds are also involved in the depletion of the ozone layer.
Examples of these gases are chlorofluorocarbons (CFC's), hydrofluorocarbons (HFC's), perfluorocarbons (PFC's), bromofluorocarbons (halocarbons) and sulfur hexafluoride (SF_6). HFC's and PFC's are the most heat-absorbent substances. All the above compounds are anthropogenic, and are generated in a variety of industrial processes.

The CFC-compounds, containing chlorine, fluorine, and carbon, are commonly employed as refrigerants, solvents, and foam blowing agents during the production of plastic foams and polystyrenes. CFC's are very stable in the troposphere, where they act as strong greenhouse gases. E.g., one CFC molecule contributes many more times to the heating of the Earth than a CO_2 molecule. Well-known examples of CFC's are CFC-11 ($CFCl_3$) and CFC-12 (CF_2Cl_2), having a GWP of approximately 5000 and 8500, respectively. The lifetime of CFC-11 is about 45 years, and of CFC-12 around 100 years; see Sections 6.4 and 6.6 for further details.
CFC's are broken down by ultraviolet radiation in the stratosphere, thereby releasing chlorine atoms. These atoms react with ozone to chlorine monoxide and molecular oxygen:

$$Cl + O_3 \rightarrow ClO + O_2 \,. \qquad (6.1\text{-}1)$$

Chlorine monoxide in turn reacts with atomic oxygen, yielding again monochlorine and molecular oxygen:

$$ClO + O \rightarrow Cl + O_2 \,. \qquad (6.1\text{-}2)$$

The chlorine atoms formed in this way are again available to react with ozone, according to the reaction (6.1-1). Therefore, due to these catalytic reactions, one CFC molecule can be responsible for the depletion of thousands of ozone molecules.

Clearly, CFC's participate in the depletion of the ozone layer during their decomposition in the stratosphere. Further, they may stay there for more than 100 years, because of their very long lifetimes.

Halons, consisting of bromine (B), fluorine (F), and carbon (C), are used as fire extinguishing agents. Like CFC's, halons are broken down by ultraviolet radiation in the stratosphere, thereby releasing bromine atoms. These bromine atoms will also support ozone depletion by catalytic reactions similar to Equations (6.1-1) and (6.1-2). Bromine is many times more effective at destroying ozone than chlorine. By reason of their harmfulness, an entirely global production stop of the substances CFC's and halons is aimed for. The effects of current CFC's and halons in the atmosphere are expected to be gone not until the second half of this century.

Hydrofluorocarbon (HFC) compounds, consisting of hydrogen, fluorine, and carbon, is a class of replacements for CFC's. Because they do not contain chlorine, they do not deplete the ozone layer. Unfortunately, several HFC's remain very powerful greenhouse gases.

The HCFC's (hydrochlorofluorocarbons) are another class of chemicals being used to replace the CFC's. They contain chlorine and thus deplete stratospheric ozone, but to a much lesser extent than CFC's.

The PFC-compounds, consisting of carbon (C) and fluorine (F), have extremely high GWP's and very long lifetimes (see Table 6.4-1). They do not deplete stratospheric ozone, but there is concern about their impact on global warming.

In addition to the compounds mentioned in the foregoing, the properties of some other important species that may pollute the environment are specified below.

Aerosols, involving extremely fine dust or droplets of microscopic size, arise from combustion processes, dust storms, and sea salt. They may cause chemical reactions, and are condensation nuclei for water vapor. These solid or liquid particles may vary from 10^{-9} to 10^{-6} m in diameter (see Figure 1.1-1).

Asbestos, fibrous particles originating from brake linings, asbestos construction plates and insulating material, gives increased risk of a tumor in the lungs after prolonged exposure.

Fly ash, solid particles in combustion gases from power generation utilities using coals. Fly ash is for 99% filtered in furnaces. The non filtered, part is called *fly dust*. Fly ash and fly dust may contain heavy metals, i.e., metals with a density greater than 4500 kg/m^3.

Hydrochloric acid HCl and hydrogen fluoride HF, mainly formed during burning of waste matter containing chlorine and fluorine.

Heavy metals, originate in the atmosphere from metal industry, and burning useless stuff to ashes. *Vanadium* (V) and *nickel* (Ni) are found in mineral oil. Leaded fuel contains *tetraethyl lead*, which may be added to petrol as an antiknock agent. *Lead* (Pb) is also applied in power lines (cable burning). *Mercury* (Hg) ends up in the environment through combustion of fuels containing Hg, such as coals and waste matter. Natural gas also contains some mercury.
Olfactory substances, such as hydrogen sulfide H_2S and ammonia NH_3, are produced in a large number of industrial processes.

6.2 TOXIC SUBSTANCES

Among the many pollutants in the atmosphere, there are a number of toxic substances, such as:

sulfur oxides	SO_x
nitrogen oxides	NO_x
ammonia	NH_3
hydrogen sulfide	H_2S
hydrochloric acid	HCl
hydrogen fluoride	HF
chlorine	Cl
carbon monoxide	CO
sulfuric acid	H_2SO_4
soot	C
benzene	C_6H_6
ozone	O_3

Sulfur oxides (SO_2 and SO_3) arise during combustion of sulfurous fuels, such as sulfurous coal and fuel oil. Mainly SO_2 is formed, which is much less toxic than SO_3. Sulfur dioxide SO_2 is an irritant gas for the respiratory system. Very high concentrations may occur in so-called *winter smog*, or *acid smog* (see below). The replacement of coal with natural gas reduces the emission of SO_2 tremendously.
The emission of nitrogen oxides NO_x has enlarged substantially, mainly due to the enormous increase in the number of motor vehicles. NO is a colorless gas, while NO_2 is colored red-brownish, and is much more toxic than NO.
The gases SO_2 and NO_x are harmful to the respiratory system.

A highly irritant substance, mainly originating from livestock industry, is gaseous ammonia NH_3. Other gases, emitted in smaller quantities, such as hydrogen sulfide H_2S, hydrochloric acid HCl, hydrogen fluoride HF, and monochlorine Cl, are also irritant gases for eyes and mucous membranes. Refuse incineration contributes significantly to the emission of HCl and HF.

A highly toxic substance is carbon monoxide CO. Threatening for our health is its great affinity to the blood pigment hemoglobin, compared with oxygen O_2. Therefore, carbon monoxide is highly poisonous as it causes obstruction of the oxygen transport in the blood and thus the oxygen supply to the body. Carbon monoxide originates from burning coal and other fuels, but not from burning natural gas. Presently, CO emissions are primarily caused by exhaust gas emissions of road traffic. Especially, during short-distance rides, the production of carbon monoxide by motorized vehicles is substantial.

When the atmosphere contains high concentrations of sulfur dioxide SO_2, some sulfur trioxide SO_3 is formed, via the reaction:

$$SO_2 + \tfrac{1}{2}O_2 \rightarrow SO_3 \,. \tag{6.2-1}$$

Sulfur trioxide readily reacts with water to form sulfuric acid H_2SO_4:

$$SO_3 + H_2O \rightarrow H_2SO_4 \,. \tag{6.2-2}$$

Sulfuric acid is harmful to the respiratory system. It may be formed in engine exhaust streams because of the simultaneous presence of sulfur trioxide and water. In addition to the obvious deleterious effects of producing sulfuric acid, SO_3 also poisons automobile three-way catalytic converters. Thus, low sulfur levels are required in gasolines (Ref. 31).

Soot consists of agglomerates of fine carbon particles to which hydrocarbons can be absorbed. Soot formation is favored by inadequate mixing of fossil fuel and air and by high flame temperatures. The soot particles are carriers of PAH (polycyclic aromatic hydrocarbons) into the respiratory system (Ref. 5).

Combination of sulfur dioxide, sulfuric acid and soot particles in the atmosphere may cause the formation of *winter* or *acid smog*. This mixture of smoke and fog, may be dangerous not only because it reduces visibility but also because the chemicals in the smoke can damage health when inhaled. This is especially a considerable problem in areas where soft, brownish-black coal (lignite) is burned for the generation of energy (see also Section 6.4).

Emissions of volatile organic compounds (VOC's) have increased enormously during the past forty years. As mentioned before, an example is benzene C_6H_6, an inflammable, colorless gas with a distinctive odor. Benzene, which is highly carcinogenic, may be used as an additive in motor fuel, and released into the atmosphere when incomplete combustion occurs.

Presently, near ground level the so-called *photochemical* or *summer smog* frequently occurs, produced by nitrogen oxides and volatile organic compounds. The combination of these substances leads to the formation of ozone O_3, under influence of ultraviolet solar radiation. The formation of tropospheric ozone is outlined in Section 6.3.
Ozone is a very strong oxidant, reacting easily with other substances. For example, rubber is oxidized very fast through ozone. Above all, ozone is very toxic, and causes irritation of the eyes and is also very harmful to our respiratory system. Ozone is a gas with an oppressive odor, and is colored blue in higher concentrations.

Of high toxicity are also the polychlorinated dibenzodioxins (PCDD's) and dibenzofurans (PCDF's). These substances are formed in refuse incineration plants, through combustion of refuse containing PVC. By incineration exhaust gases, dioxins end up in the pasturage, and might finish up in milk for consumption.

Pertaining to the indoor environment in large office buildings, the so-called *sick building syndrome* (SBS) has aroused interest during the last decades. Symptoms are, for example, fatigue, respiration problems, throat and skin irritations, and headaches. It may be caused by insufficient air-conditioning (energy saving), too many people working in too small rooms, and the presence of many heat-producing devices. Certain construction materials and coverings may also contribute to SBS. A well-known example is diphenylamine $(C_6H_5)_2NH$, used in certain floor coverings and curtains, and for the manufacture of plastics. Ethereal substances, as used in felt-tipped pens, may also cause SBS when used in poor ventilated rooms.

6.3 TROPOSPHERIC OZONE FORMATION

As mentioned earlier, stratospheric ozone is highly desirable. It protects us against harmful ultraviolet radiation from the Sun, because no solar UV-radiation with wavelengths shorter than 280 nm can reach the surface of the Earth.

Also the troposphere contains ozone to an amount of about 10% of the total ozone in the atmosphere (see the atmospheric ozone profile in Figure 5.8-1).
The occurrence of tropospheric ozone O_3 is of great significance as, contrary to stratospheric ozone, tropospheric ozone not only is a risk to us in a direct way, but also affects the climate in that it contributes considerably to the greenhouse effect. Moreover, tropospheric ozone is the photochemical precursor of the hydroxyl radical OH, which is a very important atmospheric oxidant (see the reactions hereafter).

The ozone concentrations in the troposphere are made up of ozone coming down from the stratosphere and, for the most part, ozone produced by photochemical reactions in the troposphere itself.

The formation of ozone in the troposphere mainly is due to the presence of elevated levels of photochemical oxidants. The ozone is formed as a result of photochemical oxidation of carbon compounds such as carbon monoxide CO, methane CH_4, and other hydrocarbons under the influence of solar radiation and in the presence of nitrogen oxides NO_x.
As mentioned in Section 6.1, these pollutant substances are emitted into the atmosphere by natural processes as well as by man-made activities.
Ozone formed in the troposphere can be divided into ozone in the free troposphere and ozone in the boundary layer of the atmosphere. As shown previously in Figure 1.3-2, the atmospheric boundary layer reaches from the Earth's surface up to an altitude of about 2 km. Here the ozone concentration is determined by the special local conditions, which cause photochemical smog formation.

A prerequisite to a thorough understanding of the ozone problem is knowledge of *tropospheric ozone chemistry*, i.e., the study of the complex chains of chemical reactions involved. It is certainly beyond the scope of this text to give a complete overview of the reactions leading to the formation of tropospheric ozone. The mention of the following reactions, taken from References 5, 18, 21, and 35, is merely an attempt to illustrate some effects that may occur when NO_x is emitted into the troposphere. Nitrogen oxides are of importance because current commercial airplanes cruise at altitudes between 8 and 13 km, through which their NO_x emissions are injected directly into the upper part of the troposphere (Ref. 1).

The reactions governing the formation of tropospheric ozone are primarily driven by the above-mentioned hydroxyl radical OH, which is a product of a two-step reaction initiated by ozone:

$$O_3 + hf \rightarrow O(^1D) + O_2 , \tag{6.3-1}$$

$$O(^1D) + H_2O \rightarrow 2\,OH , \tag{6.3-2}$$

where $O(^1D)$ is an exited oxygen atom.
Practically, the OH radical is capable of reacting with all pollutants, such as carbon monoxide CO, methane CH_4, and other hydrocarbons.

As examples, in the following the reactions are illustrated of the photochemical oxidation of carbon monoxide CO and methane CH_4, which show the important role of NO_x on the formation of tropospheric ozone.

For the oxidation of CO the following reaction with OH radicals will occur:

$$CO + OH \rightarrow CO_2 + H. \tag{6.3-3}$$

The H-atoms attach to oxygen molecules:

$$H + O_2 + M \rightarrow HO_2 + M, \tag{6.3-4}$$

where M is a third body molecule that absorbs the excess energy.
The hydroperoxy radicals HO_2 react with the present NO molecules, forming NO_2 and hydroxyl radicals OH:

$$HO_2 + NO \rightarrow OH + NO_2. \tag{6.3-5}$$

In the presence of sunlight, NO_2 is split up by photolysis, forming NO and oxygen radicals:

$$NO_2 + hf \rightarrow NO + O(^3P), \tag{6.3-6}$$

where $O(^3P)$ is a ground-state oxygen atom.
This atom reacts with molecular oxygen O_2 under collision with a non reacting molecule M (mostly N_2 or O_2), to form ozone:

$$O(^3P) + O_2 + M \rightarrow O_3 + M. \tag{6.3-7}$$

At the same time, ozone may be destroyed by reaction with NO, to reform NO_2:

$$O_3 + NO \rightarrow NO_2 + O_2. \quad (6.3\text{-}8)$$

The net result of the above reactions becomes:

$$CO + 2\,O_2 + hf \rightarrow CO_2 + O_3. \quad (6.3\text{-}9)$$

This reaction shows that complete oxidation of CO results in the creation of CO_2 and O_3, and that for each CO molecule that is oxidized, one ozone molecule will be formed.

The oxidation of methane CH_4 takes place via the following reaction chain:

$$CH_4 + OH \rightarrow CH_3 + H_2O, \quad (6.3\text{-}10)$$

$$CH_3 + O_2 + M \rightarrow CH_3O_2 + M, \quad (6.3\text{-}11)$$

$$CH_3O_2 + NO \rightarrow CH_3O + NO_2, \quad (6.3\text{-}12)$$

$$CH_3O + O_2 \rightarrow CH_2O + HO_2, \quad (6.3\text{-}13)$$

$$CH_2O + hf \rightarrow CO + H_2, \quad (6.3\text{-}14)$$

$$HO_2 + NO \rightarrow OH + NO_2, \quad (6.3\text{-}15)$$

$$NO_2 + hf \rightarrow NO + O\ (2\ x), \quad (6.3\text{-}16)$$

$$O + O_2 + M \rightarrow O_3 + M\ (2\ x). \quad (6.3\text{-}17)$$

The overall reaction for the oxidation of CH_4 is:

$$CH_4 + 4\,O_2 + hf \rightarrow CO + H_2 + H_2O + 2\,O_3. \quad (6.3\text{-}18)$$

Chemical depletion or formation of O_3 due to the presence of nitrogen oxides NO_x depends on the intensity of solar radiation. This is described by the so-called *photostationary equilibrium*, which concerns the equilibrium between NO_x, O_3 and O_2.
Nitrogen oxides, undoubtedly, are the most significant aero-engine emission species, typically comprising more than 90% of NO and 10% of NO_2 (Ref. 34). The NO released into the atmosphere, oxidizes further into NO_2 either by oxygen or ozone:

$$2\,NO + O_2 \rightarrow 2\,NO_2. \qquad (6.3\text{-}19)$$

Oxidation by oxygen takes two molecules of NO for one oxygen molecule. Therefore, at low NO concentrations, this reaction proceeds very slowly.

Via the reaction (6.3-6), NO_2 is broken up by photolysis into NO and O, whereupon according to the reaction (6.3-7), the oxygen radicals combine with O_2 to form O_3:

$$NO_2 + hf \rightarrow NO + O(^3P), \qquad (6.3\text{-}6)$$

$$O(^3P) + O_2 + M \rightarrow O_3 + M. \qquad (6.3\text{-}7)$$

Ozone is destroyed according to the reaction (6.3-8):

$$O_3 + NO \rightarrow NO_2 + O_2. \qquad (6.3\text{-}8)$$

The resulting photo-stationary equilibrium, comprising the reactions (6.3-6) to (6.3-8), yield the net result:

$$NO_2 + O_2 + hf \leftrightarrow NO + O_3. \qquad (6.3\text{-}20)$$

This means that due to ultraviolet radiation the chemical equilibrium shifts toward the right-hand side of the reaction (6.3-20), leading to the formation of ozone.

The time and spatial scales at which the various compounds contribute to ozone formation depend on their reactivity. Fewer reactive compounds such as CO and CH_4 affect ozone formation on a global scale, determining the long-term average ozone concentration.

As apparent from the reactions (6.3-10) to (6.3-17), the amount of the greenhouse gas methane CH_4 in the atmosphere is reduced by the emission of nitrogen dioxide. So besides the formation of the greenhouse gas ozone, NO_x also deforms a greenhouse gas (Ref. 1).

As noticed already in Section 6.2, near the ground we are faced with the formation of photochemical or summer smog, which phenomenon concerns the high peak concentrations of ozone in city centers during hot and sunny days. Although OH radicals in a polluted atmosphere react with a variety of hydrocarbons (HC) from motor vehicles, in a clean atmosphere they react mainly with CO and CH_4, according to the reactions (6.3-3) and (6.3-10):

$$CO + OH \rightarrow CO_2 + H, \tag{6.3-3}$$

$$CH_4 + OH \rightarrow H_2O + CH_3. \tag{6.3-10}$$

The H atoms and hydrocarbon radicals CH_3 attach to oxygen molecules to form HO_2 and CH_3O_2 radicals via the reactions (6.3-4) and (6.3-11):

$$H + O_2 + M \rightarrow HO_2 + M, \tag{6.3-4}$$

$$CH_3 + O_2 + M \rightarrow CH_3O_2 + M. \tag{6.3-11}$$

Next, the HO_2 and CH_3O_2 radicals react with NO, whereby NO_2 is formed and OH radicals are reformed according to the reactions (6.3-12) and (6.3-15):

$$NO + CH_3O_2 \rightarrow NO_2 + CH_3O, \tag{6.3-12}$$

$$NO + HO_2 \rightarrow NO_2 + OH. \tag{6.3-15}$$

The above reactions show that with the OH radicals, (oxidizing) HO_2 radicals are formed from VOC's, without the OH radicals being consumed in the process. Apparently, the presence of hydrocarbons leads to the formation of OH radicals, and so to the formation of ozone according to reaction (6.3-20). This

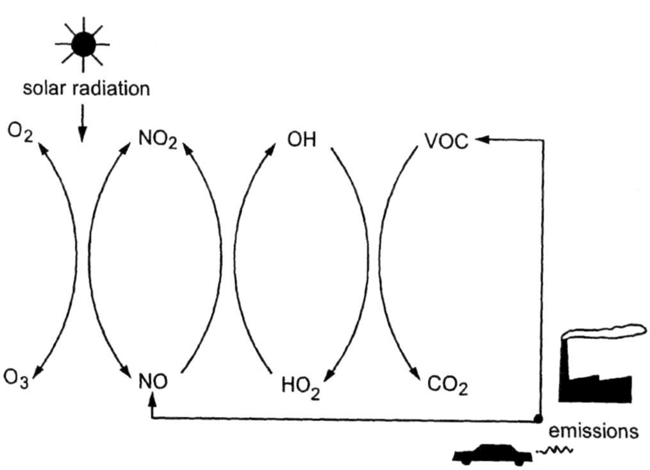

Figure 6.3-1. Diagram illustrating ozone formation near ground level.

process of ozone formation near ground level from nitrogen oxides and hydrocarbons, under the influence of solar radiation, is pictured in Figure 6.3-1. Also, we mention again the occurrence of acid or winter smog. Then the presence of chemicals, such as sulfur dioxide SO_2 and aerosols in the smoke may be responsible for negative effects on our health. When the acid smog contains high concentrations of pollutant, it may have a dirty yellow color and may be very dense ('pea-soup' smog). Then dispersion of pollutants hardly takes place.
Both winter and summer smog are found mainly in a stable atmosphere, with little or no convective motions. The effects of summer and winter smog and the more direct effects of hydrocarbons from incomplete combustion of fossil fuels, usually, are denoted as "local pollution".

With respect to the role of aviation, we should emphasize the importance of NO_x emissions from airplanes at their cruise altitudes. These emissions, when compared to the total NO_x production by combustion of fossil fuels at ground level, are estimated to produce only a very small portion of the total amount of nitrogen oxides in the atmosphere. Nevertheless, emissions from high-flying airplanes may be of primary importance since they eject the NO_x directly into the atmosphere at altitudes where NO_x concentrations normally are very low and practically no other sources are found.
It is worth noting that approximately 60% of the NO_x emissions by airplanes are concentrated in the zone confined by northern latitudes of 30 and 60 degrees, and that the lifetime of NO_x in the tropopause region is about 5-10 days, whereas at low altitudes the lifetime is less than one day (Ref. 35). Therefore, the concentration of nitrogen oxides in the atmosphere varies strongly, not only spatially but also in time.

It is beyond doubt that much attention must be devoted to the ozone chemistry in the upper region of the troposphere, since it is believed that nitrogen oxides at these altitudes contribute considerably to the enhanced greenhouse effect, and, when rising to the stratosphere, also support the depletion of the ozone layer (see Section 6.6).

6.4 THE ENHANCED GREENHOUSE EFFECT

The Earth receives energy through solar radiation. Due to thermal equilibrium, i.e., the balance between absorbed solar radiation and emitted infrared radiation, there is a global equilibrium temperature on Earth.

The majority of the energy coming from the Sun within the range of wavelengths smaller than visible light, is absorbed at high altitudes in the atmosphere. For example, under normal circumstances the short-wave UVB and UVC radiation is almost entirely filtered within the ozone layer at an altitude of about 30 km, so their harmful effects are hardly perceived near the Earth's surface. Also scattering chiefly by the air molecules and clouds, takes place. The scattered radiation creates the dispersed luminosity of the sky during daytime. The color of the sky is primarily bright blue because the shorter wavelengths are scattered more than the longer ones (see Section 5.3).
When solar radiation penetrates deeper into the atmosphere, the energy from smaller wavelengths is also absorbed, especially in the infrared spectral region. In an atmosphere of which the composition is not influenced by human activities, the absorption is mainly caused by greenhouse gases such as water vapor H_2O, carbon dioxide CO_2, methane CH_4 and nitrous oxide N_2O. Of these, the most important radiative trace substances are water vapor H_2O and carbon dioxide CO_2. Absorption of radiation energy in a specific atmospheric layer causes the local temperature to rise, due to the increase in internal energy of the molecules. Owing to the filtering effect of the atmosphere, only a limited range of wavelengths is admitted through the so-called *windows for visible light* and *telecommunication*.

The preceding remarks only relate to radiation coming directly from the Sun. Solar radiation heats up the atmosphere. If the Earth would only absorb heat, without any heat emission, our planet would be warming up constantly. However, the existence of an equilibrium temperature implies that the Earth emits radiation. This radiation is almost entirely within the infrared spectrum, due to the relatively low surface temperatures of the Earth. Especially carbon dioxide and water vapor in the atmosphere absorb a significant part of the infrared radiation emitted by the Earth. As a result, the atmosphere is heated up from underneath, as it was caught within the atmosphere. As we already know, this warming of the Earth's atmosphere due to absorption of infrared radiation is the *natural greenhouse* or *blanketing effect*, on account of which life on Earth is possible. Because of this phenomenon, the global climate is warmer than it would be otherwise. As described in Chapter 5, if the atmosphere did not contain any radiative trace substances, the average surface temperature of the Earth would be more than 30°C below the current mean temperature of 15°C. In other words, the natural greenhouse effect is essential for all forms of life on Earth.
We may explain the greenhouse effect by examining first of all the equilibrium temperature of the Earth's surface in the absence of greenhouse gases (Figure 6.4-1a). Assuming that the ground acts as a perfect blackbody, and using the

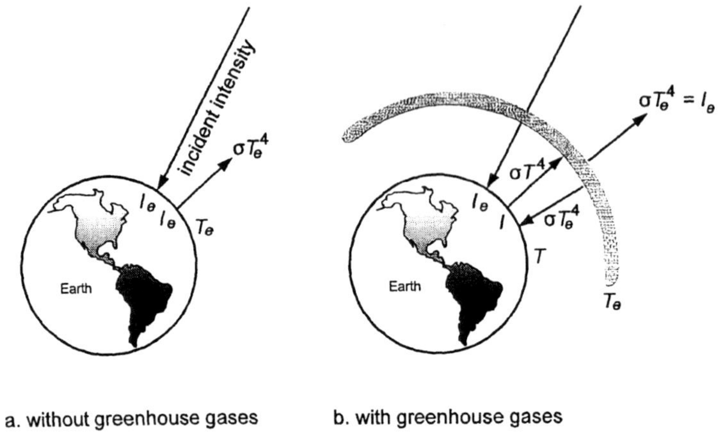

Figure 6.4-1. The greenhouse effect.

condition that the absorbed intensity I_e equals the emitted intensity, the temperature of the ground surface T_e is given by the Stefan-Boltzmann law:

$$I_e = \sigma T_e^4 \,, \tag{6.4-1}$$

where σ is the Stefan-Boltzmann constant.

Next, as represented in the previous Figure 1.1-1 by the working of a greenhouse, we consider the situation that at some height above the ground a shield of glass is present (Figure 6.4-1b). The shield will transmit the short-wave solar radiation unimpededly. On the other hand, the long-wave radiation emitted by the ground is completely impeded by the shield, which emits energy itself from its two surfaces downwards to the ground and upwards to the atmosphere. Denoting the shield's temperature T_s, the intensities emitted by the glass are given by:

$$I_s = \sigma T_s^4 \,. \tag{6.4-2}$$

Now, the ground surface receives an increased amount of energy compared to the situation without the glass shield. Accordingly, also the temperature of the ground surface will be higher ($T > T_e$).
As the upward emitted energy by the shield must be equal to the solar energy ($I_s = I_e$), according to Equation (6.4-1), we find: $T_e = T_s$. Clearly, the glass has now the temperature that the ground had before installing the shield.
From the intensity balance of the shield we have:

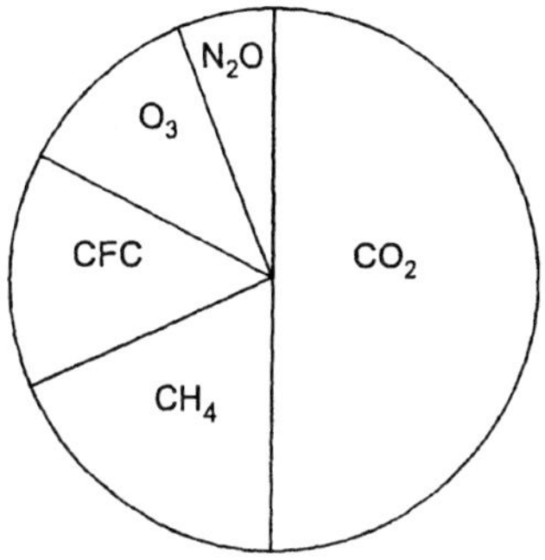

Figure 6.4-2. Relative importance of greenhouse gases to the enhanced greenhouse effect.

$$2\sigma T_e^4 = \sigma T^4 \quad \text{or} \quad T = 2^{\frac{1}{4}} T_e \, . \tag{6.4-3}$$

For $T = 15°C = 288.15$ K, we obtain $T_e = 242.30$ K or $\Delta T = T - T_e \approx 46°C$. Remembering the numerical evaluation made in Chapter 5, Section 5.4, we must conclude that the above result gives an overestimation of ΔT. This is certainly not surprising in view of the various simplifying assumptions used in the analysis. Nevertheless, the above approach to the problem may provide some insight into the appearance of the greenhouse effect. For further study, the reader is referred to Ref. 10.

The greenhouse effect might become a problem if enhanced by anthropogenic activities, which generally concern emissions of carbon dioxide CO_2, methane CH_4, nitrous oxide or laughing gas N_2O, and ozone O_3. A main contribution to the greenhouse effect stems also from aerosol spray cans, which employ as propellants chemicals such as chlorofluorocarbons (CFC's), CF_2Cl_2 and $CFCl_3$. These substances, of which large amounts were released into the atmosphere during the years 1950 to 1990, are strong greenhouse gases because they are chemically inert in the troposphere. An impression of the relative importance of the various greenhouse gases to the enhanced greenhouse effect is depicted in Figure 6.4-2, showing that anthropogenic emissions of CO_2 are responsible for at least half of the enhanced greenhouse effect.

The enhanced greenhouse effect is further aggravated by the increase of tropospheric ozone O_3, which is formed by photochemical reaction of volatile organic compounds (VOC's) with nitrogen oxides NO_x.
The enhancement of the greenhouse effect is chiefly ascribed to the increased amount of CO_2 in the atmosphere. Important anthropogenic sources of CO_2 are the combustion of fossil fuels and the deforestation, which is the conversion of forest to land for other use, of our planet. Forests absorb CO_2 by photosynthesis. Figure 6.4-3 shows data on the CO_2 concentrations in the

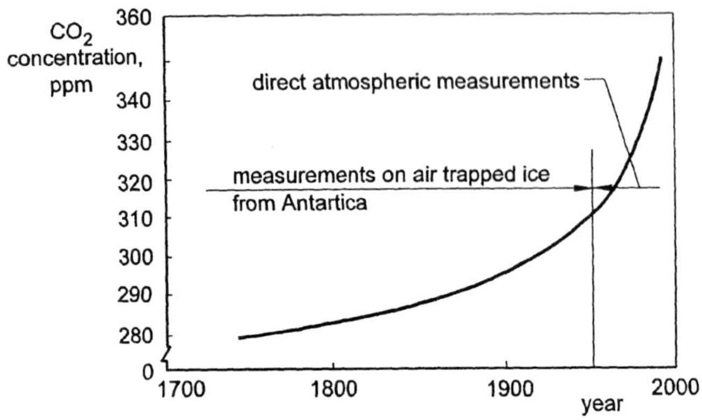

Figure 6.4-3. CO_2 concentrations in the past 250 years.

atmosphere as indicated by measurements on air trapped ice from Antarctica, and by direct atmospheric measurements at Mauna Loa, Hawaii (Ref. 18). The plot shows that 250 years ago there was a CO_2 concentration of about 280 ppm in the atmosphere, which has been increased to 360 ppm currently. If this trend continues, the concentration of CO_2 will reach a level of about 450 ppm by the year 2050. An important point is that the effective lifetime of CO_2 in the atmosphere is believed to be in excess of 100 years. In other words, today's emissions of CO_2 will go on affecting the climate for more than the next 100 years. In favor of CO_2 can be said that it has virtually no direct effect on the human state of health.

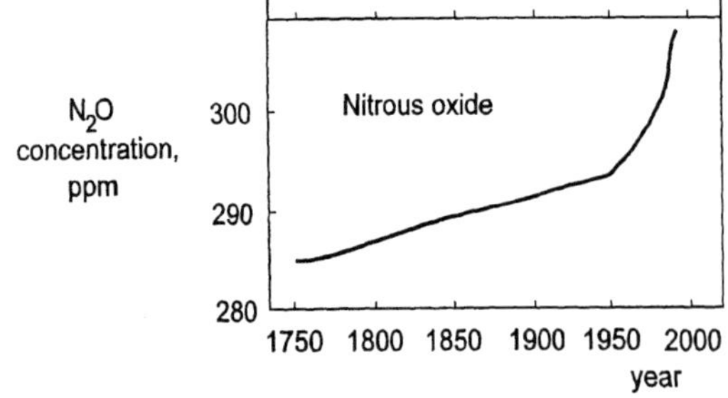

Figure 6.4-4. CH_4 and N_2O concentrations in the atmosphere in the past 250 years.

The emissions of methane CH_4 and nitrous oxide N_2O show similar trends, especially when we consider the past five decades (Figure 6.4-4). Both, the emissions of CH_4 and N_2O are relatively low. Nevertheless they contribute significantly to the greenhouse effect (see Figure 6.4-2). As mentioned before, atmospheric methane is a trace gas which affects the Earth's climate both directly and indirectly. Nitrous oxide N_2O is a very stable gas which is not broken down in the troposphere. It is only in the stratosphere that nitrous oxide may be decomposed by photolysis.
In the troposphere, the amount of ozone has doubled during the past hundred years. This is almost certainly a result of increasing emissions of carbon monoxide CO and nitrogen oxides NO_x from man-made activities (Ref. 36).

The various greenhouse gases have different radiative efficiencies and hence different impacts on the global warming. As explained earlier in Chapter 5, the index *Global Warming Potential* (GWP) may be used to compare the relative importance of the various greenhouse gases to global warming, taking into account the lifetime of the pollutant (Ref. 24). The GWP is defined as the ratio of the global warming caused by a given mass of a substance to the warming by a similar mass of carbon dioxide.
Examples of average values of GWP valid for a period of 100 years as reported by IPCC (References 24-27), are given in Table 6.4-1, showing that chlorofluorocarbons etc. contribute very strongly to the enhanced greenhouse effect. In the table also are given the estimated lifetimes of the pollutants.

The GWP values listed in Table 6.4-1 refer to the emissions of long-lived, well-

Gas	Chemical formula	GWP	Lifetime in year
Carbon dioxide	CO_2	1	100
Methane	CH_4	22	12
Nitrous oxide	N_2O	300	114
Ozone	O_3	1800	0.25
Trichlorofluoromethane	CCl_3F (CFC-11)	5000	45
Dichlorodifluoromethane	CCl_2F_2 (CFC-12)	8500	100
Fluoroform	CHF_3 (HFC-23)	12000	260
Chlorodifluoromethane	$CHClF_2$ (HCFC-22)	1700	11.9
Perfluoroethane	C_2F_6	11900	10000
Sulfur hexafluoride	SF_6	22200	3200

Table 6.4-1. Global Warming Potentials (10 year time horizons).

mixed gases, of which the presence has a direct effect on global warming. However, gases in the atmosphere also can contribute indirectly to global warming. Indirect trapping of heat occurs when chemical transformations of the original gas produce a gas or gases that are greenhouse gases, or when a gas influences the atmospheric lifetimes of other gases.
Therefore, the use of GWP is disputable, because the global warming potential of gases with an indirect influence on the heating of the Earth, such as CO, NO_x, H_2O, and SO_2 , cannot be determined truthfully.
E.g., for airplane emissions indirect climate change from tropospheric ozone produced by NO_x emissions is not linearly proportional to the amount of NO_x emitted but depends also on location, season, and especially the altitude. Also, the occurrence of contrails (H_2O) cannot be related directly to the extent to which they promote larger expanses of cirrus cloud.

Climatological investigations reveal that the anthropogenic emissions of greenhouse gases are responsible for a rise in mean global temperature. This may imply a change in the climate, which is the average weather experienced in a particular region over a long period of time. Then cold periods will become rare, and warm periods more numerously. The oceans will expand and the continental ice caps will melt, causing the surface of the sea to rise. Flood disasters may occur, along with extreme weather conditions endangering the human state of health.
It may be appreciated that a rise of sea level mainly is caused by expansion of the oceans due to an increased water temperature, because on the basis of Archimedes law, only melting of land ice delivers a (minor) contribution to Earth's water volume. On the other hand, decreased ice-mass means that less incoming radiation will be reflected, leading to an increased surface temperature on Earth.

So far, climatologists still disagree on whether the enhanced greenhouse effect already causes the climate to change or not. However, there is unanimity on the fact that emissions of greenhouse gases need to be reduced, and that deforestation has to be stopped, to prevent global warming.

6.5 ACIDIFICATION AND ACID RAIN FORMATION

Unnatural acidity of precipitation directly affects the environment. It is mainly caused by the emissions of nitrogen oxides NO_x, sulfur oxides SO_x, and ammonia NH_3. Nitrogen oxides and sulfur oxides are formed during power

generation, in heating utilities, and are present in the exhaust streams of motor vehicle engines. Ammonia mainly originates from organic fertilizer, thus from livestock industry and agricultural fertilization and also from artificial fertilizer industry.
With high sulfur concentrations in the atmosphere, acid rain may cause severe corrosion of mineral building materials. Also, acid rain may be harmful to forests and vegetation. For instance, acid rain can cause thinning of the crowns of many coniferous tree species (see also Chapter 1, Section 1.4).

Acid substances, when dissolved in water, are characterized by releasing protons H^+, which then are present as hydronium ions H_3O^+. The acidity (sour taste) of a solution is determined by its concentration of H^+-ions.
One of the strongest acids is hydrochloric acid HCl, containing 6×10^{26} H^+- ions per m^3. Pure water contains 6×10^{19} H^+-ions per m^3.
A solution containing fewer protons per m^3 than pure water is called an *alkali* or a *base*. This is a substance able to absorb H^+-ions from another substance. Our sense of taste has no alkali-receptors, which makes the term alkali not as appealing as acid.
As one kilomole of a substance always contains 6.022469×10^{26} particles, Avogadro's number, the number of protons in a solution can be expressed in terms of kmol. The concentration of hydrochloric acid is therefore approximately 1 kmol/m^3 and pure water 10^{-7} kmol/m^3. Sodium hydroxide NaOH contains about 10^{-14} kmol/m^3. Table 6.5-1 shows the large variety in proton concentration for several substances.

Substance	H^+/m^3	$kmol/m^3$	pH-value
hydrochloric acid	6×10^{26}	1	0
battery acid	6×10^{25}	10^{-1}	1
citric acid	6×10^{24}	10^{-2}	2
acid rain	6×10^{24} - 6×10^{21}	10^{-2} - 10^{-5}	2 - 5
vinegar	6×10^{23}	10^{-3}	3
natural rain	$6\times10^{17.4}$	$10^{-5.6}$	5.6
drinking-water	$6\times10^{16.5}$	$10^{-6.5}$	6.5
pure water	6×10^{19}	10^{-7}	7
sea-water	6×10^{18}	10^{-8}	8
soap	6×10^{17}	10^{-9}	9
liquid ammonia	6×10^{15}	10^{-11}	11
sodium hydroxide	6×10^{12}	10^{-14}	14

Table 6.5-1. Proton concentration and acidity for several substances.

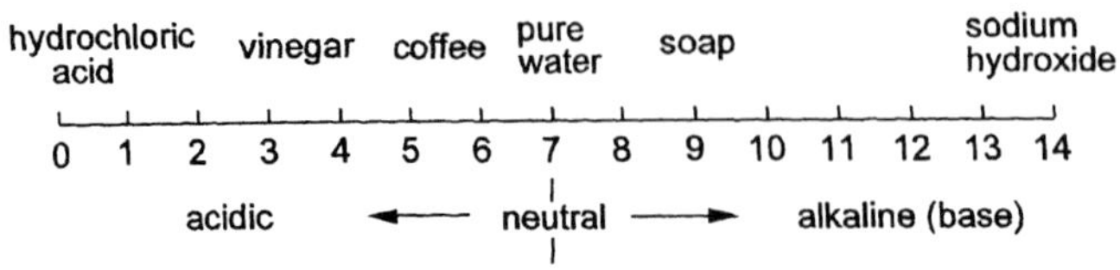

Figure 6.5-1. Acidity and the pH-scale.

For practical reasons, the acidity of a substance is measured on a logarithmic scale:

$$pH = -\log\left(\frac{H^+}{H_0^+}\right), \tag{6.5-1}$$

where *'pH'* means *potential of hydrogen*, H^+ is the concentration of protons in kmol/m^3 and $H_0{}^+$ is a reference value of 1 kmol/m^3.
When a solution has a *pH*-value of 7, like pure water, it is neither acid, nor alkaline. Thus, on the scale defined by Equation (6.5-1), $pH = 7$ represents neutrality, values higher than 7 alkalinity and values lower than 7 acidity (Figure 6.5-1). In a general sense, the lower the *pH*-value, the higher the acidity. E.g., a solution with a *pH*-value of 5 has an acidity 10 times higher than a solution with a *pH*-value of 6.

The *pH*-value of acid precipitation, i.e., precipitation like rain, snow, hail and freezing rain, is lower than for normal, natural rain. As mentioned above, acid rain is formed when the air contains nitrogen oxides, sulfur oxides, or ammonia. These gases are slightly soluble in water and, so that they pass into solution in the water droplets of clouds to form dilute nitric, sulfuric or carbonic acids.

Nitrogen oxides NO and NO_2 are formed during combustion processes using atmospheric air as an oxidizer:

$$N_2 + O_2 \rightarrow 2\,NO. \tag{6.5-2}$$

The nitrogen oxide NO released into the atmosphere is oxidized to nitrogen dioxide NO_2 according to the following reaction:

$$2NO + O_2 \rightarrow 2\,NO_2. \tag{6.5-3}$$

It appears that direct absorption of nitrogen oxides in liquid droplets hardly occurs due to their poor solubility. Actually, the NO_2 in the atmosphere readily combines with OH radicals to form nitric acid HNO_3 in the gaseous phase (Ref. 5):

$$NO_2 + OH + M \rightarrow HNO_3 + M. \tag{6.5-4}$$

Nitric acid is absorbed in the form of drops and contributes to the acidification of rain due to its strong dissociation:

$$HNO_3 + H_2O \rightarrow H_3O^+ + NO_3^-. \tag{6.5-5}$$

Sulfur oxides, especially SO_2, arise from the combustion of sulfurous fuels. When dissolving sulfur dioxide in water vapor, sulfurous acid H_2SO_3 will be formed:

$$SO_2 + H_2O \rightarrow H_2SO_3. \tag{6.5-6}$$

From the oxidation of sulfur trioxide SO_3 by water vapor H_2O, sulfuric acid H_2SO_4 is formed:

$$SO_3 + H_2O \rightarrow H_2SO_4. \tag{6.5-7}$$

If sulfuric acid is formed, the rain water acidifies due to the strong dissociation of H_2SO_4:

$$H_2SO_4 + 2\,H_2O \rightarrow 2\,H_3O^+ + SO_4^-. \tag{6.5-8}$$

When ammonia NH_3 is dissolved in water it is converted into ammonium NH_4 and hydroxyl radicals OH, through the reaction:

$$NH_3 + H_2O \rightarrow NH_4^+ + OH^-. \tag{6.5-9}$$

From the above, it will be clear that virtually all acid precipitation originates from anthropogenic sources.

6.6 STRATOSPHERIC OZONE DEPLETION

Ozone is a pollutant at the Earth's surface where it directly can lead to health problems, it is a greenhouse gas in the troposphere, and in the stratosphere it acts as a natural filter that absorbs and blocks the Sun's ultraviolet radiation ($\lambda < 280$ nm). The typical variation of ozone concentration with the altitude can be seen in the previous Figure 5.8-1. Nearly 90% of the atmospheric ozone is in the stratosphere and is referred to as the *ozone layer.*

As discussed earlier in Chapter 5, stratospheric ozone is formed by a process known as the *Chapman mechanism.* It involves two photochemical and two chemical reactions, repeated below:

$$O_2 + hf\,(\lambda < 250\text{ nm}) \rightarrow 2O, \qquad (6.6\text{-}1)$$

$$O + O_2 + M \rightarrow O_3 + M, \qquad (6.6\text{-}2)$$

$$O_3 + hf\,(\lambda < 320\text{ nm}) \rightarrow O_2 + O, \qquad (6.6\text{-}3)$$

$$O + O_3 \rightarrow 2\,O_2. \qquad (6.6\text{-}4)$$

In the reactions (6.6-1) and (6.6-2) ozone is formed, while on the contrary in the reactions (6.6-3) and (6.6-4) ozone is depleted. As a result there is more or less a constant amount of ozone in the stratosphere.

The altitude of the stratosphere extends from about 10 km to about 50 km (see the previous Figure 4.6-1). The upper part of the stratosphere gets warmer with increasing altitude, which phenomenon is caused by ozone absorbing radiation. A consequence of the positive value of the temperature gradient is that the stratosphere is very stable, so there is much less vertical mixing in this region than in the troposphere.
In the troposphere relatively fast circulation and mixing lead to air flows that spread out pollutants and bring them back to the Earth's surface. Precipitation taking place in the troposphere also makes the pollutants come back to Earth. On the contrary, in the stratosphere pollution can stay for years, because of the prevailing lapse rate conditions and the cloudless layers.

Over the past decades, there has been a considerable global stratospheric ozone loss, of which the reductions are most pronounced above the polar regions in local springtime. It is known that ozone depletion at first results from the release into the atmosphere of the man-made compounds containing chlorine, fluorine or bromine, and carbon (see Section 6.1).

Compounds such as chlorofluorocarbons and halons are called *Ozone-Depleting Substances*, ODS's. These detrimental chemicals are generally very stable in the troposphere and only degrade under intense ultraviolet radiation in the stratosphere. Being inert in the troposphere, they can survive there for many years. When these gases migrate into the stratosphere, they disintegrate due to

the ultraviolet radiation acting upon them, thereby releasing monochlorine and bromine atoms, which then deplete ozone. In this respect the quantity *Ozone Depletion Potential* (ODP) is used, which is the ratio of the impact on ozone of a given substance compared to the impact of a similar mass of CFC-11 (see Table 6.4-1).
E.g., the chlorine atoms will establish the following powerful catalytic ozone depletion reaction chain (see also the previous Equations (6.1-1) and (6.1-2)):

$$Cl + O_3 \rightarrow ClO + O_2, \qquad (6.6\text{-}5)$$

$$2\ ClO + M \rightarrow Cl_2O_2 + M, \qquad (6.6\text{-}6)$$

$$Cl_2O_2 + hf \rightarrow 2\ Cl + O_2. \qquad (6.6\text{-}7)$$

The net effect of the reaction cycle is:

$$2\ O_3 + hf \rightarrow 3\ O_2. \qquad (6.6\text{-}8)$$

Apparently, during this cycle chlorine monoxide ClO is formed, which reacts further with oxygen radicals to regenerate atomic chlorine:

$$ClO + O \rightarrow Cl + O_2. \qquad (6.6\text{-}9)$$

The chlorine radical Cl is able to react with ozone again. Thus, chlorine atoms remain participating in the process, without being consumed themselves. As stated before, as a consequence of this chain reaction, only one chlorofluorocarbon molecule entering the catalytic process can lead to the destruction of as many as thousands of ozone molecules per single atom of chlorine.

Within the framework of international measures to reduce the production and consumption of CFC's and similar species, considerable effort is put into finding appropriate substitutes.
Despite all positive achievements accomplished by banning the CFC's, it will still take many years before the effects of these hardly decomposable species, which have been emitted into the atmosphere until about 1990, will be vanishing completely. They end up in the stratosphere, where they may remain for more than a century, continuously destroying the ozone layer (see Table 6.4-1).
Therefore, about 90% of the current CFC-molecules in the atmosphere, are still in the higher layers of the atmosphere, so that a perceptible recovery of the ozone layer is not to be expected before halfway this century.

At this point, we recall that solar ultraviolet radiation is commonly split into UVA-, UVB-, and UVC radiation (see the previous Table 5.2-1), of which UVB and UVC are harmful. UVA is not absorbed by ozone, whereas UVB is mostly absorbed by ozone, although some reaches the Earth. The dangerous UVC-radiation is completely absorbed by ozone and molecular oxygen.
Obviously, depletion of the ozone layer will lead to a higher UVB level, which causes the harmful effects to life on Earth to rise.

In this respect, also the atmospheric effects of aviation-induced nitrogen oxides may be of importance. For example, 20 - 40% of the nitrogen oxides produced by subsonic aviation are injected directly into the stratosphere, where the ambient concentrations of NO_x are low. It should be noted that if a future generation of supersonic transport airplanes would be realized with cruise altitudes in the stratosphere, 100% of their nitrogen oxides would be directly emitted into the ozone layer (Ref. 37). Also note that in the stratosphere vertical spreading of pollutants is slow, while horizontal mixing is faster and notably extensive. Additionally, the lifetime of NO_x in the stratosphere is larger than in the troposphere.

Generally, two types of chemical reactions involving NO_x may occur in the stratosphere. First there is a set of neutral reactions, where formation and destruction are balanced:

$$NO + O_3 \rightarrow NO_2 + O_2, \qquad (6.6\text{-}10)$$

$$NO_2 + hf \rightarrow NO + O, \qquad (6.6\text{-}11)$$

$$O + O_2 + M \rightarrow O_3 + M. \qquad (6.6\text{-}12)$$

On the other hand, stratospheric nitrogen oxides may act as catalytic destructors of ozone (Ref. 36), according to the following set of reactions in which the nitrogen oxide itself is maintained in the process (Reference 32):

$$NO + O_3 \rightarrow NO_2 + O_2, \qquad (6.6\text{-}13)$$

$$NO_2 + O \rightarrow NO + O_2, \qquad (6.6\text{-}14)$$

$$O_3 + hf \rightarrow O + O_2. \qquad (6.6\text{-}15)$$

The net effect of these reactions is:

$$2\,O_3 + hf \rightarrow 3O_2. \qquad (6.6\text{-}16)$$

Apparently, the net effect of these reaction cycles, which are important in the altitude region where most ozone is located, results in ozone depletion, and so contributes to the disruption of the natural radiation balance.

In favor of the presence of NO_x in the stratosphere can be said that it also reacts with chlorine monoxide:

$$NO_2 + ClO + hf \rightarrow ClONO_2, \qquad (6.6\text{-}17)$$

which partially nullifies the effect of CFC on the ozone layer.

Greenhouse gases such as CO_2, methane CH_4, and N_2O, can rise to very large altitudes, where they may affect the chemistry of the stratosphere, directly by chemical reactions and indirectly through temperature changes. Rising concentrations of greenhouse gases may reduce the temperature of the stratosphere, resulting in an increase of the ozone column.
An increase in N_2O concentration, for example, will decrease the ozone column as nitrous oxide is decomposed almost entirely in the stratosphere, leading to the formation of ozone depleting NO radicals:

$$N_2O + O(^1D) \rightarrow 2\,NO, \qquad (6.6\text{-}18)$$

where the oxygen atom has to be in an excited state.

The effects of methane may involve the formations of water H_2O and OH radicals. Then, the presence of the hydroxyl radical OH contributes to stratospheric ozone depletion, as is shown by the following catalytic cycle:

$$OH + O_3 \rightarrow HO_2 + O_2, \qquad (6.6\text{-}19)$$

$$HO_2 + O \rightarrow OH + O_2, \qquad (6.6\text{-}20)$$

$$O_3 + hf \rightarrow O + O_2. \qquad (6.6\text{-}21)$$

The net effect of the chain reaction results in ozone depletion:

$$2\,O_3 + hf \rightarrow 3\,O_2. \qquad (6.6\text{-}22)$$

With regard to the above, it seems appropriate to presume that there is still insufficient clear knowledge involving the complex relationship between the enhanced greenhouse effect and the depletion of the ozone layer.

As a result of the atmospheric global circulation (see Figure 2.4-1), pollutant species such as CFC's and NO_x in the upper layers of the atmosphere will move to the northern and southern polar regions. This is the cause of the so-called *ozone hole* above the South Pole, which is a very large reduction of the ozone column, far below the common level. Since the 1980's, the hole in the ozone layer arises every year in the Antarctic springtime.

During the *polar night*, a night during from March to September at the continental ice cap of the southern polar region, temperatures of −85°C may occur (see Table 2.2-1). This extreme coldness is of significance to the ozone deficiency. In this region NW-winds prevail, while the surface roughness of the land is negligible. Therefore, powerful air flows rage, maintaining dramatically low temperatures, because moderate subtropical influences are excluded. In combination with the high pressure zone at the South Pole, the extremely low temperatures prevent the supply of fresh ozone-rich air during the polar night.

Another effect of the extreme coldness is the formation of polar stratospheric clouds, which cause that the nitrogen oxides become captured within ice crystals. Therefore, in the beginning of September, when the polar spring arises and the first sunlight penetrates, the chlorine atoms can deplete the ozone in an unobstructed way.

Halfway October, when the Sun is high enough above the South Pole to restore the ozone layer, it has been destroyed for about 60%. In November the ozone layer is warmed up sufficiently to let fresh air, containing ozone formed in other regions, enter the southern polar region. From the beginning of December, when the ozone hole has vanished, it takes to the next September to start the depletion process over again. Obviously the hole in the ozone layer is dependent on the season.

Also in the northern polar region an ozone hole has been established, which, however, is smaller-scaled, as the temperatures over the North Pole are not as low as over the South Pole. The temperature required to capture the nitrogen oxides is therefore not always achieved.

The shrinking of the ozone layer not only occurs in the polar regions, but on all geographical altitudes, except for the tropics.

Presently, the hole in the ozone layer above the South Pole may to be at its maximum. In all probability the situation will become better in the course of this century, as a result of the banning of CFC's and halons.

7 FUNDAMENTALS OF JET PROPULSION

7.1 TYPES OF JET PROPULSION SYSTEMS

The two main types of jet propulsion systems are the *turbojet* or *straight jet* and the *bypass jet engine* or *turbofan*. Both engine types are so-called *air breathing engines* because they use the oxygen from the atmospheric air to burn the kerosine fuel. The generation of thrust is based on Newton's third law of motion. According to the principle "action = reaction", the engines accelerate air backwards with respect to the aircraft.

The basic layout of the turbojet is sketched in Figure 7.1-1. The turbojet consists of five distinct components; air intake, compressor, combustor, turbine and exhaust nozzle. The core of the engine, comprising compressor, combustor and turbine is termed the *gas generator*. In Figure 7.1-1 also are shown the typical variations in pressure, temperature and velocity of the flow when passing through the engine. The changes in intake and nozzle are only effectuated by way of the form of the passages through which the air or gas flows.
Due to external compression, the air enters the intake at a velocity which is somewhat lower than the flight speed of the aircraft. In the intake the air is slowed down further, causing a pressure and temperature rise of the air. Two

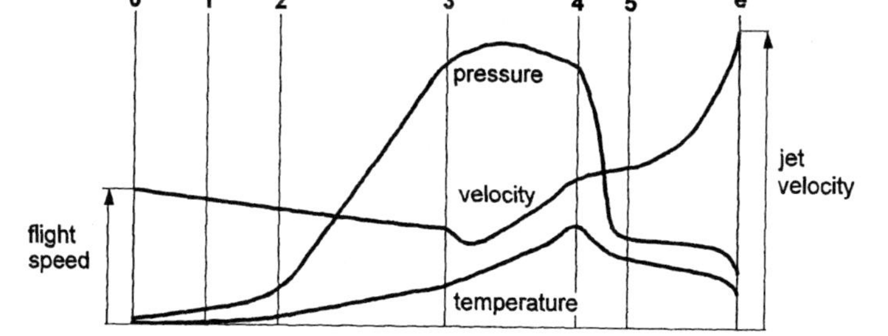

Figure 7.1-1. Basic layout of turbojet engine.

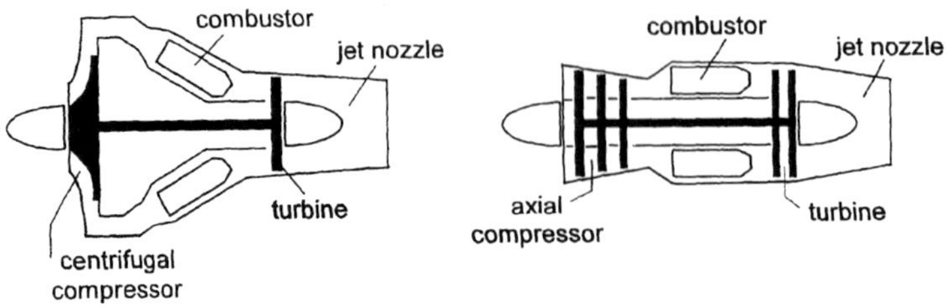

Figure 7.1-2. Compressor types.

compressor types can be distinguished, centrifugal and axial compressors (Figure 7.1-2). In the centrifugal compressor the air is taken in near the center of an impeller, which flings the air in the direction of the circumference. This gives a high pressure rise at the compressor exit.

In modern gas turbine engines, mostly the axial flow compressor is incorporated, which consists of a series of rotating many-bladed fans or rotors. These fans rotate between sets of fixed stator blades. Each set of rotating and fixed blades forms a compressor stage, where the flow is alternately accelerated and decelerated. As a result, in the axial flow compressor, which is usually composed of a large number of stages, the passing air obtains a high pressure and temperature rise. A further discussion of the characteristics of overall compressor performance will follow in Section 8.2 of Chapter 8.

In the combustor heat energy from burning fuel is added to the air flow, providing a very high temperature at the combustor exit.

The turbine, which is a series of rotating blades, extracts energy from the gas stream to drive the compressor. The *turbine entry temperature* equals the combustor exit temperature. This temperature may be restricted, because of the critical thermal load of the turbine blades (see Section 8.3).

The remaining energy from the gas generator is expanded in the propelling nozzle and ejected as high-velocity gas stream.

As will be explained in Chapter 8, the turbojet is only an effective propulsion device at very high flight speeds. Moreover, due to its high jet velocity, the turbojet not only is wasteful of fuel but also tends to be very noisy. These problems can be solved through extracting a larger part of the power from the hot gases by means of a stretched turbine, and supplying it to a ducted multi bladed fan. This arrangement, by which also cold air is accelerated, is called a *turbofan* (Figure 7.1-3). A turbofan accelerates a much larger air mass flow rate than the turbojet, through which a higher propulsive efficiency in the high-subsonic speed range is obtained. Part of the air impelled by the fan passes through the bypass duct and is discharged as a cold jet, delivering a large contribution to the engine thrust. The remainder of the fan air goes through the engine core and becomes a hot flow.

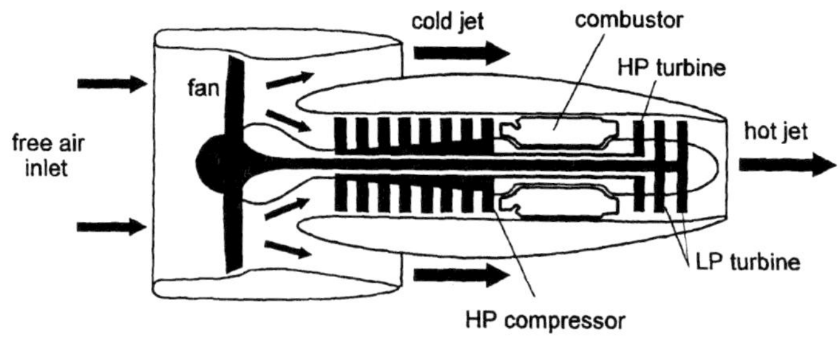

A spool is a rotating assembly in a jet engine, consisting of a turbine, a drive shaft, and a compressor or fan.

Figure 7.1-3. Two-spool turbofan layout.

A characteristic design figure for a turbofan is its *bypass ratio*, which is the ratio of the mass flow rate of the cold (secondary) flow passing through the bypass duct to the mass flow rate of the hot (primary) flow passing through the gas generator. In formula:

$$B = \frac{m_c}{m_h} , \tag{7.1-1}$$

where B is the bypass ratio and m_c and m_h are the mass flow rates of the cold and hot stream, respectively. Since the total mass flow is $m_c + m_h = m$, it follows that:

$$\frac{m_c}{m} = \frac{B}{B+1} \quad \text{and} \quad \frac{m_h}{m} = \frac{1}{B+1} . \tag{7.1-2}$$

The bypass ratio ranges from unity on small airplanes to five or more on civil airliners. Advanced propulsion systems with a bypass ratio beyond the current values are called ultra high bypass ratio (UHBR) engines.

In comparison with the turbojet, vital reductions in fuel consumption and noise generation are obtained.
It may be remarked that the early engines with a low bypass ratio were initially referred to as *bypass engines*. The name turbofan was firstly employed only for high bypass ratio engines but is presently used for most bypass engines.

On the basis of the turbojet, in this chapter the fundamentals of jet propulsion will be considered. After that, in Chapter 8, especially, the turbofan engine will be examined and in Chapter 9 propeller propulsion.

7.2 DEFINITION OF THRUST

The generation of thrust requires that the time rate of change of linear momentum for the hot gases leaving the nozzle exit is higher than for the air flow entering the intake of the engine.

To explain the calculation of thrust, we consider an engine on a pylon under a wing (Figure 7.2-1). Around the engine a control volume is specified, which extends far upstream where the pressure and velocity have their free stream values p_0 and V_0. The side-boundaries are parallel to the velocity V_0 and are sufficiently far removed from the engine that the local pressures equal the free stream pressure p_0. At the rear, the control surface coincides with the nozzle exit plane, because engine manufacturers, normally, define the thrust as the rate of change of linear momentum between the state points 0 and e in Figure 7.2-1. Most of the air entering the control volume passes around the engine and only a small part flows through the core of the engine.

Assuming one-dimensional, steady flow and neglecting the fuel mass flow rate $m_f (m_f \ll m)$, the condition for continuity is:

$$m + (S - A_0)\,\rho_0\,V_0 + \Delta m = m + (S - A_e)\,\rho_0\,V_0\,, \tag{7.2-1}$$

where $m = \rho_0\,V_0\,A_0$ is the mass flow rate passing through the engine, S is the cross-sectional area of the control volume, Δm the air mass flow rate entering the control volume through the side-boundaries, and A_e the nozzle exit area.

From Equation (7.2-1) we find:

$$\Delta m = (A_0 - A_e)\,\rho_0\,V_0\,. \tag{7.2-2}$$

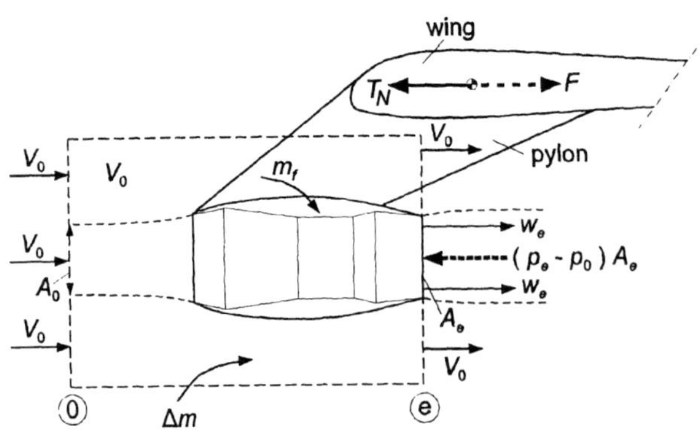

Figure 7.2-1. Definition of thrust.

If we consider relative pressures (relative to p_0) around the control surface, the momentum equation gives:

$$F - (p_e - p_0) A_e =$$
$$= m (w_e - V_0) - \rho_0 V_0^2 (S - A_0) - \rho_0 V_0^2 (A_0 - A_e) + \rho_0 V_0^2 (S - A_e) \ . \quad (7.2\text{-}3)$$

In Equation (7.2-3), $\rho_0 V_0^2 (A_0 - A_e)$ is the time rate of change of linear momentum of the air entering the side-boundaries, and F is the force exerted by the engine on the flow.
Rearranging Equation (7.2-3) yields:

$$F = m (w_e - V_0) + (p_e - p_0) A_e \ . \quad (7.2\text{-}4)$$

According to Newton's third law of motion, the *net thrust,* T_N, available for the propulsion of the airplane is also given by:

$$T_N = m (w_e - V_0) + (p_e - p_0) A_e \ . \quad (7.2\text{-}5)$$

The first term of the right-hand side of Equation (7.2-5) is called the *momentum thrust* and the second term is indicated as *pressure thrust.*
The thrust produced at $V_0 = 0$ is known as the *gross thrust,* T_G, which follows from Equation (7.2-5) as:

$$T_G = m\, w_e + (p_e - p_0) A_e \ . \quad (7.2\text{-}6)$$

A convenient form for the thrust equation is obtained when we introduce the equivalent jet velocity, $(w_e)_{eq}$, defined by:

$$(w_e)_{eq} = w_e + \frac{A_e}{m} (p_e - p_0) \ . \quad (7.2\text{-}7)$$

Using this definition we get:

$$T_N = m [(w_e)_{eq} - V_0] \ . \quad (7.2\text{-}8)$$

An important performance parameter for the relationship between the net thrust and the size of the engine is the *specific thrust* (net thrust per unit weight flow rate of air):

$$\psi_T = \frac{T_N}{m g} = \frac{m (w_e - V_0) + (p_e - p_0) A e}{m g} \ . \quad (7.2\text{-}9)$$

In the special case the nozzle exit pressure equals the ambient pressure, the latter equation reduces to:

$$\psi_T = \frac{T_N}{m g} = \frac{(w_e - V_0)}{g} \ . \quad (7.2\text{-}10)$$

Expressing the thrust as the product of specific thrust and weight flow rate of air passing through the fan will be useful because it provides an indication of the frontal area of the engine and corresponding drag at given thrust.

Before reviewing the performance calculation of the turbojet, it may be worthwhile to proceed with a brief summary of the fundamentals of thermodynamics in the next two sections.

7.3 THE FIRST AND SECOND LAW OF THERMODYNAMICS

The *first law of thermodynamics* is a formal statement of the conservation of energy in a thermodynamic system, which says that the energy supplied to the system must equal the sum of the increase in the internal energy and the energy leaving the system as work. Obviously, a system requires the use of a control surface enclosing the process. The differential form of the first law for a fixed mass m of a gas is usually written as:

$$\mathrm{d}Q = \mathrm{d}U + \mathrm{d}W , \tag{7.3-1}$$

where $\mathrm{d}Q$ is the amount of energy added to the system, $\mathrm{d}W$ the amount of work done by the system, and $\mathrm{d}U$ the corresponding change in internal energy of the system.
It is convenient to employ lowercase letters to denote the values of extensive variables per unit mass. Then, Equation (7.3-1) is expressed as:

$$\mathrm{d}q = \mathrm{d}u + \mathrm{d}w . \tag{7.3-2}$$

When the gas expands quasi-statically, i.e., when the process is carried out extremely slowly, the work done by the gas is given by $p\mathrm{d}v$, where p is the (static) pressure and $\mathrm{d}v$ is the change in volume. Then, Equation (7.3-2) becomes the form:

$$\mathrm{d}q = \mathrm{d}u + p\mathrm{d}v = \mathrm{d}u + p\mathrm{d}(1/\rho) , \tag{7.3-3}$$

where $1/\rho = v$ is the specific volume, being the volume per unit mass, and hence is the inverse of the density.
When we use the concept of enthalpy, h, defined as:

$$h = u + \frac{p}{\rho} , \tag{7.3-4}$$

we have for an infinitesimal change of state:

$$dh = du + d\left(\frac{p}{\rho}\right) = du + p\,d\left(\frac{1}{\rho}\right) + \left(\frac{1}{\rho}\right)dp \;. \tag{7.3-5}$$

Eliminating u from Equations (7.3-3) and (7.3-5) leads to:

$$dq = dh - (1/\rho)\,dp \;. \tag{7.3-6}$$

The *specific heat* c of a substance is the heat energy required to change the temperature of unit mass of substance by one degree kelvin. In formula:

$$Q = mc\Delta T \;, \tag{7.3-7}$$

where Q is the heat energy given off or absorbed, ΔT the increase or decline of temperature, and m the mass of the substance.
Contrary to solids and gases, the values of the specific heat for gases, depend on how the process is carried out. Two important processes are those in which either the volume or the pressure is kept constant.

According to the above definition, the *specific heat at constant volume,* c_v, of a gas, is the amount of heat energy that must be supplied to raise the temperature of 1 kg of the gas by one degree kelvin at constant volume.
Therefore, from Equation (7.3-3):

$$c_v = \left(\frac{dq}{dT}\right)_v = \left(\frac{\partial u}{\partial T}\right)_v \;. \tag{7.3-8}$$

When we are dealing with an ideal gas, we have the perfect gas law:

$$\frac{p}{\rho} = \frac{R_a}{M}T = RT \;, \tag{7.3-9}$$

where R_a is the universal gas constant, M is the molecular mass of the gas, and R is the specific gas constant (gas constant per unit mass). Furthermore, we have the simplifying condition that for an ideal gas the internal energy depends only on temperature:

$$u = u(T) \;. \tag{7.3-10}$$

Then, according to Equation (7.3-8), the specific heat at constant volume can be written as:

$$c_v = \frac{du}{dT} \;. \tag{7.3-11}$$

When one kilogram of an ideal gas is heated at constant pressure, we find from Equations (7.3-3) and (7.3-11):

$$dq = c_p dT = du + p d(1/\rho) = c_v dT + p d(1/\rho) \;, \tag{7.3-12}$$

where c_p is the *specific heat at constant pressure*, i.e., the amount of heat energy that must be supplied to raise the temperature of 1 kg of the gas by one degree kelvin at constant pressure.

Using the perfect gas law, Equation (7.3-12) can be modified into the form:

$$c_p dT = c_v dT + R dT \;. \tag{7.3-13}$$

Hence,

$$R = c_p - c_v \;. \tag{7.3-14}$$

Expressing the specific heat at constant pressure for an ideal gas in terms of enthalpy, we obtain from Equation (7.3-6):

$$c_p = \left(\frac{dq}{dT}\right)_p = \frac{dh}{dT} \;. \tag{7.3-15}$$

The above equations learn us that for a gas obeying the perfect gas law, the specific heats c_p and c_v are functions of temperature only.
In this respect, it is worth noting that the internal energy of a gas is composed of the translational kinetic energy of the molecules, and the energy associated with molecular rotation and vibration, and electronic excitation. However, when the molecules of the gas are dissociating into atoms, the specific heats not only are a function of temperature but also of pressure (Ref. 38).

A quantity of considerable importance in thermodynamic calculations is the ratio of specific heats:

$$\gamma = \frac{c_p}{c_v} \;. \tag{7.3-16}$$

From Equations (7.3-14) and (7.3-16) follows:

$$c_v = \frac{R}{\gamma - 1} \;, \tag{7.3-17}$$

and

$$c_p = R\frac{\gamma}{\gamma - 1} \;. \tag{7.3-18}$$

The *second law of thermodynamics* makes use of the quantity *entropy*, S, defined such that the change in entropy dS during an infinitesimal part of a reversible process is given by the relation:

$$\mathrm{d}s = \frac{\mathrm{d}q}{T} , \tag{7.3-19}$$

where s is the *specific entropy*, i.e., the entropy per unit mass of gas.
The quantity entropy is useful because the entropy of an isolated thermodynamic system never decreases. It remains constant for reversible (isentropic) processes and increases for irreversible processes.

By inserting Equation (7.3-5) into the first law of thermodynamics for a unit mass of gas at rest in the form (7.3-3), we easily obtain:

$$\mathrm{d}s = \frac{\mathrm{d}h}{T} - \left(\frac{1}{\rho T}\right) \mathrm{d}p . \tag{7.3-20}$$

Using the relation $\mathrm{d}h = c_p \, \mathrm{d}T$ (Equation (7.3-15)) and the perfect gas law (7.3-9), Equation (7.3-20) can be written as:

$$\mathrm{d}s = c_p \frac{\mathrm{d}T}{T} - R \frac{\mathrm{d}p}{p} . \tag{7.3-21}$$

This expression can be integrated, giving the entropy change per unit mass as:

$$s - s_0 = c_p \ln \frac{T}{T_0} - R \ln \frac{p}{p_0} , \tag{7.3-22}$$

where the subscript "0" denotes a reference condition of temperature and pressure.
It should be appreciated that entropy is a state variable, which can be used to describe the state of a thermodynamic system, along with p, T, ρ, and h. Its value depends only on the state of the system, and not on the history of the process or the past history of how it got there. Therefore, as visualized in Figure 7.3-1, entropy can be expressed in terms of two other quantities, e.g., temperature and pressure:

$$s = s(T,p) . \tag{7.3-23}$$

The isobars in Figure 7.3-1 can be determined by applying Equation (7.3-22) and the corresponding expression for the enthalpy: $h - h_0 = c_p \, (T - T_0)$. The reference values used for the graph are: $T_0 = 288.15$ K and $p_0 = 101325$ N/m^2 (0 m ISA).

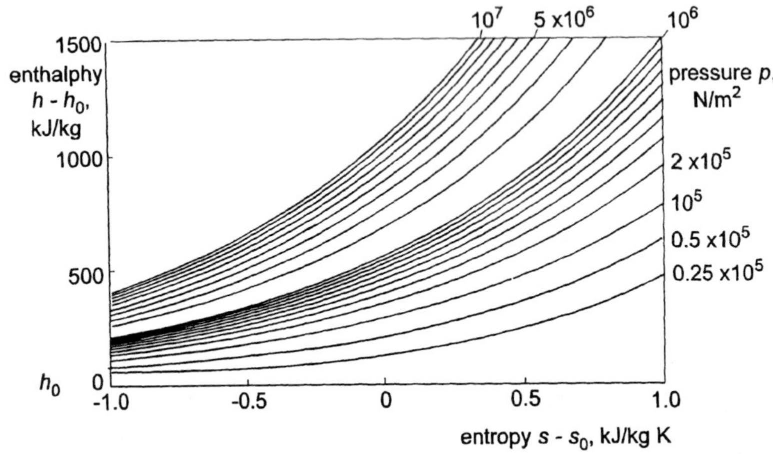

Figure 7.3-1. Enthalpy - entropy diagram for air.

For an isentropic (reversible and adiabatic) process where no heat is supplied to the gas we have $dq = 0$, $ds = 0$ and $s = s_0$. Then, Equation (7.3-22) leads to:

$$0 = \ln\frac{T}{T_0} - \frac{R}{c_p}\ln\frac{p}{p_0}\ . \qquad (7.3\text{-}24)$$

Using $R = c_p - c_v$ and $\gamma = c_p / c_v$, Equation (7.3-24) can be changed into the form:

$$\frac{T}{p^{\frac{\gamma - 1}{\gamma}}} = \text{constant}. \qquad (7.3\text{-}25)$$

Using the perfect gas law, Equation (7.3-25) can also be expressed as:

$$\frac{p}{\rho^{\gamma}} = \text{constant}, \qquad (7.3\text{-}26)$$

and

$$\frac{T}{\rho^{\gamma - 1}} = \text{constant}. \qquad (7.3\text{-}27)$$

Equations (7.3-25) to (7.3-27) are the well-known *Poisson relations* which provide information on the variations of p, T, and ρ in an isentropic process.

7.4 THE FLOW ENERGY EQUATION

To derive the flow energy equation, we consider a gas flow through the diverging channel shown in Figure 7.4-1, and define the system as the amount of gas between the cross-sections 1 and 2. Now the differential form of the first law of thermodynamics may be written as:

$$dq = de + dw \, , \tag{7.4-1}$$

where dq is the amount of energy added to the system, de is the change in internal energy, and dw is the work done by the system.
The internal energy e is all the energy inside the control surface. Ignoring potential energy, we have in a gas flow:

$$e = u + \frac{V^2}{2} \, , \tag{7.4-2}$$

where u denotes the internal molecular energy, and $V^2/2$ represents the kinetic energy of the system per unit mass.

The increment of the work per unit mass of gas flow dw can be divided into two parts:

$$dw = dw_m + p\,d\frac{1}{\rho} \, , \tag{7.4-3}$$

where again the term $pd(1/\rho)$ is the increment of work done by the gas on the control surface of the system and dw_m comprises all other forms of work extracted from the gas.
After a time dt, in Figure 7.4-1 the system will have advanced to the position 1'- 2'.

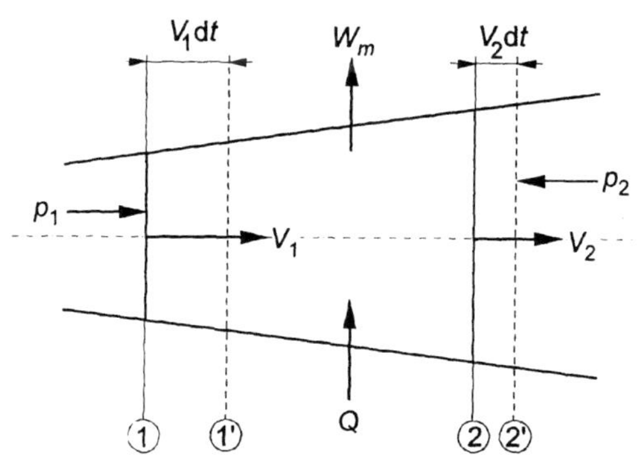

Figure 7.4-1. Derivation of flow energy equation.

Then, the principle of conservation of energy requires that the net energy transferred to the system equals the energy in region 1' - 2' at time $t + dt$ minus the energy in region 1 - 2 at time t.
Assuming steady, one-dimensional, frictionless flow, the internal energy in region 1' - 2 remains, so that the net energy supplied to the system is also given by the energy in region 2 - 2' at time $t + dt$ minus the energy in region 1 - 1' at time t.
If the properties of the masses in the regions 2 – 2' and 1 – 1' can be assumed to be the same as the properties of the gases when they pass their respective entry-sections 1 and 2, we derive the energy balance for a flowing gas as:

$$m(q - w_m) - (p_2 V_2 A_2 - p_1 V_1 A_1)]\mathrm{d}t = m\left[(u_2 - u_1) + \left(\frac{V_2^2}{2} - \frac{V_1^2}{2}\right)\right]\mathrm{d}t \ , \quad (7.4\text{-}4)$$

where m is the mass flow rate, A is the cross-sectional area of the channel, and the quantity pVA represents the work per unit time done by the system on its surroundings.
Equation (7.4-4) expresses that in the time dt, heat is supplied and energy in the form of work is extracted at rates q and w_m units of energy per unit mass.
Now, we use the continuity equation which expresses the physical principle that mass can be neither created nor destroyed. Referring to the steady flow in Figure 7.4-1, there will be no variation with time of the mass of the gas confined inside the channel between the cross-sections 1 and 2. As expressed by the equation below, the same rate of mass flow will cross each section of the channel:

$$m = \rho V A = \text{constant} . \quad (7.4\text{-}5)$$

Using the equality $m = \rho_1 V_1 A_1 = \rho_2 V_2 A_2$, we obtain the following form for Equation (7.4-4):

$$q - w_m = \left(\frac{p_2}{\rho_2} - \frac{p_1}{\rho_1}\right) + (u_2 - u_1) + \left(\frac{V_2^2}{2} - \frac{V_1^2}{2}\right) , \quad (7.4\text{-}6)$$

or in differential form:

$$\mathrm{d}q - \mathrm{d}w_m = \mathrm{d}\frac{p}{\rho} + \mathrm{d}u + V\mathrm{d}V . \quad (7.4\text{-}7)$$

When we use the concept of enthalpy, h, according to Equation (7.3-5), then, for an infinitesimal process:

$$\mathrm{d}q - \mathrm{d}w_m = \mathrm{d}h + V\mathrm{d}V = \mathrm{d}h_t , \quad (7.4\text{-}8)$$

where h_t is referred to as the *total* or *stagnation* enthalpy, defined by:

$$h_t = h + \frac{V^2}{2} . \qquad (7.4\text{-}9)$$

From Equations (7.4-8) and (7.3-15) we get:

$$dh_t = c_p dT + V dV = c_p dT_t , \qquad (7.4\text{-}10)$$

where T and T_t are the *static* and *total temperature,* respectively. Apparently, the decrease in kinetic energy is equal to the increase in enthalpy.
From Equation (7.4-10) we find that the relation between T and T_t is given by:

$$T_t = T + \frac{V^2}{2\, c_p} . \qquad (7.4\text{-}11)$$

With the *Mach number* $M = V/c$, named after the Austrian professor Ernst Mach (1838-1916), where c is speed of sound, $c = \sqrt{\gamma R T}$, and $\gamma = c_p / c_v$, we can transform Equation (7.4-11) into:

$$T_t = T\left(1 + \frac{(\gamma - 1)}{2} M^2 \right) . \qquad (7.4\text{-}12)$$

Physically, total enthalpy h_t and total temperature T_t are the enthalpy and temperature, respectively, which a gas flow of enthalpy h or static temperature T and velocity V would have when brought to rest adiabatically.

With Equations (7.3-25) and (7.3-27) we also find for an isentropic process:

$$p_t = p\left(1 + \frac{(\gamma - 1)}{2} M^2 \right)^{\frac{\gamma}{\gamma - 1}} \qquad (7.4\text{-}13)$$

and

$$\rho_t = \rho\left(1 + \frac{(\gamma - 1)}{2} M^2 \right)^{\frac{1}{\gamma - 1}} , \qquad (7.4\text{-}14)$$

where p_t and ρ_t is the total pressure and density, respectively. Their values occur when the flow is slowed down isentropically to zero velocity.

Equations (7.4-11) to (7.4-14) show that static temperature, pressure, density, and velocity of a moving gas are equivalent in terms of energy to a total temperature, total pressure and total density. This will be useful in studying the gas flow through a jet engine.

By using the relations: $M = V/c$, $c = \sqrt{\gamma R T}$ and $\gamma = c_p / c_v$, Equation (7.4-13) can be written as:

$$p_t = p\left(1 + \frac{(\gamma - 1)}{2\gamma}\frac{\rho}{p}V^2\right)^{\frac{\gamma}{\gamma - 1}} . \tag{7.4-15}$$

The latter form is known as *Bernoulli's equation for compressible flow.*

Combining Equations (7.4-8) and (7.4-10), the steady flow energy equation can be written in the in the following differential form:

$$dq - dw_m = c_p dT_t \, . \tag{7.4-16}$$

In terms of finite quantities we have:

$$Q - W_m = mc_p \Delta T_t \, , \tag{7.4-17}$$

where Q and W_m are the heat and mechanical work transfers per unit time, and m is the rate of mass flow.

The steady flow energy equation in the form (7.4-17) will appear to be very useful in determining the performance of jet engines.

It should be appreciated that the entropy change of a flowing gas mass can be determined by replacing T and p by T_t and p_t in Equation (7.3-22):

$$s - s_0 = c_p \ln\frac{T_t}{T_{t_0}} - R\ln\frac{p_t}{p_{t_0}} \, , \tag{7.4-18}$$

where the subscript "0" denotes the reference condition of total temperature and total pressure.

For example, the relation between the entropy rise and the decrease in total pressure $(p_{t_2} < p_{t_0})$ in a one-dimensional, steady, adiabatic flow ($T_t = T_{t_0}$ = constant) follows from Equation (7.4-18) as:

$$\Delta s = s_2 - s_0 = R\ln\frac{p_{t_0}}{p_{t_2}} \, , \tag{7.4-19}$$

where the subscripts "0" and "2" denote the condition at the entry and exit of the control volume (see the intake flow in Figure 7.1-1).

We conclude this section with considering the speed of sound. This quantity, by definition, is the rate at which infinitesimal small disturbances propagate through a medium.

In order to derive an expression for the speed of sound, we examine a disturbance front or *sound wave*, moving with speed c through a perfect gas at rest in a

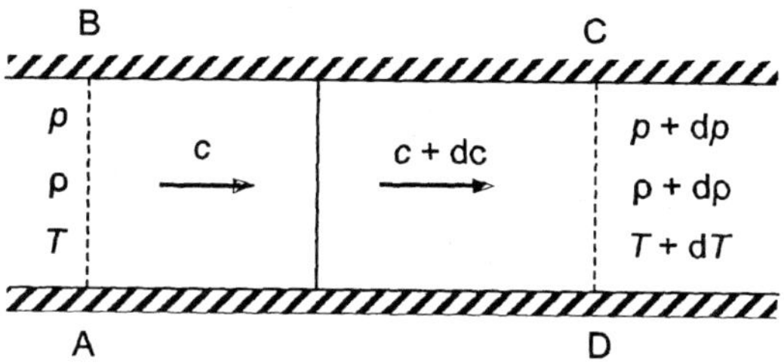

Figure 7.4-2. Propagation of a sound wave.

constant area duct (Figure 7.4-2).

If the coordinate system is fixed to the wavefront and the disturbance is moving to the left, then the gas will approach the front with speed c. The pressure, density and temperature are p, ρ and T, respectively. Behind the front, the speed of the gas flow is $c + \mathrm{d}c$, and the pressure, density and temperature are $p + \mathrm{d}p$, $\rho + \mathrm{d}\rho$ and $T + \mathrm{d}T$, respectively.

Applying the continuity equation, Equation (7.4-5), to the control volume $ABCD$ in Figure 7.4-2 requires that:

$$\mathrm{d}(\rho c) = 0 \quad \text{or} \quad \rho \mathrm{d}c + c\mathrm{d}\rho = 0 \;, \tag{7.4-20}$$

hence,

$$c = -\rho \frac{\mathrm{d}c}{\mathrm{d}\rho} \;. \tag{7.4-21}$$

The conservation of the momentum theorem for the control volume states that the net force acting on the gas along the boundaries of the control volume must be equal to the variation of the momentum per unit time, i.e.,

$$-\mathrm{d}p = \mathrm{d}(\rho c^2) = c^2\mathrm{d}\rho + 2\rho c \mathrm{d}c \;. \tag{7.4-22}$$

Eliminating $\mathrm{d}c$ from Equations (7.4-21) and (7.4-22), we find the speed of sound is:

$$c^2 = \frac{\mathrm{d}p}{\mathrm{d}\rho} \;. \tag{7.4-23}$$

Assuming that the flow through the sound wave is isentropic, the relationship between pressure and density is given by Equation (7.3-26):

$$\frac{p}{\rho^\gamma} = \text{constant } (C) \;. \tag{7.4-24}$$

Hence,

$$\frac{dp}{d\rho} = \frac{d(C\rho^{\gamma})}{d\rho} = \gamma C\rho^{(\gamma - 1)} = \gamma \frac{p}{\rho} . \qquad (7.4\text{-}25)$$

Combination of Equations (7.4-23) and (7.4-25) results in the following expression for the speed of sound:

$$c = \sqrt{\gamma \frac{p}{\rho}} . \qquad (7.4\text{-}26)$$

Substitution of the perfect gas law, Equation (7.3-9), into Equation (7.4-26) yields:

$$c = \sqrt{\gamma R T} . \qquad (7.4\text{-}27)$$

This equation shows that the speed of sound in a perfect gas depends only on the absolute temperature of the gas.

7.5 THE IDEAL TURBOJET CYCLE

In examining the performance of a turbojet engine, we may consider the *ideal* or *Brayton cycle*. The idealization of the various processes includes the following assumptions:

a. The working fluid behaves as a perfect gas with constant specific heats and has the same composition throughout the cycle.

b. The compression and expansion processes occur isentropically.

c. Combustion takes place at constant total pressure.

d. The mass flow rate of the working fluid is constant throughout the cycle ($m_f \ll m$).

e. In the convergent exhaust nozzle the flow is expanded to the free stream static pressure.

The ideal turbojet cycle is depicted in Figure 7.5-1 in terms of pressure p versus specific volume $v = 1/\rho$. The various engine components are assigned by numerical stations as defined in Figure 7.1-1. The analogous representation of the cycle in the *Mollier* or *h - s diagram* is given in Figure 7.5-2.

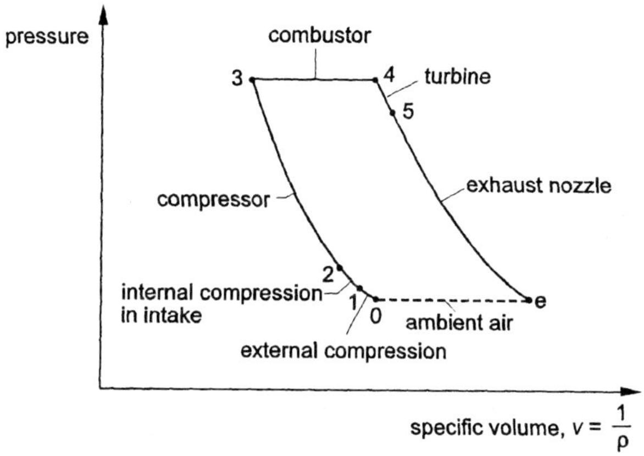

Figure 7.5-1. Ideal turbojet cycle.

As discussed in Section 7.4, static temperature, static pressure, and velocity of a gas flow are equivalent to total temperature and pressure. Hence, in studying the conditions of the flow passing the engine, kinetic energy is taken into account implicitly by using the total properties T_t and p_t according to Equations (7.4-12) and (7.4-13).
Starting from the free stream conditions p_0, T_0, and V_0 (or M_0) in point 0 in Figures 7.5-1 and 7.5-2, the air flows through the various engine components.

1. The intake (0 – 2). Due to the external and internal deceleration of the air

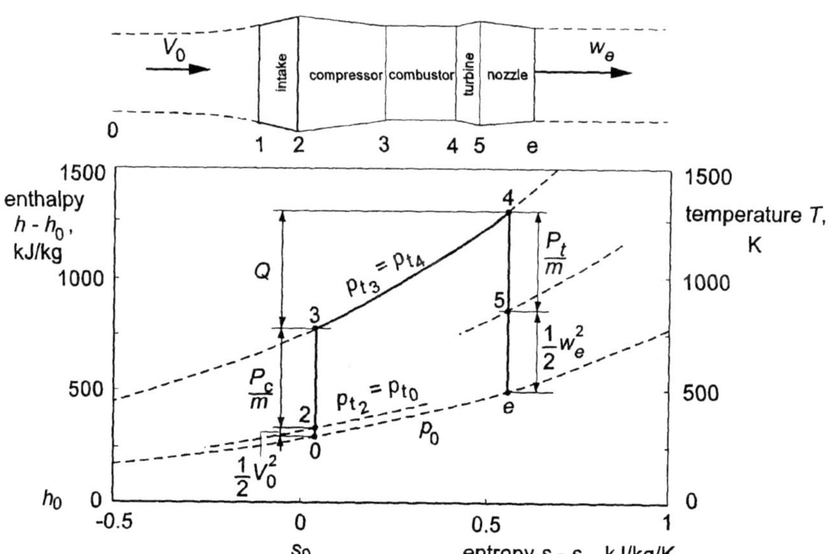

Figure 7.5-2. H-s diagram for ideal turbojet cycle.

relative to the engine, the air pressure rises from p_0 to $p_{t_2} = p_{t_0}$ at the inlet of the compressor:

$$p_{t_2} = p_0 \left(1 + \frac{(\gamma - 1)}{2} M_0^2 \right)^{\frac{\gamma}{\gamma - 1}} . \tag{7.5-1}$$

Likewise, the temperature at point 2 equals the free stream total temperature;

$$T_{t_2} = T_0 \left(1 + \frac{(\gamma - 1)}{2} M_0^2 \right) . \tag{7.5-2}$$

2. The compressor (2 – 3). From Equations (7.4-12) and (7.4-13) we obtain:

$$\frac{T_t}{T} = \left(\frac{p_t}{p} \right)^{\frac{\gamma}{\gamma - 1}} . \tag{7.5-3}$$

Further, according to Equation (7.3-25) we have $T / p^{\{(\gamma - 1) / \gamma\}}$ = constant for an isentropic process, so that Equation (7.5-3) becomes:

$$\frac{T_t}{p_t^{\frac{\gamma - 1}{\gamma}}} = \text{constant} . \tag{7.5-4}$$

Applying Equation (7.5-4) to the compression process, we find the compressor outlet temperature from:

$$T_{t_3} = T_{t_2} (\varepsilon_c)^{\frac{\gamma - 1}{\gamma}} , \tag{7.5-5}$$

where $\varepsilon_c = p_{t_3} / p_{t_2}$ is the pressure ratio of the compressor.

3. The combustor (3 – 4). Because the combustion takes place at constant total pressure, we have:

$$p_{t_4} = p_{t_3} \cdot \tag{7.5-6}$$

From the flow energy equation for steady, one-dimensional, frictionless flow, Equation (7.4-17), the heat added to the flow per unit time is given by:

$$Q = m_f H = m c_p (T_{t_4} - T_{t_3}) , \tag{7.5-7}$$

where H is the heating value of the fuel (see the previous Table 1.5-1), T_{t_4} is the *turbine entry temperature*, and c_p is the specific heat of air at constant pressure.

4. The turbine (4 – 5). The temperature at the turbine exit is found from the condition that turbine power (P_t) equals compressor power (P_c):

$$m c_p (T_{t_4} - T_{t_5}) = m c_p (T_{t_3} - T_{t_2}) \ , \tag{7.5-8}$$

or

$$T_{t_4} - T_{t_5} = T_{t_3} - T_{t_2} \ . \tag{7.5-9}$$

The pressure at the turbine exit follows from:

$$p_{t_5} = p_{t_4} \left(\frac{T_{t_5}}{T_{t_4}} \right)^{\frac{\gamma}{\gamma - 1}} \ . \tag{7.5-10}$$

5. The exhaust nozzle (5 – e). Since the high pressure, high temperature gases are expanded down isentropically to ambient pressure p_0, the Mach number M_e at the nozzle exit is given by:

$$\frac{p_{t_e}}{p_e} = \frac{p_{t_5}}{p_0} = \left(1 + \frac{\gamma - 1}{2} M_e^2 \right)^{\frac{\gamma}{\gamma - 1}} \ , \tag{7.5-11}$$

or

$$M_e = \sqrt{\frac{2}{\gamma - 1} \left[\left(\frac{p_{t_e}}{p_0} \right)^{\frac{\gamma - 1}{\gamma}} - 1 \right]} \ . \tag{7.5-12}$$

Referring to Equations (7.4-12) and (7.4-13), we see that the temperature T_e is obtained from:

$$T_e = \frac{T_{t_e}}{1 + \frac{\gamma - 1}{2} M_e^2} = \frac{T_{t_e}}{\left(\frac{p_{t_e}}{p_0} \right)^{\frac{\gamma - 1}{\gamma}}} \ . \tag{7.5-13}$$

Using the expression for the speed of sound, $c = \sqrt{\gamma R T}$, the jet velocity can be written as:

$$w_e = M_e\sqrt{\gamma R T_e} = \sqrt{\frac{2\gamma R}{\gamma - 1} T_{t_e}\left[1 - \left(\frac{p_0}{p_{t_e}}\right)^{\frac{\gamma-1}{\gamma}}\right]} . \tag{7.5-14}$$

Substituting Equation (7.5-14) into Equation (7.2-8) yields the specific thrust:

$$\psi_T = \frac{1}{g}\left[\sqrt{\frac{2\gamma R}{\gamma - 1} T_{t_e}\left[1 - \left(\frac{p_0}{p_{t_e}}\right)^{\frac{\gamma-1}{\gamma}}\right]} - V_0\right] , \tag{7.5-15}$$

where ψ_T is the net thrust T_N per unit weight flow rate of air through the engine.

The efficiency of energy conversion within the engine, is called the *thermal efficiency*, η_{th} of the cycle. This quantity is defined as the ratio of the increase in kinetic energy of the air mass flow rate through the engine divided by the heat energy, added to the core flow in unit time:

$$\eta_{th} = \frac{P_j}{m_f H} = \frac{\frac{1}{2} m (w_e^2 - V_0^2)}{m_f H} , \tag{7.5-16}$$

where $P_j = \frac{1}{2}\, m\, (w_e^2 - V_0^2)$ is the net power output of the engine, i.e., the useful power for propulsion.
On the basis of Equation (7.4-11) we can express the power output P_j as:

$$P_j = m c_p [(T_{t_e} - T_e) - (T_{t_0} - T_0)] , \tag{7.5-17}$$

and under Equation (7.4-17), the heat supplied to the flow as:

$$m_f H = m c_p (T_{t_4} - T_{t_3}) . \tag{7.5-18}$$

The thermal efficiency can now be written as:

$$\eta_{th} = \frac{(T_{t_e} - T_e) - (T_{t_0} - T_0)}{(T_{t_4} - T_{t_3})} = \frac{(T_{t_e} - T_{t_0}) - (T_e - T_0)}{(T_{t_4} - T_{t_3})} . \tag{7.5-19}$$

Using Equation (7.5-9) and the equalities: $T_{t_e} = T_{t_5}$ and $T_{t_2} = T_{t_0}$, we have:

$$T_{t_4} - T_{t_e} = T_{t_3} - T_{t_0} \text{ or } T_{t_e} - T_{t_0} = T_{t_4} - T_{t_3} . \tag{7.5-20}$$

From Equations (7.5-19) and (7.5-20) we find the expression:

$$\eta_{th} = 1 - \frac{T_e - T_0}{T_{t_4} - T_{t_3}} . \tag{7.5-21}$$

Introducing the overall pressure ratio $\varepsilon = \dfrac{p_{t_3}}{p_0}$ of the working cycle:

$$\varepsilon = \frac{p_{t_3}}{p_0} = \frac{p_{t_3}}{p_{t_2}} \frac{p_{t_2}}{p_0} = \varepsilon_c \left(1 + \frac{\gamma - 1}{2} M_0^2 \right)^{\frac{\gamma}{\gamma - 1}} . \tag{7.5-22}$$

We can write:

$$\frac{T_{t_3}}{T_0} = \frac{p_{t_3}}{p_0}^{\frac{\gamma - 1}{\gamma}} = \varepsilon^{\frac{\gamma - 1}{\gamma}} , \tag{7.5-23}$$

and

$$\frac{T_{t_4}}{T_e} = \frac{p_{t_4}}{p_e}^{\frac{\gamma - 1}{\gamma}} = \frac{p_{t_3}}{p_0}^{\frac{\gamma - 1}{\gamma}} = \varepsilon^{\frac{\gamma - 1}{\gamma}} . \tag{7.5-24}$$

From Equations (7.5-23) and (7.5-24) we find:

$$\frac{T_{t_3}}{T_0} = \frac{T_{t_4}}{T_e} = \varepsilon^{\frac{\gamma - 1}{\gamma}} . \tag{7.5-25}$$

Finally, by combination of the Equations (7.5-25) and (7.5-21), the expression for thermal efficiency becomes:

$$\eta_{th} = 1 - \frac{1}{\varepsilon^{\frac{\gamma - 1}{\gamma}}} . \tag{7.5-26}$$

A graphic representation of Equation (7.5-26) is given in Figure 7.5-3, emphasizing the importance of a high flight Mach number and a high compressor pressure ratio. The latter requirement, in turn, means a high thrust setting Γ, i.e., a high compressor rpm (revolutions per minute) or engine speed.

The optimum thermal efficiency as follows from Equation (7.5-26) is independent of the temperature to which the air flow is raised in the combustor. However, it is worth pointing out that the compression of the air causes an

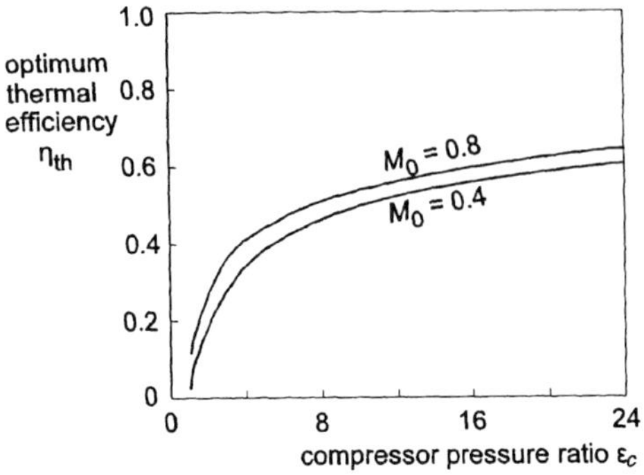

Figure 7.5-3. Optimum thermal efficiency.

increase in temperature at the entry of the combustor. As a consequence, the higher the overall pressure ratio, the lower the amount of fuel which can be burned without going beyond the limit of the turbine entry temperature (see Chapter 10 for a more detailed discussion). Certainly, the maximum overall pressure ratio increases with increasing allowable turbine entry temperature and with that the maximum achievable thermal efficiency.

Next we consider the *propulsive* or *Froude efficiency*, which is defined as the ratio of the power directly available for the propulsion of the airplane to the increase in kinetic energy of the air mass flow rate through the engine:

$$\eta_j = \frac{P_a}{P_j} = \frac{T_N V_0}{\frac{1}{2} m \left(w_e^2 - V_0^2 \right)} , \qquad (7.5\text{-}27)$$

where $P_a = T_N V$ is called *power available*.
Substitution of Equation (7.2-5) into Equation (7.5-26) yields with $p_e = p_0$:

$$\eta_j = \frac{m\,(w_e - V_0)\,V_0}{\frac{1}{2} m \left(w_e^2 - V_0^2 \right)} = \frac{2}{1 + \frac{w_e}{V_0}} = \frac{2}{2 + \frac{T_N}{m V_0}} . \qquad (7.5\text{-}28)$$

Although approximate, Equation (7.5-28) shows the essential properties associated with propulsion. If the jet velocity is nearly equal to the flight speed, the kinetic energy of the jet is used very efficiently and the propulsive efficiency tends to unity. Unfortunately, according to $T_N = m\,(w_e - V_0)$, the net thrust goes to zero as w_e tends to V_0.
The *overall* or *total efficiency* of the propulsion process is the ratio of the power available for propulsion and the heat energy supplied to the flow in unit time:

$$\eta_{tot} = \frac{T_N V_0}{m_f H} \ . \qquad (7.5\text{-}29)$$

Examining Equations (7.5-16) and (7.5-27), we observe that:

$$\eta_{tot} = \eta_{th}\,\eta_j \ . \qquad (7.5\text{-}30)$$

Another important engine characteristic is the amount of fuel required to generate a given net thrust. This quantity may be expressed by the fuel weight flow rate F, divided by the net thrust:

$$c_T = \frac{F}{T_N} \ , \qquad (7.5\text{-}31)$$

where c_T is the *thrust specific fuel consumption* (TSFC), or shortly *specific fuel consumption* (SFC), which has the dimensions N/Ns or N/Nh.
From the definitions given above, it follows that the specific fuel consumption is inversely proportional to overall efficiency and heating value of the fuel, and proportional to flight speed V_0:

$$c_T = \frac{g\,V_0}{H\,\eta_{tot}} \ . \qquad (7.5\text{-}32)$$

In Figure 7.5-4 are shown typical results of "design point" calculations, i.e., the performance calculated for specified values of flight altitude, airspeed, turbine entry temperature, and compressor pressure ratio. Considered is an ideal turbojet cycle, applying to a flight at an altitude of H = 11 km (ISA) and a flight Mach number M_0 = 0.8. Plotted is specific fuel consumption versus specific thrust for a number of compressor pressure ratios and turbine entry temperatures. The curves

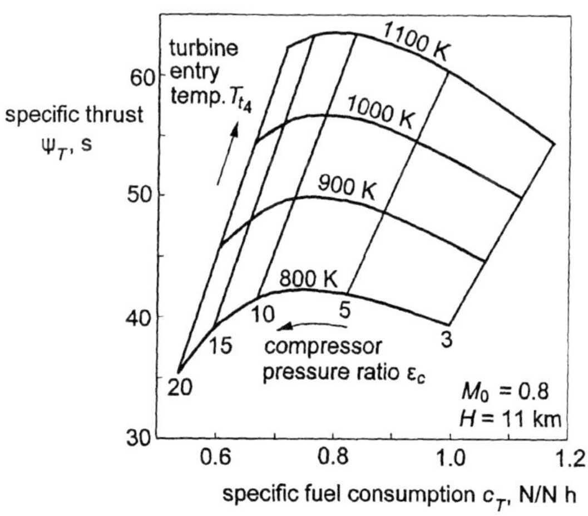

Figure 7.5-4. Specific thrust and specific fuel consumption for ideal turbojet cycle.

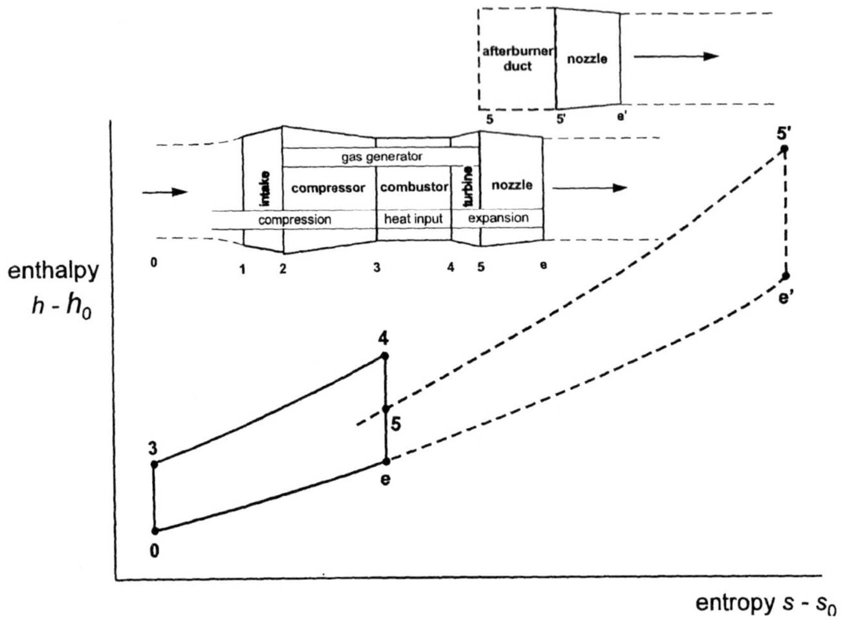

Figure 7.5-5. Cycle of ideal turbojet with afterburner.

show that specific thrust and specific fuel consumption increase with increasing turbine entry temperature. At constant T_{t4}, however, an increase in pressure ratio ε_c causes a reduction in specific fuel consumption.
Turbojet engines may be equipped with a reheat system or afterburner. Then supplementary fuel is burned downstream of the turbine and upstream of the exhaust nozzle. Figure 7.5-5 shows the reheat cycle of the ideal turbojet with an afterburner. Burning of extra fuel is possible because the exhaust gas entering the nozzle comprises an excess of unburned oxygen. Afterburning is a method to achieve a temporary increase in thrust for takeoff and climb or for transition from subsonic to supersonic flight velocities.
When the engine is equipped with an afterburner, it is essential that the cross-sectional area of the nozzle outlet is variable, because of the large change in density of the gas stream resulting from the increased gas temperature (see also Section 7.8).

Finally, it should be emphasized that when the exhaust gases are not expanded fully to p_0 in the nozzle, the pressure p_e in the exit plane will be greater than p_0. Consequently, there will be a pressure thrust as expressed by Equation (7.2-5); see also Section 7.8. Then, thermal and propulsive efficiency must be determined by employing the original definitions of η_{th} and η_j, according to the Equations (7.5-16) and (7.5-27).

7.6 COMPONENT EFFICIENCIES

From the calculations in Section 7.5 we conclude that for given flight altitude and airspeed, both specific thrust and specific fuel consumption for the ideal turbojet cycle are completely determined by compressor pressure ratio and turbine entry temperature.
The actual performance, however, will also depend on the efficiencies of the engine components. Deviations from the ideal behavior arise mainly from the effects of friction and turbulence in the flow, through which for example we have non isentropic compression in the intake and compressor, and non isentropic expansion in the turbine and nozzle. In the following a short description of the definitions of most of the imperfections is given.

In the *intake* the air is decelerated to a very low velocity before it flows into the compressor. Because there is no heat transfer, the total temperature remains constant as the flow velocity reduces from the flight speed ahead of the intake to a low velocity at the compressor inlet. At subsonic airspeeds, viscous effects result in a reduction of total pressure, whilst for supersonic flight speeds a further cause for total pressure loss is the occurrence of shock waves. To describe the efficiency of the intake, the ratio of total pressure at the compressor inlet to the free stream total pressure may be used:

$$\eta_R = \frac{p_{t_2}}{p_{t_0}} . \tag{7.6-1}$$

The quantity η_R is termed the *pressure recovery factor* of the intake. At subsonic airspeeds η_R may have a value of about 0.95 (Ref. 8).
The pressure ratio (7.6-1) is represented on the *h-s* diagram in Figure 7.6-1. The

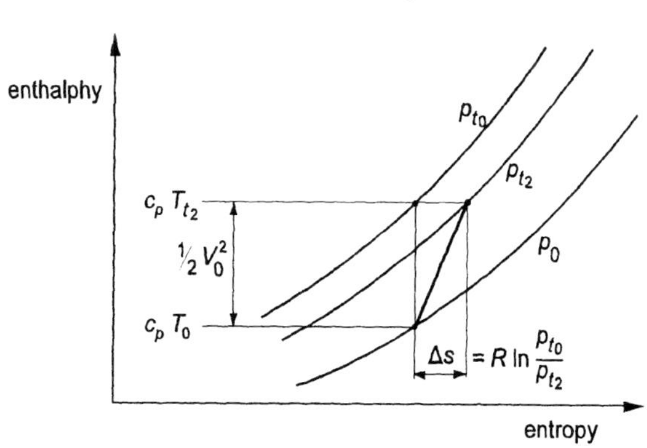

Figure 7.6-1. Representation of intake efficiency.

change in entropy Δs is given by the previous equation (7.4-19).

For intakes designed for subsonic airspeeds, η_R is experienced to be virtually constant with flight Mach number. At supersonic airspeeds it is most common to specify the intake losses by means of the intake pressure recovery factor as a function of flight Mach number M_0 .

The general form of the enthalpy - entropy diagram for an actual turbojet cycle is shown in Figure 7.6-2, aside from the assumption that the gases in the propelling nozzle expand to the free stream pressure ($p_e = p_0$).
The efficiency of the *compressor*, η_c, may be expressed in terms of the compressor isentropic efficiency, which is the ratio of the ideal and actual compressor powers:

$$\eta_c = \frac{m\, c_p\,((T_{t_3})_{\text{is}} - T_{t_2})}{m\, c_p\,(T_{t_3} - T_{t_2})} = \frac{(T_{t_3})_{\text{is}} - T_{t_2}}{T_{t_3} - T_{t_2}} \; . \tag{7.6-2}$$

Equation (7.6-2) expresses that for a given compressor pressure ratio, the actual total temperature rise exceeds the isentropic value so that the actual compression

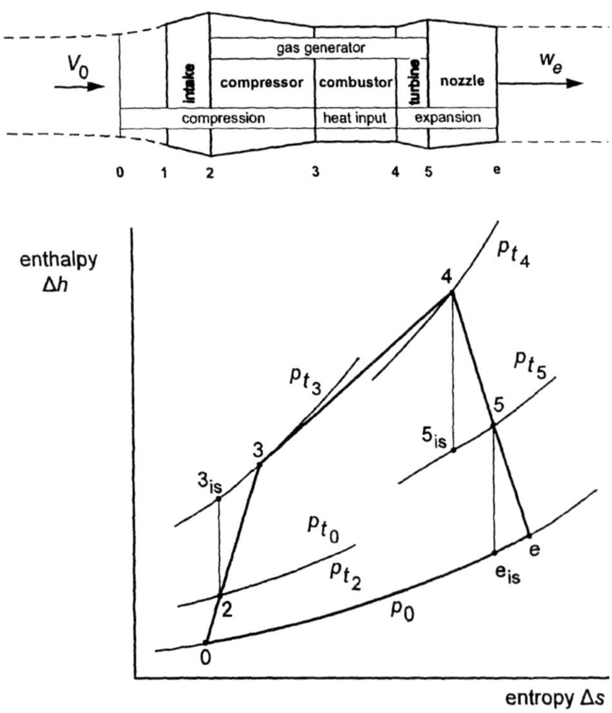

Figure 7.6-2. Actual turbojet cycle ($p_e = p_0$).

process requires more power than the ideal process.

The rise of temperature for a given compressor pressure ratio ε_c, from Equation (7.6-2), is found as follows:

$$T_{t_3} - T_{t_2} = \frac{T_{t_2}}{\eta_c}\left(\varepsilon_c^{\frac{\gamma-1}{\gamma}} - 1\right) . \tag{7.6-3}$$

A drawback of the isentropic efficiency is the fact that it is dependent on the magnitude of the compressor pressure ratio. For this reason, also the concept of *polytropic efficiency is used.* This is the isentropic efficiency of an infinitely small pressure change such that it has a fixed value throughout the entire process. Then:

$$\eta_{\text{pol}} = \frac{(\mathrm{d}T_t)_{\text{is}}}{\mathrm{d}T_t} = \text{constant} . \tag{7.6-4}$$

Also, from Equation (7.5-4), we have for an isentropic process:

$$\frac{(\mathrm{d}T_t)_{\text{is}}}{T_t} = \frac{\gamma-1}{\gamma}\frac{\mathrm{d}p_t}{p_t} . \tag{7.6-5}$$

Combination of Equations (7.6-4) and (7.6-5) gives:

$$\eta_{\text{pol}}\frac{(\mathrm{d}T_t)_{\text{is}}}{T_t} = \frac{\gamma-1}{\gamma}\frac{\mathrm{d}p_t}{p_t} . \tag{7.6-6}$$

Integrating between the limits 2 and 3 furnish:

$$\eta_{\text{pol}} = \frac{\ln\left(\frac{p_{t_3}}{p_{t_2}}\right)^{\frac{\gamma-1}{\gamma}}}{\ln\left(\frac{p_{t_3}}{p_{t_2}}\right)} , \tag{7.6-7}$$

or

$$\frac{T_{t_3}}{T_{t_2}} = \varepsilon_c^{\frac{\gamma-1}{\eta_{\text{pol}}\gamma}} . \tag{7.6-8}$$

Substitution of Equation (7.6-8) into Equation (7.6-3) produces the following relation between the polytropic and the isentropic efficiency of the compression process:

$$\eta_c = \frac{\varepsilon_c^{\frac{\gamma - 1}{\gamma}} - 1}{\varepsilon_c^{\frac{\gamma - 1}{\eta_{pol}\gamma}} - 1} . \tag{7.6-9}$$

Similarly, non isentropic expansion in the turbine causes that for a given turbine pressure ratio the gas comes out at a higher temperature than in the ideal case (see Figure 7.6-2). This behavior may be expressed by the *turbine isentropic efficiency*, which is defined as the ratio of the actual and ideal turbine powers:

$$\eta_t = \frac{m\, c_p\, (T_{t_4} - T_{t_5})}{m\, c_p\, (T_{t_4} - (T_{t_5})_{is})} = \frac{T_{t_4} - T_{t_5}}{T_{t_4} - (T_{t_5})_{is}} . \tag{7.6-10}$$

Combining Equations (7.5-10) and (7.6-10) results in the following expression for the turbine pressure ratio at a given temperature drop:

$$\frac{p_{t_5}}{p_{t_4}} = \left(1 - \frac{T_{t_4} - T_{t_5}}{\eta_t\, T_{t_4}} \right)^{\frac{\gamma}{\gamma - 1}} . \tag{7.6-11}$$

For an expansion we have: $\eta_{pol} = dT_t / (dT_t)_{is}$, yielding:

$$\frac{p_{t_5}}{p_{t_4}} = \left(\frac{T_{t_5}}{T_{t_4}} \right)^{\frac{y}{\eta_{pol}(\gamma - 1)}} . \tag{7.6-12}$$

Consequently,

$$\eta_t = \frac{1 - \left(\frac{T_{t_5}}{T_{t_4}} \right)^{\frac{\eta_{pol}(\gamma - 1)}{\gamma}}}{1 - \left(\frac{T_{t_5}}{T_{t_4}} \right)^{\frac{\gamma - 1}{\gamma}}} . \tag{7.6-13}$$

Typical variations of η_c and η_t as functions of compressor pressure ratio are given in Figure 7.6-3 for a polytropic efficiency of 0.85. The curves show that η_c decreases and η_t increases with increasing pressure ratio.

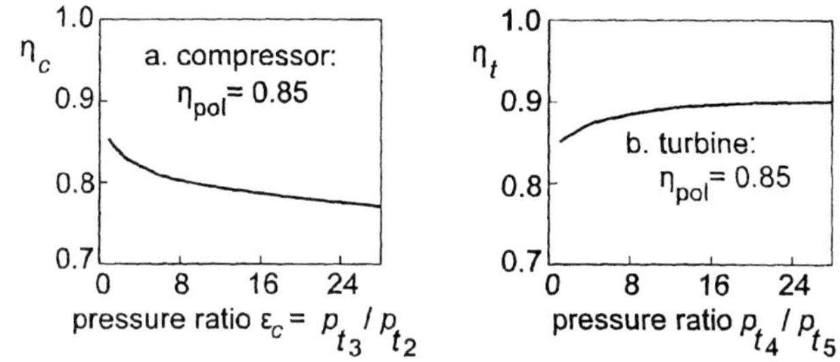

Figure 7.6-3. Typical variations of η_c and η_t.

Another factor that affects the output of the actual turbojet cycle is the variation of specific heat c_p with the conditions of the gas. Under normal working conditions, generally, c_p is a function of temperature only.
Figure 7.6-4 shows the variations of c_p and γ with temperature for air and typical combustion gases (Ref. 40). The curves indicate that c_p rises somewhat with increasing temperature and with increasing fuel-air ratio. The opposite is true of the specific heat ratio , $\gamma = c_p / c_v$, since this quantity is given by (see Chapter 3):

$$\frac{1}{\gamma} = 1 - \frac{R_a}{M c_p} , \tag{7.6-14}$$

where R_a is the universal gas constant and M is the molecular mass of the gas.
In Figure 7.6-4 use is made of the fact that the molecular mass of combustion gases is approximately equal to that of air so that γ is related to c_p by Equation (7.6-14) with $R = R_a / M = 287$ m^2 /s^2 K.

For jet engine design calculations, it is common practice to use particular mean

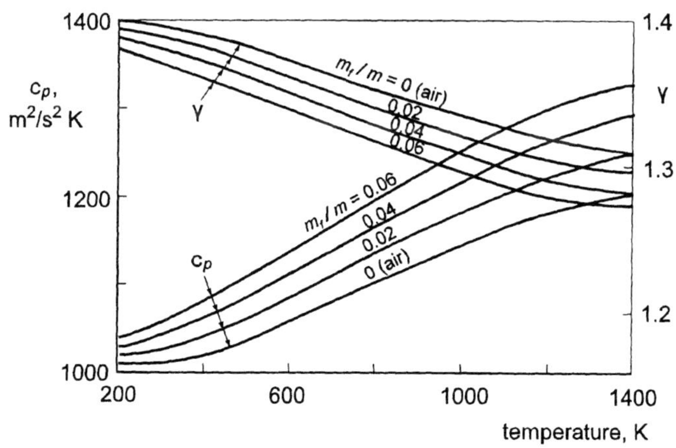

Figure 7.6-4. c_p and γ versus temperature.

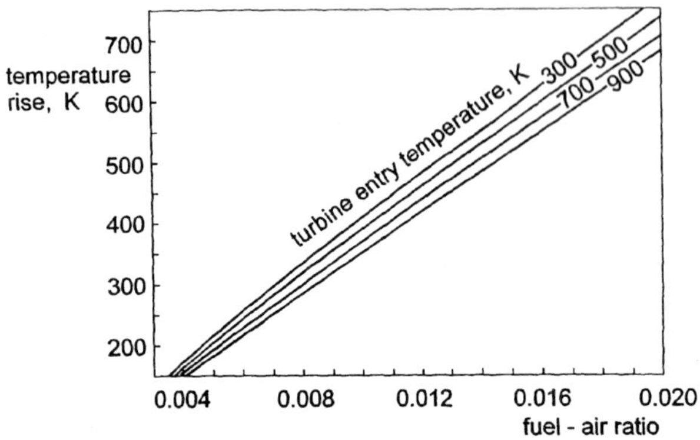

Figure 7.6-5. Temperature rise by combustion.

values of c_p and γ for the compression and expansion processes, namely for intake and compressor (air): c_p = 1005 m^2/s^2 K, γ = 1.4, and for turbine and exhaust nozzle (gas): c_{pg} = 1147 m^2/s^2 K, γ_g = 1.333.
In order to establish the fuel consumption for given values of T_{t_3} and T_{t_4}, it is essential to match the heat energy, that is, to balance the sum of the heat energy in the air mass flow rate at the inlet of the combustion chamber and the fuel mass flow rate at the fuel temperature to that in the combustion gas at the turbine entry temperature. The required computations may be facilitated by applying the curves given in Figure 7.6-5, which show the increase of temperature $(T_{t_4} - T_{t_3})$ as a function of fuel-air ratio for a number of initial temperatures T_{t_3} (Ref. 40). These data concern complete burning of a reference fuel. When making use of Figure 7.6-5, the fuel mass flow rate is easily obtained from the product of fuel-air ratio and air mass flow rate.

The *combustion* process involves two losses, a loss due to imperfect conversion of fuel into heat energy and a drop in total pressure. A few remarks with regard to these subjects will be made in Chapter 10. A thorough discussion of combustion losses, however, is left to the more advanced studies of propulsion.

Finally, we mention that driving the compressor by the turbine is coupled with losses due to bearing friction. For this purpose the mechanical efficiency, η_m, is introduced, which may be defined as the ratio between the power of compressor and turbine:

$$\eta_m = \frac{[c_p(T_{t_3} - T_{t_2})]}{[c_{pg}(T_{t_5} - T_{t_4})]}\,, \tag{7.6-15}$$

where η_m, typically, has a value of about 0.95 (Ref. 41).

For the turbojet cycle in Figure 7.6-2, it is assumed that there is also some loss of total pressure when the gases are expanded in the propelling nozzle from p_{t_5} to $p_e = p_0$.

To take account of exhaust nozzle loss in cycle calculations, the nozzle isentropic efficiency, η_n, may be used, which is defined as the ratio of the actual kinetic energy of the gas stream at the nozzle exit to that which would have obtained from isentropic flow when in both cases the expansion to the free stream pressure p_0 is completed in the nozzle:

$$\eta_n = \frac{w_e^2}{(w_e)^2_{\text{is}}}\,. \tag{7.6-16}$$

Using the flow energy equation (7.4-11), we have for adiabatic flow where $T_{t_5} = T_{t_e}$:

$$\eta_n = \frac{T_{t_5} - T_e}{T_{t_5} - (T_e)_{\text{is}}}\,. \tag{7.6-17}$$

On applying the Poisson relation (7.3-25), we obtain the actual temperature drop in a convenient form for use in Figure 7.6-2:

$$T_{t_5} - T_e = \eta_n T_{t_5}\left[1 - \left(\frac{1}{p_{t_5}/p_0}\right)^{\frac{\gamma - 1}{\gamma}}\right]. \tag{7.6-18}$$

The achieved efficiencies η_n are about 0.95 (Ref. 41).

Some refinement of turbojet performance prediction may be obtained by taking into consideration the design of the exhaust nozzle. To explain this subject, first of all, it is necessary to consider the properties of a flow through a channel of varying cross section.

7.7 ISENTROPIC FLOW THROUGH A CHANNEL OF VARYING CROSS SECTION

For steady flow the mass flow rate crossing a section of the passage, follows from the continuity equation:

$$m = \rho V A \ . \tag{7.7-1}$$

Assuming perfect gas conditions, Equation (7.7-1) can be written in the form:

$$m = \frac{p}{R\,T} M \sqrt{\gamma R T}\, A = \frac{p}{\sqrt{T}} \sqrt{\frac{\gamma}{R}}\, M A \ . \tag{7.7-2}$$

By inserting Equations (7.4-12) and (7.4-13) into Equation (7.7-2) the mass flow rate can be expressed in terms of total properties:

$$m = \frac{p_t}{\sqrt{T_t}} \sqrt{\frac{\gamma}{R}}\, M \left(1 + \frac{\gamma - 1}{2} M^2 \right)^{-\frac{\gamma + 1}{2(\gamma - 1)}} A \ . \tag{7.7-3}$$

Since for steady isentropic flow the quantities m, T_t and p_t are constants, we find by differentiating Equation (7.7-3) that:

$$\frac{dA}{A} + \frac{dM}{M} - \frac{\frac{\gamma + 1}{2} M^2}{1 + \frac{\gamma - 1}{2} M^2} \frac{dM}{M} = 0 \ . \tag{7.7-4}$$

Rearranging, we get for the variation of the Mach number with cross-sectional area:

$$\frac{dM}{M} = \frac{1 + \frac{\gamma - 1}{2} M^2}{1 - M^2} \frac{dA}{A} \ . \tag{7.7-5}$$

The related variations of Mach number, velocity, temperature, pressure and density with area follow from Equation (7.7-5) in combination with the Equations (7.4-11) to (7.4-14) of Section 7.4.

The effects of area change in isentropic flow are displayed in Figure 7.7-1. From this illustration and Equation (7.7-5) it can be seen that the effects of area change are opposite for subsonic flow ($M < 1$) and supersonic flow ($M > 1$).

Considering a converging-diverging channel, the minimum cross-sectional area is called the *throat* ($dA = 0$).There, according to Equation (7.7-5), the flow must

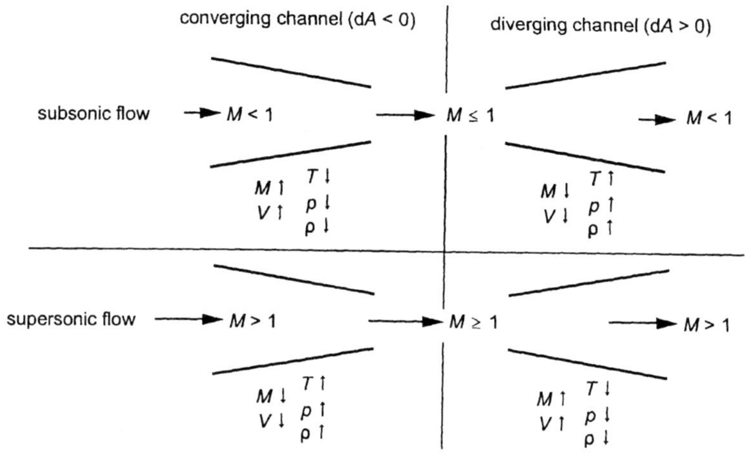

Figure 7.7-1. Convergent and divergent channel flow.

satisfy the condition that the Mach number either shows an extreme ($dM = 0$) or is equal to unity (Figure 7.7-2).

Under the first-mentioned condition, in the channel subsonic expansion takes place followed by subsonic diffusion or for supersonic flow a decreasing Mach number in the converging section but not reaching the Mach number unity, and in the diverging section an increasing Mach number.

When $M = 1$ we have the condition that the flow velocity at the throat equals the local speed of sound. Now, the flow is said to be *choked* because the mass flow rate is the maximum which the throat can cross with the given total temperature and pressure. The mass flow rate in a choked flow, from Equation (7.7-3), is given by:

$$m^{*} = \frac{p_t}{\sqrt{T_t}} \sqrt{\frac{\gamma}{R}\left(\frac{\gamma + 1}{2}\right)^{-\frac{\gamma + 1}{2(\gamma - 1)}}} A^{*} . \qquad (7.7\text{-}6)$$

In this equation an asterisk is used to mark the properties of the flow at the throat. Equating the Equations (7.7-6) and (7.7-3) yields:

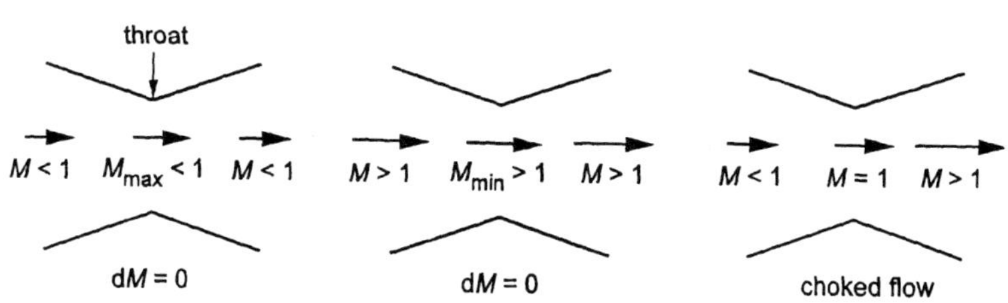

Figure 7.7-2. Flow in convergent - divergent channel.

$$\frac{A}{A^*} = \frac{1}{M}\left[\frac{2}{\gamma+1}\left(1+\frac{\gamma-1}{2}M^2\right)\right]^{\frac{\gamma+1}{2(\gamma-1)}} . \qquad (7.7\text{-}7)$$

The Mach number at a particular section of a channel with a given shape, follows directly from the ratio A/A^* in Equation (7.7-7). Then, with known total conditions, Equations (7.4-12) to (7.4-14) furnish the corresponding values of T, p, and ρ.

The conditions at the throat are found by setting $M = 1$ in Equations (7.4-12) to (7.4-14). For air ($\gamma = 1.4$) we obtain:

$$\frac{T_t}{T^*} = \frac{\gamma+1}{2} = 1.2, \qquad (7.7\text{-}8)$$

$$\frac{p_t}{p^*} = \left(\frac{\gamma+1}{2}\right)^{\frac{\gamma}{\gamma-1}} = 1.893, \qquad (7.7\text{-}9)$$

$$\frac{\rho_t}{\rho^*} = \left(\frac{\gamma+1}{2}\right)^{\frac{1}{\gamma-1}} = 1.577. \qquad (7.7\text{-}10)$$

In order to establish in a convergent-divergent channel a subsonic expansion in the convergent part with a Mach number of unity at the throat section, followed by a supersonic expansion in the divergent part of the channel, the pressure at the throat must be greater than the ambient pressure at the outlet p_0. Thus, the pressure ratio required for obtaining the condition $M^* = 1$ is given by:

$$\frac{p_t}{p_0} \geq \left(\frac{\gamma+1}{2}\right)^{\frac{\gamma}{\gamma-1}} = 1.893. \qquad (7.7\text{-}11)$$

To conclude, once again, in the case of a converging-diverging channel, the presence of a throat does not necessarily imply $M^* = 1$ because for values of p_t / p_0 lower than given by Equation (7.7-9) subsonic flow is found throughout the channel (unchoked flow).

7.8 FLOW CONDITIONS AT THE NOZZLE EXIT

Aircraft gas turbine engines designed for subsonic flight speeds employ a convergent exhaust nozzle (Figures 7.8-1a and b).

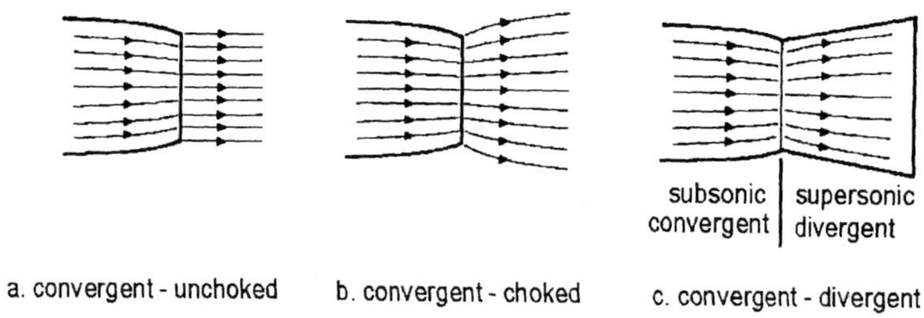

Figure 7.8-1. Propelling nozzle flows.

Since the expansion process in the exhaust nozzle is nearly isentropic, we can make use of Equation (7.4-13) to write:

$$p_{t_5} = p_{t_e} = p_e \left(1 + \frac{(\gamma - 1)}{2} M_e^2 \right)^{\frac{\gamma}{\gamma - 1}} . \tag{7.8-1}$$

According to Section 7.7, in the case of a convergent nozzle a critical pressure ratio is distinguished, which furnishes sonic velocity at the nozzle exit.
Insertion of $M_e = 1$ into Equation (7.8-1) yields:

$$\left(\frac{p_{t_e}}{p_e} \right)_{cr} = \left(\frac{\gamma + 1}{2} \right)^{\frac{\gamma}{\gamma - 1}} . \tag{7.8-2}$$

For air, using $\gamma = 1.4$, the critical pressure ratio is 1.893, while for gas, taking γ_g = 1.333, $(p_{t_e}/p_e)_{cr}$ equals 1.851.

These values apply of course only to the case of isentropic expansion in the nozzle. For pressure ratios below the critical value, the pressure p_e in Equation (7.8-1) may be taken equal to p_0. Then:

$$\frac{p_{t_e}}{p_0} < \left(\frac{\gamma + 1}{2} \right)^{\frac{\gamma}{\gamma - 1}} . \tag{7.8-3}$$

When the latter condition exists, we have *unchoked flow*. This implies that $M_e < 1$ and $p_e = p_0$ so that the jet flows out as a cylindrical stream (Figure 7.8-1a). Because of the equality $p_e = p_0$ the pressure term in Equation (7.2-5) for the thrust is zero, and the nozzle exit Mach number is given by Equation (7.5-12).
When the actual nozzle pressure ratio is greater than the critical value, we have:

$$\frac{p_{t_e}}{p_0} > \left(\frac{\gamma + 1}{2} \right)^{\frac{\gamma}{\gamma - 1}} . \tag{7.8-4}$$

Now we speak of *choked flow* since the mass flow rate has its maximum value.

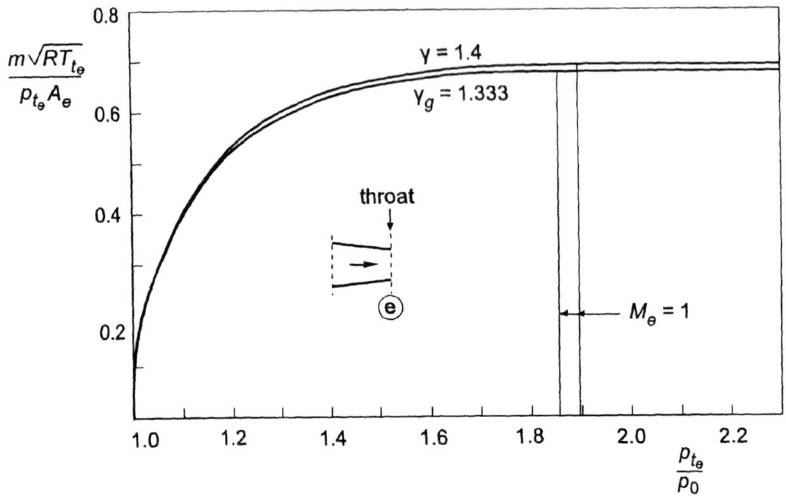

Figure 7.8-2. Nozzle characteristics for isentropic flow.

The Mach number is unity at the exit plane, and the static pressure at the nozzle exit becomes:

$$\frac{p_{t_e}}{p_e} = \left(\frac{\gamma + 1}{2}\right)^{\frac{\gamma}{\gamma - 1}}, \tag{7.8-5}$$

where $p_e > p_0$.

For a so-called *choked nozzle* of given area A_e, the mass flow rate of a nondimensional flow can readily be found from Equation (7.7-6) as:

$$\frac{m\sqrt{R\,T_{t_e}}}{p_{t_e} A_e} = \sqrt{\gamma}\left(\frac{2}{\gamma + 1}\right)^{\frac{\gamma + 1}{\gamma - 1}}. \tag{7.8-6}$$

As shown in figure 7.8-2, at given values of p_{t_e} and T_{t_e}, the (maximum) mass flow rate is not dependent on the critical pressure ratio.
It should be noted that also at some point in the turbine a maximum value of $\sqrt{R\,T_t}/p_t A$ is reached at a pressure ratio which produces choking conditions.

For most thrust settings and flight conditions of interest to gas turbine engines, the flow at the nozzle exit will be choked so that some expansion of the jet occurs downstream of the nozzle (Figure 7.8-1b).

To achieve maximum engine thrust under such conditions, clearly, the nozzle would have to be convergent-divergent (Figure 7.8-1c). Indeed, the Olympus

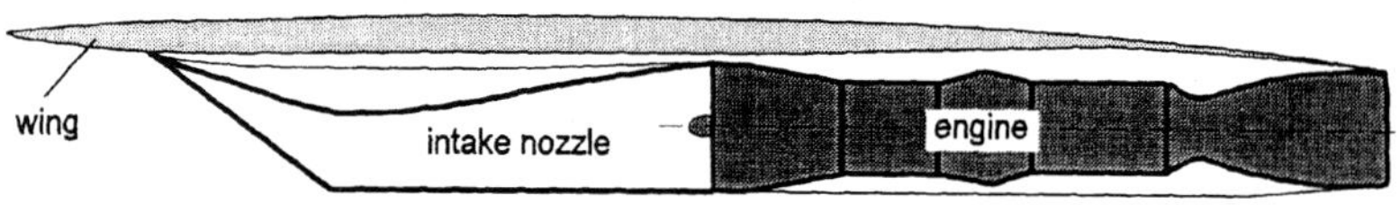

Figure 7.8-3. Concorde's Olympus two-spool turbojet engine.

engines, of the in 2003 decommissioned supersonic aircraft the "Concorde", where very high pressure ratios p_{t_e}/p_0 occurred, made use of convergent-divergent nozzles in order that the pressure p_e matches the ambient pressure p_0 (Figure 7.8-3). These so-called *condi-nozzles* could be varied in shape and throat area to match the operating conditions.
For the generation of thrust at subsonic airspeeds, however, it is not profitable incorporating a condi-nozzle, since its advantages will not balance the increases in engine weight and airplane drag.

Continuing the performance prediction of a subsonic turbojet with a simple convergent nozzle, Figure 7.8-4 presents the thermal efficiencies of actual cycles, using typical efficiency values for intake, compressor, turbine, and nozzle. As also assumed in Figure 7.5-4, the flight Mach number M_0 is 0.8 and the flight altitude H is 11 km (ISA). To simplify matters, it is adopted that the working fluid behaves throughout the engine as a perfect gas with fixed thermodynamic properties, which are equal to those of air. Also, the assumptions are made of perfect combustion and isentropic expansion of the flow in the exhaust nozzle. It is seen from Figure 7.8-4 that as the compressor pressure ratio is enlarged for a given turbine entry temperature, the thermal efficiency is increased until a peak value is achieved. Further increase in pressure ratio then diminishes the thermal efficiency. Apparently, there is a maximum pressure ratio corresponding to a given turbine entry temperature.

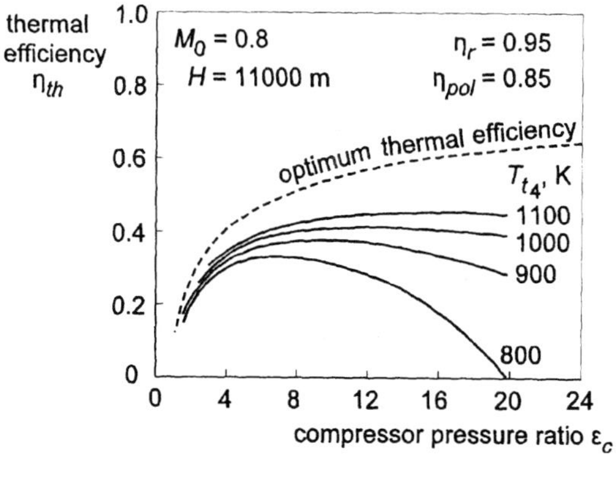

Figure 7.8-4. Thermal efficiency.

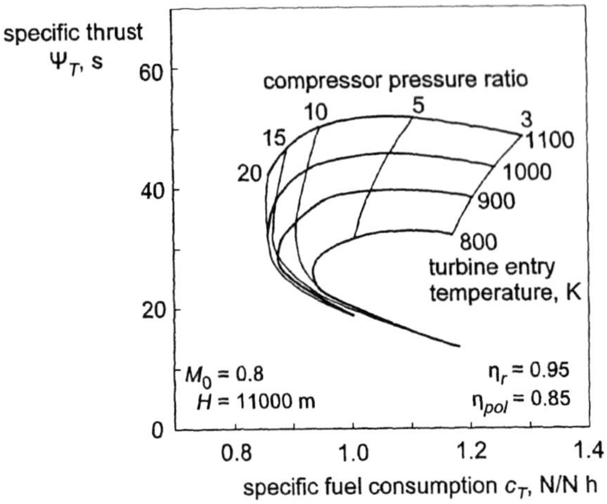

Figure 7.8-5. Actual turbojet cycle performance.

In order to demonstrate how specific fuel consumption and specific thrust are affected by the efficiencies of intake, compressor and turbine, results of design point calculations are presented in Figure 7.8-5, using the same basic figures and assumptions as in Figure 7.8-4. Looking at the actual and ideal performance in Figure 7.8-5 and 7.5-4, we see that the picture is largely the same, except that noticeable performance penalties arise from the imperfections.

Returning to the turbojet with afterburner, we stated earlier that it is essential to have a variable nozzle outlet area. To explain this requirement, we assume that a choked convergent nozzle is used, so that the mass flow rate crossing the nozzle exit is given by:

$$m = \rho_e c_e A_e \ . \tag{7.8-6}$$

By introducing the relations: $p_e / \rho_e = R\, T_e$ and $c_e = \sqrt{\gamma R T_e}$ into Equation (7.8-6), the area A_e can be written as:

$$A_e = \frac{m\sqrt{T_e}}{p_e\sqrt{\dfrac{\gamma}{R}}} \ . \tag{7.8-7}$$

Clearly, the magnitude of the cross-sectional area is proportional to $\sqrt{T_e}$. Hence, the variation of T_e when applying afterburning, demands for an adjustable area A_e. Other disadvantages of the application of afterburning are increased specific fuel consumption, enlarged jet noise levels, and increased engine weight.

8 PERFORMANCE OF TURBOFAN AND TURBOJET ENGINES

8.1 TURBOFAN ENGINE DESIGN POINT PERFORMANCE

In the turbofan engine the available energy from the gas generator is converted to one of greater mass flow rate and lower exhaust velocity, than it was directly expanded in the hot nozzle. In the two-spool layout of the previous Figure 7.1-3 the fan and the high pressure compressor are driven by separate turbine assemblies through coaxial shafts. This configuration is suitable for mediate compressor pressure ratios and bypass ratios. At very high values of these parameters, the three-spool system of Figure 8.1-1 with its moderate compression ratios may be of advantage to ensure that each compressor will operate efficiently over the entire engine speed range.
To understand the operation of gas turbine engines, it will be useful to follow a numerical example of a design point performance calculation of a hypothetical two-spool turbofan engine, using the engine stations as numbered in Figure 8.1-2. The following simplifications will be used:

a. The working fluid behaves as a perfect gas with fixed mean thermodynamic properties. This assumption is appropriate for all the propulsion systems considered in this book, since the high combustion temperatures generally result in sufficiently low densities for the ideal-gas behavior to be reasonable approximations. Values of c_p = 1005 m^2/s^2 K, γ =1.4 and R = 287 m^2/s^2K, are used for the air flowing through intake and compressor, and c_{p_g} = 1147 m^2/s^2K, γ_g = 1.333 and $R_g = R$ = 287 m^2/s^2K for the combustion gases.
- The fuel mass flow rate entering the combustion chamber is neglected with respect to the air mass flow rate. Furthermore, no bleed air is taken from the

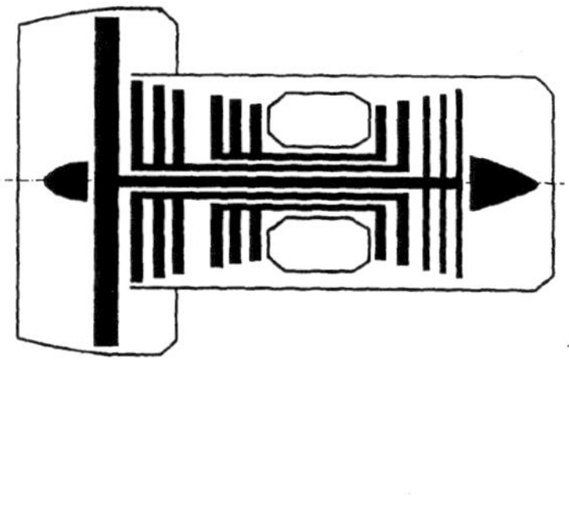

Figure 8.1-1. Three-spool turbofan arrangement.

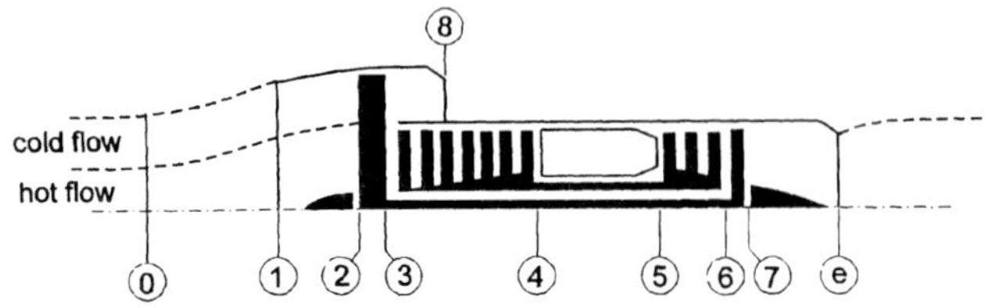

Figure 8.1-2. Station numbering.

compressor. This means that the compressor mass flow rate equals the turbine mass flow rate.

- The fan and core nozzle are convergent and have fixed exit areas. No mixing of hot and cold flow takes place. Both expansion processes occur isentropically.

The engine is assumed to be designed to operate at a flight Mach number $M_0 =$ 0.8 and at an altitude $H = 10{,}000$ m (ISA). In addition, the following design parameters are adopted:

- Bypass ratio: $B = 5$.
- Overall compressor pressure ratio: $p_{t_4}/p_{t_2} = 30$.
- Fan pressure ratio: $p_{t_3}/p_{t_2} = 1.6$.
- Turbine entry temperature: $T_{t_5} = 1600$ K.

For a flight at H = 10,000 m (ISA), the ambient static temperature is $T_0 = 223$ K. The total temperature at the fan inlet comes from:

$$T_{t_2} = T_{t_0} = T_0\left[1 + \frac{\gamma - 1}{2} M_0^2\right] = 223\,[\,1 + 0.2 \times 0.64\,] = 252\,\text{K}.$$

If the pressure recovery factor of the intake $\eta_R = 0.95$, the pressure ratio at station 2 is:

$$\frac{p_{t_2}}{p_0} = \eta_R\left[1 + \frac{\gamma - 1}{2} M_0^2\right]^{\frac{\gamma}{\gamma - 1}} = 0.95\,[\,1 + 0.2 \times 0.64\,]^{3.5} = 1.448.$$

The temperature at the fan exit is given by:

$$T_{t_3} = T_{t_2}\left[\frac{p_{t_3}}{p_{t_2}}\right]^{\frac{\gamma-1}{\gamma\,\eta_{pol}}} = 252\,[\,1.6\,]^{0.336} = 295\ \text{K},$$

where for the polytropic efficiency a value $\eta_{pol} = 0.85$ is used.
Then the cold nozzle overall pressure ratio is found to be;

$$\frac{p_{t_3}}{p_0} = \frac{p_{t_3}}{p_{t_2}}\frac{p_{t_2}}{p_0} = 1.6 \times 1.448 = 2.317.$$

When the expansion processes in the nozzles are assumed to be 100% efficient, the critical pressure ratio, from Equation (7.8-2) is:

$$\frac{p_{t_3}}{p_8} = \left[\frac{\gamma+1}{2}\right]^{\frac{\gamma}{\gamma-1}} = [\,1.2\,]^{3.5} = 1.893.$$

Since $p_{t_3}/p_0 > p_{t_3}/p_8$, the cold nozzle is choked so that $M_8 = 1.0$ and $p_8 > p_0$.
Hence,

$$T_8 = T_{t_3}\left[\frac{2}{\gamma+1}\right] = 295\left[\frac{2}{2.4}\right] = 246\,\text{K}$$

and

$$\frac{p_8}{p_0} = \frac{p_8}{p_{t_3}}\frac{p_{t_3}}{p_0} = \frac{2.317}{1.893} = 1.224.$$

The cold nozzle outlet velocity is:

$$w_8 = M_8\sqrt{\gamma R T_8} = 1\sqrt{1.4 \times 287 \times 246} = 314\ \text{m/s}.$$

The contribution of the cold jet to the specific thrust of the engine is given by:

$$\frac{T_c}{m\,g} = \frac{m_c\,(w_8 - V_0)}{m\,g} + \frac{A_8\,(p_8 - p_0)}{m\,g},$$

where

$$V_0 = M_0\sqrt{\gamma R T_0} = 0.8\sqrt{1.4 \times 287 \times 223} = 239\ \text{m/s}.$$

Using the familiar expressions that $m_c = \rho_8\, w_8\, A_8$, $\rho_8 = p_8/(RT_8)$ and $m_c/m = B/(B+1)$, it readily follows that:

$$\frac{T_c}{m g} = \frac{B}{B+1} \frac{1}{g} \left[(w_8 - V_0) + \sqrt{\frac{R T_8}{\gamma}} \left(1 - \frac{p_0}{p_8} \right) \right].$$

Putting in numerical values yields:

$$\frac{T_c}{m g} = \frac{5}{6} \frac{1}{9.81} \left[(314 - 239) + \sqrt{\frac{287 \times 246}{1.4}} \left(1 - \frac{1}{1.224} \right) \right] = 9.9 \text{ s}.$$

The pressure ratio of the high-pressure compressor is:

$$\frac{p_{t_4}}{p_{t_3}} = \frac{p_{t_4}}{p_{t_2}} \frac{p_{t_2}}{p_{t_3}} = \frac{30}{1.6} = 18.75.$$

The total temperature at the outlet from the high-pressure compressor comes from:

$$T_{t_4} = T_{t_3} \left[\frac{p_{t_4}}{p_{t_3}} \right]^{\frac{\gamma - 1}{\gamma \eta_{pol}}} = 295 [18.75]^{0.336} = 790 \text{ K},$$

where again is adopted: $\eta_{pol} = 0.85$.

The high-pressure turbine temperature drop follows from the equality of compressor and turbine powers;

$$\eta_m m_h c_{p_g} \left[T_{t_5} - T_{t_6} \right] = m_h c_p \left[T_{t_4} - T_{t_3} \right],$$

where η_m is the mechanical efficiency. This quantity takes into account the losses which occur when power is transmitted from the turbine to the compressor. Setting $\eta_m = 0.99$, we find the temperature behind the high-pressure turbine to be:

$$T_{t_6} = T_{t_5} - \frac{1}{\eta_m} \frac{c_p}{c_{p_g}} \left[T_{t_4} - T_{t_3} \right] = 1600 - \frac{1}{0.99} \times \frac{1005}{1147} \left[790 - 295 \right] = 1162 \text{ K}$$

The low-pressure turbine temperature drop follows from the power demand of the low-pressure compressor:

$$\eta_m m_h c_{p_g} \left[T_{t_6} - T_{t_7} \right] = m c_p \left[T_{t_3} - T_{t_2} \right].$$

Remembering that $m / m_h = B + 1$, and using $\eta_m = 0.99$, we find:

$$T_{t_7} = T_{t_6} - \frac{B+1}{\eta_m} \frac{c_p}{c_{p_g}} \left[T_{t_3} - T_{t_2} \right] = 1162 - \frac{6}{0.99} \times \frac{1005}{1147} [295 - 252] = 934$$

To calculate the hot nozzle pressure ratio we write:

$$\frac{p_{t_7}}{p_0} = \frac{p_{t_7}}{p_{t_5}} \frac{p_{t_5}}{p_{t_4}} \frac{p_{t_4}}{p_{t_3}} \frac{p_{t_3}}{p_0} .$$

The overall turbine pressure ratio is given by:

$$\frac{p_{t_7}}{p_{t_5}} = \left[\frac{T_{t_7}}{T_{t_5}} \right]^{\frac{\gamma_g}{\eta_{pol} (\gamma_g - 1)}} .$$

With a turbine polytropic efficiency $\eta_{pol} = 0.85$ we get:

$$\frac{p_{t_7}}{p_{t_5}} = \left[\frac{934}{1600} \right]^{4.71} = 0.07924.$$

The pressure loss in the combustor is assumed to be 5% of the compressor delivery pressure, so that:

$$\frac{p_{t_5}}{p_{t_4}} = 0.95 .$$

Substitution of these numbers, yields for the nozzle pressure ratio:

$$\frac{p_{t_7}}{p_0} = 0.07924 \times 0.95 \times 18.75 \times 2.317 = 3.27.$$

Apparently, also the hot nozzle is choked, so that $M_e = 1.0$ and $p_e > p_0$. Hence,

$$T_e = T_{t_7} \left[\frac{2}{\gamma_g + 1} \right] = 934 \left[\frac{2}{2.333} \right] = 801 \text{ K} , \qquad \text{and}$$

$$\frac{p_{t_7}}{p_e} = \left[\frac{\gamma_g + 1}{2} \right]^{\frac{\gamma_g}{\gamma_g - 1}} = 1.851.$$

The pressure ratio p_e/p_0 can now be obtained, giving:

$$\frac{p_e}{p_0} = \frac{p_e}{p_{t_7}} \frac{p_{t_7}}{P_0} = \frac{3.27}{1.851} = 1.77\,.$$

The velocity of the hot jet is:

$$w_e = M_e \sqrt{\gamma_g R_g T_e} = 1 \sqrt{1.333 \times 287 \times 801} = 554 \text{ m/s}.$$

The contribution of the hot jet to the specific thrust of the engine is given by:

$$\frac{T_h}{m g} = \frac{m_h (w_e - V_0)}{m g} + \frac{A_e (p_e - p_0)}{m g}\,.$$

Recalling that $m_h / m = 1 / (B + 1)$, we find:

$$\frac{T_h}{m g} = \frac{B}{B+1} \frac{1}{g} \left[(w_e - V_0) + \sqrt{\frac{R_g T_e}{\gamma_g}} \left(1 - \frac{p_0}{p_e} \right) \right] =$$

$$= \frac{1}{6} \frac{1}{9.81} \left[(554 - 239) + \sqrt{\frac{287 \times 801}{1.333}} \left(1 - \frac{1}{1.77} \right) \right] = 8.4 \text{ s}.$$

The total specific thrust is:

$$\psi_T = \frac{T_N}{m g} = \frac{T_c}{m g} + \frac{T_h}{m g} = 9.9 + 8.4 = 18.3 \text{ s}.$$

Determination of the design net thrust requires knowledge of the total air mass flow rate, which sizes the engine, e.g., exhaust nozzle area A_8 and A_e.

The specific fuel consumption comes by dividing the fuel weight flow rate by the thrust:

$$c_T = \frac{m_f g}{T_N}\,.$$

The fuel flow equation may be written as:

$$\eta_b m_f H = m_h c_{p_g} \left[T_{t_5} - T_{t_4} \right],$$

where H is the heating value of the fuel ($H = 4.31 \times 10^7$ J/kg). The combustion efficiency η_b is introduced to take into account incomplete burning of the fuel.

Combining the latter two equations leads to:

$$c_T = \frac{1}{B+1}\,\frac{c_{p_g}}{H\eta_b}\,\frac{\left[T_{t_5} - T_{t_4}\right]}{\psi_T}\,.$$

Assuming $\eta_b = 0.95$, we obtain:

$$c_T = \frac{1147\,(1600 - 790)\,3600}{6 \times 4.31 \times 10^7 \times 0.95 \times 18.3} = 0.74\ \mathrm{N/N\,h}\ ,$$

and according to Equation (7.5-23):

$$\eta_{tot} = \frac{g\,V_0}{H\,c_T} = \frac{9.81 \times 239}{4.31 \times 10^7 \times 0.74} = 0.27\,.$$

In accordance with the above calculations, Figure 8.1-3 presents a sketch of the working cycles for this high bypass ratio engine in terms of enthalpy and entropy.

Comparing the above calculations with the turbojet performance data in Figure 7.8-5, indicates that the specific thrust of the turbofan engine strongly falls off. On the other hand, the figure obtained for the specific fuel consumption illustrates that the efficiency of the turbofan is markedly better than that of the turbojet engine.

Finally, it should be realized that the calculations made in this section concern the bare engine. Airplane performance equations, of course, require the use of the installed thrust and specific fuel consumption, which include both the

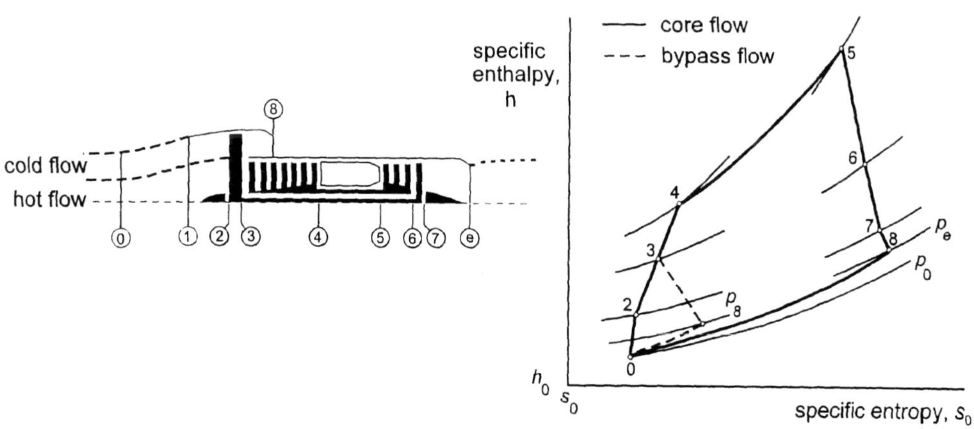

Figure 8.1-3. Enthalpy-entropy diagram for turbofan engine.

detrimental effects associated with the integration of engine and airplane structure and the losses owing to bleeding compressed air and power extraction from the turbine. However, a detailed analysis of these effects is certainly beyond the scope of this text.

8.2 COMPRESSOR CHARACTERISTICS

In the preceding section we determined the performance of the turbofan engine without giving notice to the internal details of the gas generator. Therefore, in this section and Section 8.3 short discussions are devoted to the functioning of compressor and turbine. Attention to the working of the combustion chamber will be given in Chapter 10.

As mentioned earlier, both centrifugal and axial flow compressors are employed for use in turbo-engines; see the previous Figure 7.1-2. However, in virtually all today's propulsion systems for transport airplanes, axial flow compressors are used because of their ability to produce higher pressure ratios at higher efficiencies, and for a lower frontal area at a given thrust. These features lead to a lower specific fuel consumption and decreased airplane drag, and so to improved flight performance. For this reason, we limit our discussion to axial flow compressors and axial flow turbines.

As shown in Figure 8.2-1, the axial flow compressor consists of several similar stages, each stage comprising a row of rotor blades and a row of stator blades. The rotor blades are attached to the circumference of a rotating body and the stator blades are joined to the casing of the compressor. Often, compressors have in front an additional row of stator blades, known as inlet guide vanes (IGV's), to give the air swirl in the direction of rotation.

First of all, it is worth to remember that in the inlet of the engine the air has been slowed down to a low subsonic velocity of about $M \approx 0.4$ before it enters the compressor. It is also of importance to emphasize that in compressors the airflow is always subject to negative pressure gradients and that the rise in total pressure results from power supply to the airflow by the moving rotor blades.

In Figure 8.2-1 also the changes in pressure and velocity are sketched which occur in the airflow through the compressor. The air is initially accelerated in the rotor part, thereby increasing the total temperature and total pressure. Next, the airflow is decelerated in the stator blade passages wherein the flow kinetic energy is converted to static pressure by the creation of a *diffusing flow*. The process is repeated in as many stages as are necessary to yield the required compressor pressure ratio ε_c.

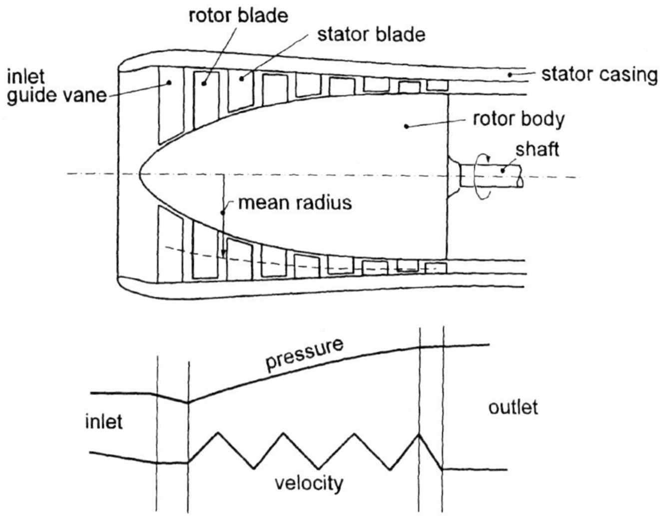

Figure 8.2-1. Axial flow compressor.

As it is desirable to keep the axial flow velocity approximately constant throughout the compressor, and due to the fact that the density increases throughout the length of the compressor, there is a marked reduction of the annular area between rotor body and stator casing in a high pressure ratio compressor. Therefore, it is necessary to reduce the blade height from the front to the rear of the compressor.

Because of the adverse pressure gradient, only a small pressure rise across each stage is gained, so that an axial flow compressor requires a large number of stages to produce a high compression ratio. However, as the number of stages increases, it becomes more difficult to ensure that each stage will operate more efficiently over a wide rotational speed range. A solution to the problem has been achieved by designing multi spool compressors, which are composed of two or more rotor assemblies, each driven by separate turbines through coaxial shafts.

The fan of the twin-spool engine in the previous Figure 7.1-3, as well as the fan of the triple-spool engine in Figure 8.1-1 is driven by a separate low pressure turbine. In both examples, the cold or bypass airflow is formed by the air accelerated by the outer portion of the fan blades as the fan drives the entire airflow into and around the core. The core or hot flow through the inner portion of the fan blades is heated in the combustion chamber. It may be noted that for low-noise emission, it is favorable that the desired fan pressure ratio of the modern high-pressure ratio turbofan engines are achieved by using only one stage of blades on the fan.

A compressor stage and its companion enthalpy-entropy diagram are sketched in Figure 8.2-2. Applying the steady flow energy equation, Equation (7.4-17), to the rotor part of the stage, and assuming that the process is adiabatic, the rotor power input, P_{cr}, results in a rise of the total temperature, given by:

$$P_{cr} = m\, c_p\, (T_{t_2} - T_{t_1}) \,, \tag{8.2-1}$$

where m is the mass flow rate passing through the stage, and c_p is the specific heat of air at constant pressure.

Because the flow through the rotor blade row will involve losses due to friction, turbulence, and air leakage across the blade tips, the total pressure rise will be less than would be obtained with an isentropic compression at the same power input. Therefore, the total pressure ratio may be expressed as:

$$\frac{P_{t_2}}{P_{t_1}} = \left[1 + \eta_r \frac{T_{t_2} - T_{t_1}}{T_{t_1}} \right]^{\frac{\gamma}{\gamma - 1}} , \tag{8.2-2}$$

where η_r is the isentropic efficiency of the process in the rotor.

The increases in total temperature and total pressure are accomplished completely within the rotor. Of course, no power is supplied to the air in the stator part, wherein merely kinetic energy is transformed to a higher static pressure at a constant total temperature($T_{t_2} - T_{t_3}$). As depicted in the h-s diagram in Figure

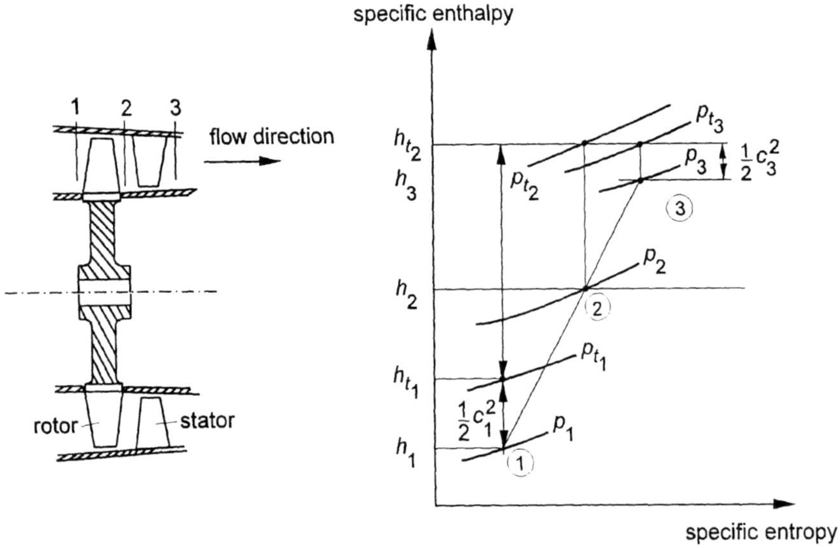

Figure 8.2-2. Compressor stage and h-s diagram.

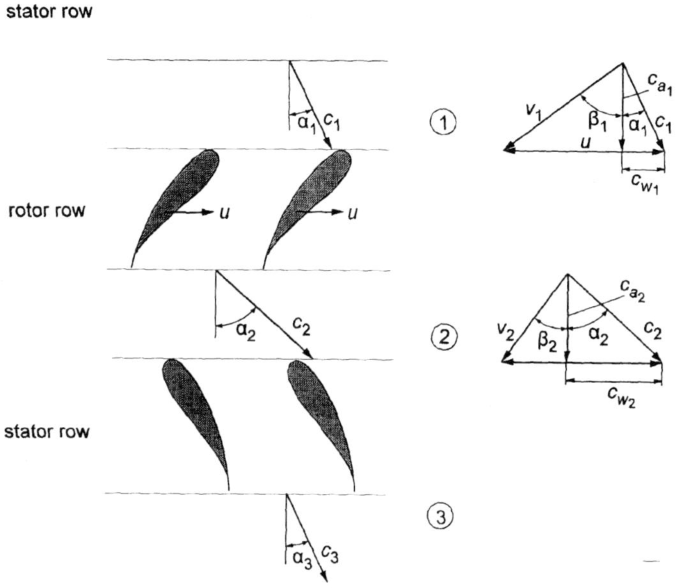

8.2-3. Compressor velocity triangles.

8.2-2, also in the stator there will be some decrease in total pressure.

To illustrate the factors determining the stage total pressure rise, in Figure 8.2-3 the velocities for the inlet and outlet of the rotor row are represented in magnitude and direction by the sides of a triangle.
To simplify the problem, considered is the airflow at the mean height of a blade where the rotor peripheral speed is u. Assuming that the flow occurs in a tangential plane at this mean radius implies that the flow velocity has only an axial component, c_a, and a tangential component, c_w (whirl velocity).

In Figure 8.2-3 the air leaves the previous row of stator blades at a speed c_1 and at an angle α_1 to the axial direction.
This speed is made up with the peripheral speed u so that the air approaches the rotor with a relative velocity v_1 at an angle β_1 from the axial direction. From the velocity triangle we directly find the axial component c_{a_1} and the tangential velocity c_{w_1} of the airflow.

The air leaves the rotor blades with a relative velocity v_2 at an angle β_2 determined by the rotor blade outlet angle.
If the velocity triangle at the rotor outlet is drawn, we see that combining the air velocity relative to the blades v_2 with the speed of the rotor blades u gives an increased absolute velocity at the outlet of c_2 at an angle α_2.

The stator blades are set at a particular angle to pick up the air approaching them at angle α_2. Then, in the passages between the stator blades the air is diffused to a lower velocity c_3 and deflected to an angle α_3.
Finally, the air is turned by the rotor speed u to a well-matched direction for entering the next row of rotor blades.

By considering the change in angular momentum of the air in passing through the rotor and assuming that the axial velocity, c_a, is constant in both axial and radial directions, the following approximate expression for the power input to the stage is obtained:

$$P_{cr} = m\,u\,(c_{w_2} - c_{w_1}) = m\,u\,c_a\,(\tan\alpha_2 - \tan\alpha_1)\ . \qquad (8.2\text{-}3)$$

The following two basic equations are readily derived from the triangles in Figure 8.2-3:

$$u = c_a\,(\tan\alpha_1 + \tan\beta_1)\ , \qquad (8.2\text{-}4)$$

and

$$u = c_a\,(\tan\alpha_2 + \tan\beta_2)\ . \qquad (8.2\text{-}5)$$

Clearly, Equation (8.2-3) can also be written as:

$$P_{cr} = m\,u\,c_a\,(\tan\beta_1 - \tan\beta_2)\ . \qquad (8.2\text{-}6)$$

The input power will be absorbed in raising the total pressure of the air and in overcoming the various losses. The whole of this input will reveal itself as a rise in total temperature of the air.
Equating (8.2-1) and (8.2-6), the rise of total temperature in the stage, ΔT_{ts}, is given by:

$$\Delta T_{ts} = T_{t_3} - T_{t_1} = T_{t_2} - T_{t_1} = u\,\frac{c_a}{c_p}\,(\tan\beta_1 - \tan\beta_2)\ . \qquad (8.2\text{-}7)$$

This gives a stage pressure ratio:

$$\varepsilon_s = \frac{p_{t_3}}{p_{t_1}} = \left[1 + \eta_s\,\frac{\Delta T_{ts}}{T_{t_1}}\right]^{\frac{\gamma}{\gamma-1}}\ , \qquad (8.2\text{-}8)$$

where η_s is the isentropic efficiency of the stage.
It can be seen from Equations (8.2-7) and (8.2-8) that a high total pressure rise in a stage, requires a high rotor blade speed u, a high axial velocity c_a, and a

high flow deflection (β_1 - β_2) in the rotor blades.
However, blade stresses will limit the rotor blade speed, and a combination of aerodynamic considerations and the occurring adverse pressure gradient will restrict the other two requirements.

Obviously, for an efficient compressor, the flow characteristics of each stage must be carefully matched. This is a relatively simple requirement to satisfy for one set of conditions, such as the design conditions (given mass flow rate, pressure ratio and rotational speed), but is much more difficult to realize when the compressor is operating at deviating (off-design) conditions.
To represent the overall performance of a compressor, i.e., the performance over a wide range of rotational speeds and inlet conditions, plots are available which give the compressor pressure ratio ε_c versus nondimensional mass flow $m\sqrt{T_{t_2}}/p_{t_2}$ for fixed values of nondimensional compressor rpm $n/\sqrt{T_{t_2}}$ and isentropic efficiency η_c (see also Section 8.6).
Note that now the subscript "2" designates the intake conditions of the compressor. The general form of these *compressor characteristics* is shown in Figure 8.2-4. It should be noted that the quantities $m\sqrt{T_{t_2}}/p_{t_2}$ and $n/\sqrt{T_{t_2}}$ are not true dimensionless as they should also include an area and a length, respectively. As the area and length both are constant for a given compressor, they may be omitted from the graph.

Figure 8.2-4 indicates that as the nondimensional mass flow is reduced at a constant nondimensional rpm the pressure ratio increases. Of course, it is important that the compressor operates near the point of maximum efficiency at

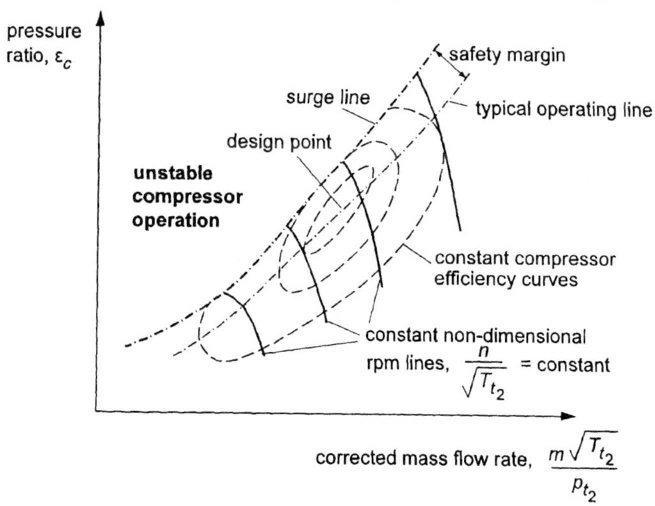

8.2-4. Compressor characteristics.

the design point.
A special feature of the characteristics of a compressor is the *surge line* on Figure 8.2-4. This line defines the upper limit for stable operation at any given engine speed, which must be established by testing of the compressor.
When surging occurs, there is uncontrollable stalling in the flow throughout the compressor. The phenomenon is attended by strong vibrations and a rise in gas temperature. As a result great damage to the compressor can occur.
The values of nondimensional airflow and compressor speed at which surge occurs define the 'surge point'. The line which joins all the surge points forms the above-mentioned surge line. To avoid a compressor surge, a typical operating line for an engine is defined.

8.3 THE AXIAL FLOW TURBINE

The turbine depends for its operation on the extraction of energy from the hot gases released from the combustion chamber and expanding them to a lower total pressure and total temperature. On impact with the blades and during the subsequent reaction through the blades, energy is absorbed, causing the turbine to rotate at high speed and so provide the power for driving the compressor(s) and accessories.
Therefore, it is truthful up to a point to say that axial flow turbines are like axial flow compressors operating in reverse. In the axial flow compressor, power is supplied to the air through which, in conformance with the energy flow equation, an increase in total pressure and total temperature is achieved. In the turbine, by contrast, the hot, high pressure gases entering the turbine expand from a higher to a lower total pressure and total temperature with an accompanying release of power.
To generate the driving power, the turbine consists of a number of stages, each

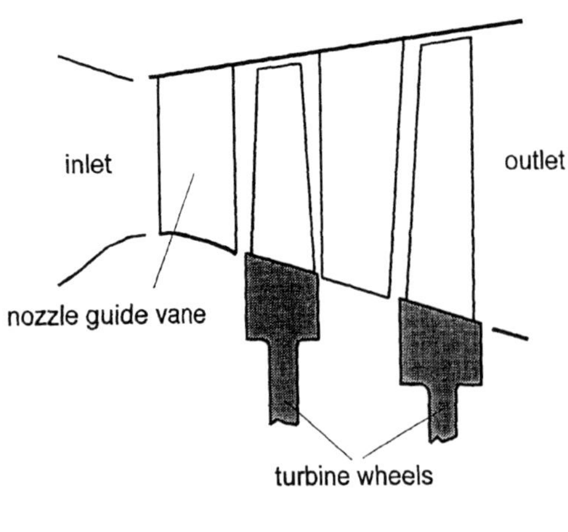

Figure 8.3-1. Axial flow turbine stages.

employing one row of stationary stator blades and one row of moving rotor blades (Figure 8.3-1).
Turbine stator blades show convergent blade passage shapes, by which the gas stream is accelerated to a high velocity. At the same time the gas flow is given a whirl in the direction of rotation of the rotor blades. Because of this behavior, there is an extended limit on the diffusion in each turbine stage. This means that a single turbine stage can provide a relatively high pressure ratio, and certainly very much more than achieved by a compressor with its unfavorable pressure gradient. This is why a single turbine stage can drive a large number of compressor stages.

Considering the expansion process, Figure 8.3-2 shows a view of the turbine stage passages with velocity triangles.
In the converging passages between the stator blades the gases are expanded to a lower static pressure, resulting in a large increase in flow velocity.
The gases leave the stator blades at a velocity c_2, directed onto the rotor blades which are rotating at mean speed u. The velocity of the gases relative to the blades becomes v_2, which is a lower velocity than the absolute velocity c_2 because the blades are moving away.
The rotor blades pick up the hot gas stream and deflect it through a considerable angle, while at the same time the gas velocity relative to the blades is increased. In this way a force is generated acting on the turbine blades.

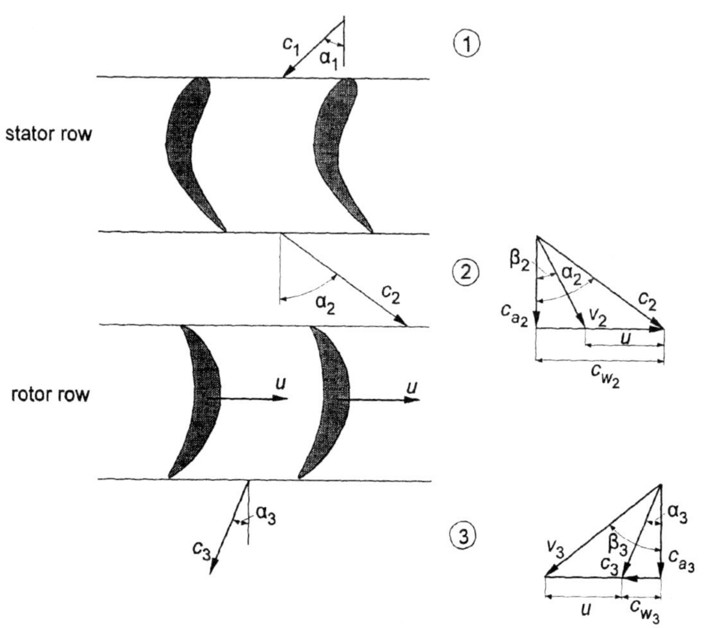

8.3-2. Turbine velocity triangles.

Naturally, in common with the axial flow compressor, only changes in gas velocity in the direction of movement of the blades will give useful forces on the turbine blades.
The power delivered by the rotor blades, P_{ts}, is given by the mass flow rate times the change in whirl velocity. Assuming c_a is constant in both axial and radial direction, we find from Figure 8.3-2:

$$P_{ts} = m u (c_{w_3} - c_{w_2}) = m u c_a (\tan \alpha_3 + \tan \alpha_2) . \tag{8.3-1}$$

Once the stage power is known, the stage total temperature drop can be found from the steady flow energy equation (cf. Equation (8.2-1)):

$$T_{t_1} - T_{t_3} = \frac{P_{ts}}{m c_{pg}} = \frac{u c_a (\tan \alpha_3 + \tan \alpha_2)}{c_{pg}} , \tag{8.3-2}$$

where c_{pg} is the specific heat at constant pressure of the combustion gases.
The stage total pressure ratio becomes:

$$\frac{p_{t_1}}{p_{t_3}} = \frac{1}{\left[1 - \frac{T_{t_1} - T_{t_3}}{\eta_s T_{t_1}} \right]^{\frac{\gamma}{\gamma - 1}}} , \tag{8.3-3}$$

where η_s is the isentropic efficiency of the stage.
The losses which prevent the efficiency from being 100 per cent, of course, are due to the same effects as occur in the compressor.

Inserting Equation (8.3-2) into Equation (8.3-3) yields:

$$\frac{p_{t_1}}{p_{t_3}} = \frac{1}{\left[\frac{1 - u c_a (\tan \alpha_3 + \tan \alpha_2)}{c_{pg} \eta_s T_{t_1}} \right]^{\frac{\gamma}{\gamma - 1}}} . \tag{8.3-4}$$

If in the turbine the energy is absorbed efficiently, the whirl will be removed from the gas stream so that the flow at the exit from the turbine will be substantially ‘straightened-out’ to give an axial flow into the exhaust nozzle of the engine.

Mention has already been made of the importance of high turbine entry temperatures (TET) in turbine engine performance. There is every incentive to the designer to operate at the highest possible temperatures as a high the designer to operate at the highest possible temperatures as a high temperature turbine engine has a lower specific fuel consumption and a higher specific thrust than a lower temperature engine. Therefore, we end this section with mentioning the never-ending search for materials which are able to withstand higher turbine entry temperatures. This quest for higher operating temperatures has led to many new alloys being developed. An additional way of increasing engine operating temperatures without exceeding material limitations is to apply turbine blade cooling, and protective and heat resistant coatings.

8.4 OFF-DESIGN PERFORMANCE

The reader is reminded that in Chapter 7 and Section 8.1 we have been considering the so-called *design point performance*.

This term indicates the performance when a given engine is running at the particular fan and compressor pressure ratio, turbine entry temperature, atmospheric conditions and forward speed for which these engine components are designed.

The remaining problem is then to provide the off-design performance, being the performance of the engine over the entire operating range of control settings, air speeds and atmospheric conditions.

Thus, off-design performance calculations involve the determination of the engine characteristics at a turbine entry temperature and intake conditions different from the on-design conditions. An example of a simplified theoretical approach of attaining the latter performance data will be reviewed in Section 8.5. Here we will merely illustrate the typical effects of the operational variables on the performance of turbojet and turbofan engines.

The engine rotational speed (thrust setting) is controlled by the fuel mass flow rate. However, at a given supply of fuel, thrust T_N, specific fuel consumption c_T, and air mass flow rate m, are dependent on the conditions of the air entering the intake. This means that for a given engine running at a given thrust setting, T_N, c_T, and m are dependent on five variables: (1) airspeed V_0 or flight Mach number M_0, (2) ambient pressure p_0 or geopotential pressure altitude H_p, (3) ambient temperature T_0, (4) humidity q, and (5) fuel mass flow rate m_f.

E.g., for net thrust and specific fuel consumption:

$$T_N = T_N(V_0, H_p, T_0, q, m_f) , \tag{8.4-1}$$

and

$$c_T = c_T(V_0, H_p, T_0, q, m_f) . \tag{8.4-2}$$

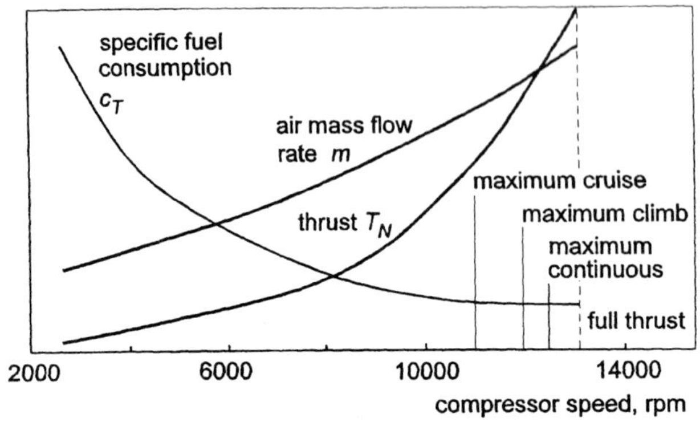

8.4-1. Typical static performance of jet engine.

When reference atmospheric conditions are assumed, the state of the incoming air is completely defined by flight speed V_0 and geopotential altitude H, through which the relations (8.4-1) and (8.4-2) reduce to:

$$T_N = T_N(V_0, H, m_f) \ , \tag{8.4-3}$$

$$c_T = c_T(V_0, H, m_f) \ . \tag{8.4-4}$$

Figure 8.4-1 gives for the jet engine a picture of the typical variations of thrust, specific fuel consumption, and air mass flow rate with thrust setting (compressor speed, rpm) at static conditions. The three performance quantities show a strong dependence on engine control setting, due to the increase of the compressor pressure ratio with increasing compressor rpm.

It should be noted that the full thrust rating in Figure 8.4-1 can be applied only for limited duration (for example, five minutes) in order to avoid a risk of damage to the engine. Therefore, the manufacturer designates a number of ratings at which the engine may be running for short periods and at which the engine may be operated continuously. E.g., the maximum thrust permitted for continuous operation, which is often named *maximum except takeoff* (*METO*) *thrust*. The maximum thrust setting permitted in cruise flight may be called *maximum cruise*. Typical variations in thrust and specific fuel consumption with flight Mach number at sea level are shown in Figure 8.4-2. It is seen that as Mach number increases from static, thrust at constant altitude and thrust setting decreases because the increase of jet velocity with airspeed is relatively small. Hence, the difference between w_e and V_0 goes down as airspeed increases so that according to Equation (7.2-8) the specific thrust will become less.

At higher airspeeds, a beneficial effect of an increasing air mass flow rate through the system more than balances, and thrust starts to rise. As a result, in

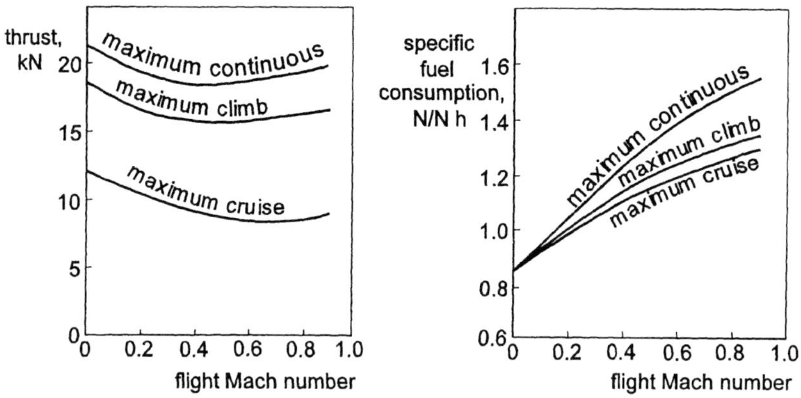

Figure 8.4-2. Typical sea level performance of turbojet engine.

the subsonic speed regime turbojet net thrust more or less is constant with airspeed.

Figure 8.4-2 also shows how specific fuel consumption varies with Mach number. Apparently, some change for the worse occurs with increasing airspeed.

With respect to the observation that the thrust is approximately independent of airspeed, it must be noted that at supersonic airspeeds the air mass flow rate becomes so large that a substantial increase in thrust takes place. The curves in

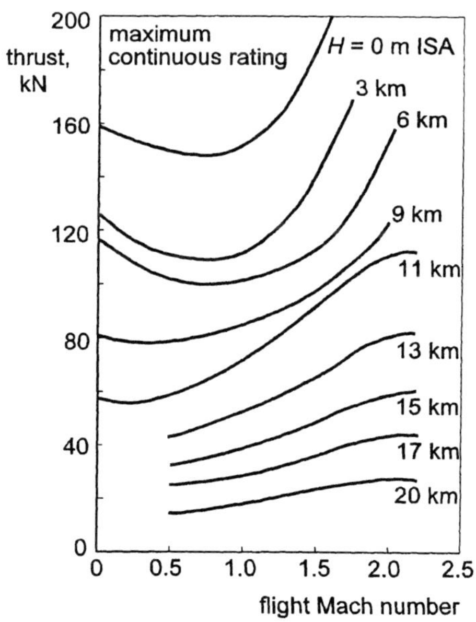

Figure 8.4-3. Typical performance of turbojet designed for use at supersonic airspeeds.

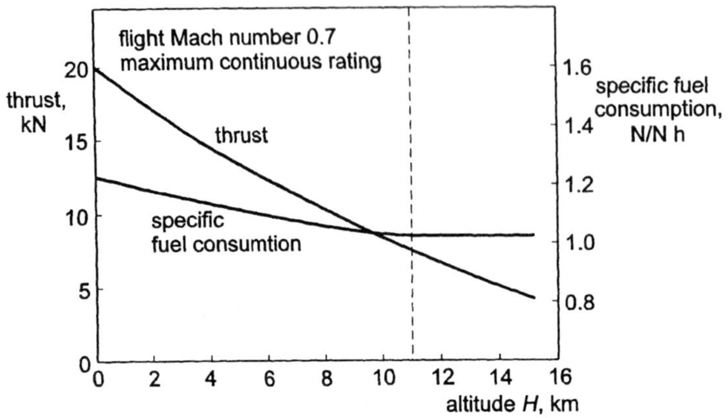

Figure 8.4-4. Typical altitude performance of turbojet engine.

Figure 8.4-3 illustrate the strong dependence of turbojet thrust on airspeed at Mach numbers greater than one. It is also seen from Figure 8.4-3 that at a given flight Mach number thrust decreases markedly with increasing altitude. As shown in Figure 8.4-4, the same influence of the altitude on thrust is found for the subsonic turbojet. Specific fuel consumption, however, manifests some improvement as the airplane gains altitude. The minimum value of the specific fuel consumption occurs in the stratosphere (ISA).

The effect of the altitude on thrust can be explained by considering the separate influences of atmospheric temperature and pressure. Looking back to the cycle analysis in Section 7.5, it may be understood that specific thrust is not dependent on the magnitude of the ambient pressure, but only on pressure ratios and the ambient temperature. Thereby, it appears that the specific thrust increases as height is gained in the troposphere.

To explain this behavior, we note that at constant airspeed, in accordance with the relation: $M_0 = V_0 / c$, the flight Mach number increases with increasing altitude in the troposphere (ISA), due to the reduction of ambient temperature. This gives, on the basis of the Poisson relation (7.5-1), a higher pressure ratio p_{t_2}/p_0 and so at constant compressor rpm, a higher nozzle pressure ratio p_{te}/p_0.

The result is that at constant airspeed and thrust setting, the specific thrust increases somewhat with increasing altitude.

On the contrary, at given airspeed and engine thrust setting, the air mass flow rate decreases with increasing altitude as the air density becomes less. Owing to these opposing influences, the variation of net thrust with altitude may be written as:

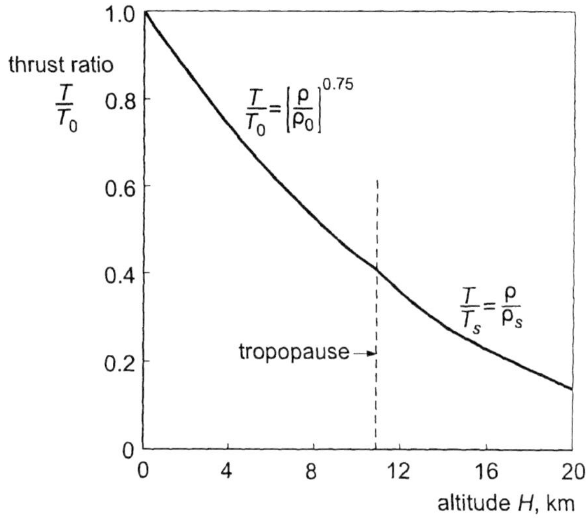

Figure 8.4-5. Variation of turbojet thrust with altitude (ISA).

$$\frac{T_N}{T_{N_0}} = \left[\frac{\rho}{\rho_0}\right]^n , \tag{8.4-5}$$

where the subscript "0" designates sea-level condition. In the troposphere (ISA) the power n is about 0.75, giving a turbojet thrust at the tropopause which is approximately 40% of its sea-level value (Figure 8.4-5).

In the stratosphere (ISA), where the air temperature is constant, the thrust varies in proportion to the density of the air;

$$\frac{T_N}{T_{N_s}} = \frac{\rho}{\rho_s} , \tag{8.4-6}$$

where the subscript "s" denotes the condition at the tropopause.
At a given geopotential pressure altitude, a significant variable in the thrust equations is the air temperature. From previous discussions we know that when the temperature increases, air mass flow rate and specific thrust decrease. Consequently, less thrust will be delivered by the engine. In Figure 8.4-6 calculation results for the ratio of turbojet and turbofan thrust in standard and off-standard atmosphere are plotted against air temperature.

Clearly, during warm days it may be necessary to restore the takeoff thrust. A temporary increase in thrust can be obtained by *liquid injection*, through which the air entering the engine is cooled by means of vaporization of a mixture of

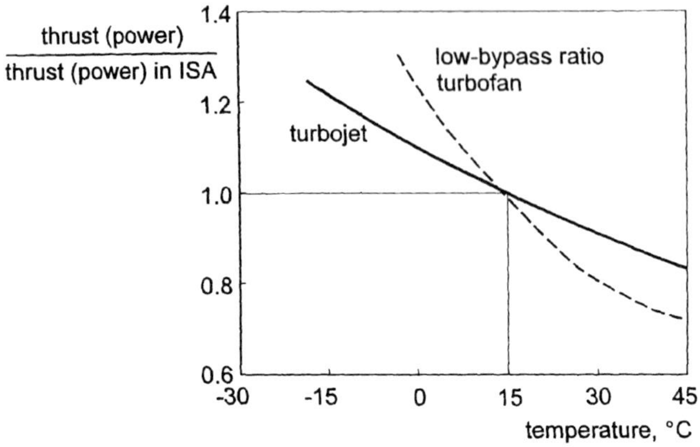

Figure 8.4-6. Effect of temperature on jet engine thrust (0 m ISA).

water and methanol. Injection of the coolant into the flow is usually achieved by spraying the liquid into the compressor inlet or directly into the combustion chamber. Nowadays, thrust augmentation is hardly applied due to the availability of the modern high-performance aircraft.

Also, the effect of the presence of humidity in the atmosphere must be mentioned. Investigations in Ref. 42 indicate, however, that the effect of humidity on turbojet performance is generally small. The test results, for which the static performance data of a turbojet at sea level were determined at constant engine speed and air temperature, show a decrease of about 5 percent in thrust for the large variation in specific humidity from zero to 0.04 (Figure 8.4-7). In order to illustrate the humidity that might be encountered in the operation of airplanes, the calculated variation of saturation specific humidity with altitude is also presented in Figure 8.4-7.

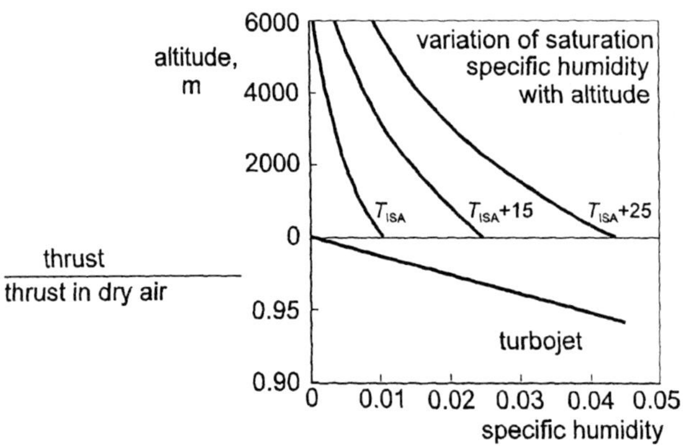

Figure 8.4-7. Effect of humidity on engine performance.

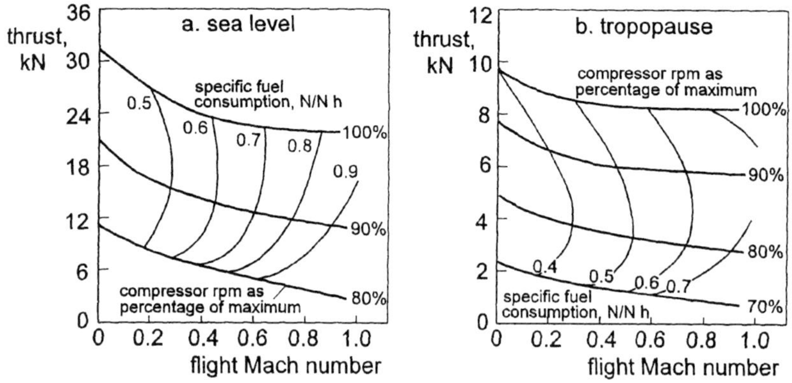

Figure 8.4-8. Typical turbofan performance.

Turning to turbofan off-design performance, the typical variation of thrust and specific fuel consumption with flight Mach number and thrust setting for a moderate bypass ratio, is shown by the curves in Figure 8.4-8 for two altitudes. Compressor speed, expressed as percentage of maximum rotational speed, is used to specify the engine thrust setting. From the plot we may see that in comparison with the turbojet there are vital reductions in specific fuel consumption (cf. Figure 8.4-2). On the other hand, the thrust of the turbofan shows a significant decrease with airspeed. Due to the large mass flow rates involved also the decrease of thrust with height is greater than for a turbojet.

Consistent with the turbojet, the variation of the turbofan thrust with altitude in the troposphere (ISA) may be expressed by Equation (8.4-5). Now, depending on bypass ratio, the power n is roughly 0.50, giving a thrust at the tropopause which is about 55% of its sea-level value.

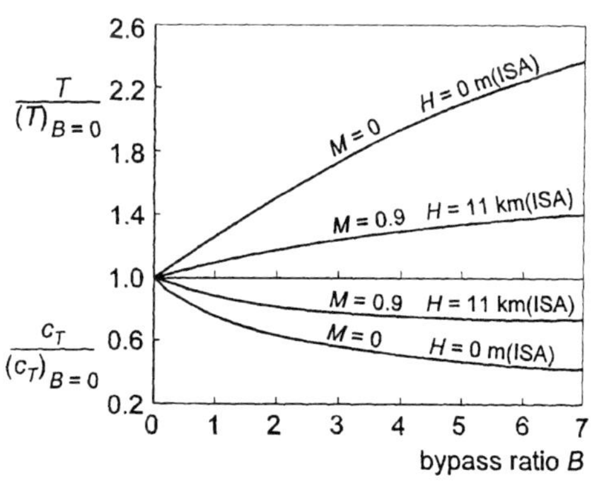

Figure 8.4-9. Effect of bypass ratio on turbofan performance.

The way in which for a turbofan engine with a given gas generator, thrust and specific fuel consumption vary with bypass ratio, is illustrated in Figure 8.4-9. Two flight conditions, takeoff at sea level and cruise flight at the tropopause (ISA) are shown. The main features are that thrust increases and specific fuel consumption decreases by the selection of a greater bypass ratio.

To explain the principal flight regimes for application of a specific engine type, particularly, the propulsive efficiency may be used.
According to Equation (7.5-27), the propulsive efficiency is given by the ratio of power available to the increase in kinetic energy of the flow. In terms of the equivalent jet velocity, we can write:

$$\eta_j = \frac{2}{1 + \dfrac{(w_e)_{eq}}{V_0}} = \frac{2}{2 + \dfrac{T_N}{m\, V_0}} \quad . \tag{8.4-7}$$

The curves in Figure 8.4-10 show that for both turbojet and turbofan, propulsive efficiency strongly increases with increasing flight Mach number. This behavior results from the fact that over the greatest part of the operation flight speed ranges, the nozzle flows are choked, which lead to almost constant equivalent jet velocities. Since turbojet thrust is produced at very high jet velocities, the higher the flight speed, the higher the propulsive efficiency. From Equation (8.4-7) and Figure 8.4-10 it also is clear that due to the increased mass flow rate and reduced mean exhaust velocity of the turbofan, this engine type is best in the high-subsonic speed range. As will be shown in Chapter 9, at low-subsonic flight speeds, the propeller remains much to be preferred to the other forms of

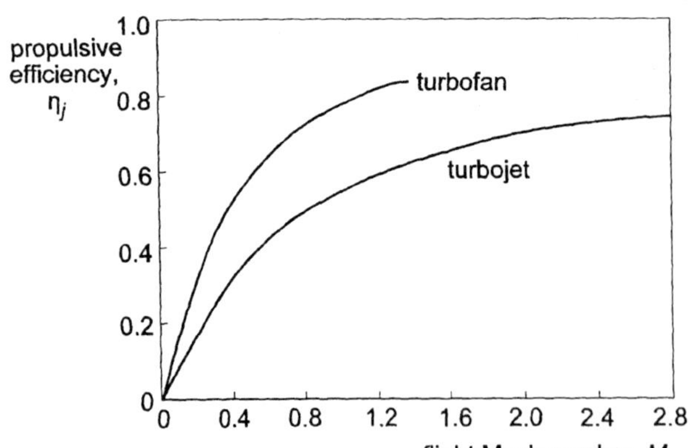

Figure 8.4-10. Typical trends of propulsive efficiency.

propulsion.
Of course, the ultimate basis on which the engine types must be compared is the overall efficiency, which follows from the product of propulsive efficiency and thermal efficiency (see the previous Equations (7.5-29) and (7.5-30)). The thermal efficiency is determined by the parameters defining the thermodynamic cycle of the flow, through which it is mainly within the control of the engine designer. Obviously, measures to improve the propulsive efficiency are only valuable if there is a resultant increase in overall efficiency.

8.5 OFF-DESIGN PERFORMANCE PREDICTION

For the design point performance calculation of the two-spool turbofan engine in Section 8.1, the four principal engine characteristics, i.e., pressure ratio of fan and compressor, bypass ratio, and turbine entry temperature, could be chosen as independent variables for the design flight condition. Under off-design conditions, however, pressure ratio of fan and compressor, and bypass ratio, are related to the turbine entry temperature. Their mutual relations follow from the compatibility of power and mass flow rate for both the low-pressure and the high-pressure spool.

Besides empirical methods, several analytical approaches for off-design performance prediction exist, some of them requiring detailed information on fan, compressor and turbine characteristics. A full analysis of this subject is given in Ref. 43. In Ref. 44 closed form solutions for the off-design performance prediction of turbo-engines using gas dynamic relationships only, are derived. In the following, as an example of the latter approach, the performance of a two-spool turbofan engine, will be outlined.
The method referred to is based on the assumption that the exhaust nozzles and low-pressure turbine stators remain choked over the range of operating conditions considered. These assumptions lead to the occurrence of aerodynamic throats in the engine, determining the working points of the engine components upstream in the engine. The adoption of choking conditions is realistic since the exhaust nozzles and low-pressure turbine may only be unchoked at very low thrust settings and flight Mach numbers.

The derivation concerns the two-spool turbofan engine arrangement shown in Figure 8.5-1, using again the station numbering according to Figure 8.1-2. Also the same assumptions as made in Section 8.1 hold, along with the following additional simplifications:

- The non isentropic processes in fan, compressor and turbines occur at constant polytropic efficiencies.

- Inlet pressure loss, mechanical efficiency, and combustion pressure loss are constant.

Figure 8.5-1 shows the compatibility between the corrected mass flow rates of the core nozzle, the low-pressure turbine (LPT) and the high-pressure turbine (HPT). The corrected turbine mass flow rates concern the values at the entry and exit planes of the turbines.

The high-pressure spool of a two-spool engine can be considered an equivalent of a single-spool engine with a fixed nozzle, with an exit area defined by the throat of the low-pressure turbine stators. For a single-spool engine with a choked exhaust nozzle, the operating point of the turbine is independent of the exhaust nozzle pressure ratio (see also Ref. 38). Hence, the turbine pressure ratio is fixed. Similarly, for a twin-spool engine with a choked low-pressure turbine, the pressure ratio of the high-pressure turbine is fixed.

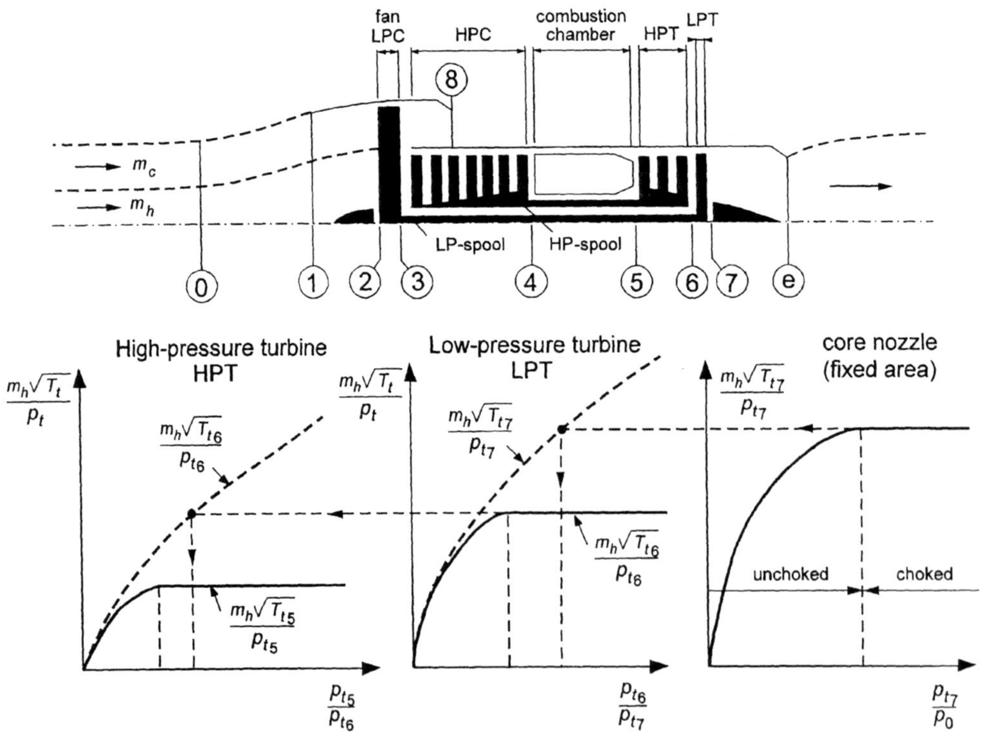

Figure 8.5-1. Mass flow rate compatibility for twin-spool turbofan engine.

For the *low-pressure spool*, the fan and the turbine are matched by the power equation:

$$m\, c_p\, [T_{t_3} - T_{t_2}] = \eta_m\, m_h\, c_{p_g}\, [T_{t_6} - T_{t_7}] \ , \tag{8.5-1}$$

where $m = m_c + m_h$.

Using the bypass ratio $B = m_c / m_h$, the equality (8.5-1) can be modified into:

$$(B + 1)\left(\frac{T_{t_3}}{T_{t_2}} - 1 \right) = \eta_m \frac{c_{p_g}}{c_p} \frac{T_{t_5}}{T_{t_2}} \frac{T_{t_6}}{T_{t_5}} \left(1 - \frac{T_{t_7}}{T_{t_6}} \right) . \tag{8.5-2}$$

By substitution of the fan pressure ratio:

$$\varepsilon_f = \left(\frac{T_{t_3}}{T_{t_2}} \right)^{\frac{\gamma\, \eta_{\text{pol}}}{\gamma - 1}} , \tag{8.5-3}$$

the power balance, Equation (8.5-2), can be written as:

$$(B + 1)\left[\left(\varepsilon_f\right)^{\frac{\gamma - 1}{\gamma\, \eta_{\text{pol}}}} - 1 \right] = \eta_m \frac{c_{p_g}}{c_p} \frac{T_{t_5}}{T_{t_2}} \frac{T_{t_6}}{T_{t_5}} \left(1 - \frac{T_{t_7}}{T_{t_6}} \right) . \tag{8.5-4}$$

According to the assumption that the low-pressure turbine and the exhaust nozzle flow remain choked over the operating range considered, the turbine pressure ratio and hence the temperature ratios T_{t_6} / T_{t_5} and T_{t_7} / T_{t_6} are invariable.

Taking the design point condition as a reference, the power balance of the low-pressure spool establishes the following relationship between the fan pressure ratio, the bypass ratio and the turbine entry temperature:

$$\frac{(B + 1)\left[\left(\varepsilon_f\right)^{\frac{\gamma - 1}{\gamma\, \eta_{\text{pol}}}} - 1 \right]}{(B_d + 1)\left[\left(\varepsilon_{f_d}\right)^{\frac{\gamma - 1}{\gamma\, \eta_{\text{pol}}}} - 1 \right]} = \varphi \ , \tag{8.5-5}$$

where

$$\varphi = \frac{[T_{t_5} / T_{t_2}]}{[T_{t_5} / T_{t_2}]_d} . \tag{8.5-6}$$

In these equations the subscript "*d*" denotes the design point conditions.

The ratio of the turbine entry temperature T_{t_5} and the inlet stagnation temperature T_{t_2} is named the *engine temperature ratio (ETR)*, which quantity is used to characterize the operating condition of the engine. Clearly, φ represents the engine temperature ratio with respect to the design condition.

A relation similar to (8.5-6) can be derived for the *high-pressure spool.* Now the equation for the power balance reads:

$$m_h c_p [T_{t_4} - T_{t_3}] = \eta_m m_h c_{p_g} [T_{t_5} - T_{t_6}] . \tag{8.5-7}$$

This equality can be written in the form:

$$\frac{T_{t_3}}{T_{t_2}} \left(\frac{T_{t_4}}{T_{t_3}} - 1 \right) = \eta_m \frac{c_{p_g}}{c_p} \frac{T_{t_5}}{T_{t_2}} \left(1 - \frac{T_{t_6}}{T_{t_5}} \right) . \tag{8.5-8}$$

Substitution of the fan pressure ratio, according to Equation (8.5-3) and the compressor pressure ratio:

$$\varepsilon_c = \left(\frac{T_{t_4}}{T_{t_3}} \right)^{\frac{\gamma \eta_{pol}}{\gamma - 1}} , \tag{8.5-9}$$

furnishes:

$$\left(\varepsilon_f\right)^{\frac{\gamma - 1}{\gamma \eta_{pol}}} \left[\left(\varepsilon_c\right)^{\frac{\gamma - 1}{\gamma \eta_{pol}}} - 1 \right] = \eta_m \frac{c_{p_g}}{c_p} \frac{T_{t_5}}{T_{t_2}} \left(1 - \frac{T_{t_6}}{T_{t_5}} \right) . \tag{8.5-10}$$

Again, taking the design point condition as a reference, and remembering the assumptions made earlier, the power balance of the high-pressure spool creates a relationship between the fan and compressor pressure ratio and the turbine entry temperature:

$$\frac{\left(\varepsilon_f\right)^{\frac{\gamma-1}{\gamma\eta_{pol}}}\left[\left(\varepsilon_c\right)^{\frac{\gamma-1}{\gamma\eta_{pol}}}-1\right]}{\left(\varepsilon_{f_d}\right)^{\frac{\gamma-1}{\gamma\eta_{pol}}}\left[\left(\varepsilon_{c_d}\right)^{\frac{\gamma-1}{\gamma\eta_{pol}}}-1\right]}=\varphi\ , \tag{8.5-11}$$

where φ is given by Equation (8.5-6).

Besides power matching for the low-pressure and high-pressure spool, also mass flow rate compatibility must be satisfied between the fan and fan exhaust nozzle, between the high-pressure compressor and the high-pressure turbine, and between the low-pressure turbine and the core nozzle.

For a choked cold flow at the fan exhaust nozzle exit, the corrected air mass flow rate through the fan is constant:

$$m_c\frac{\sqrt{T_{t_3}}}{p_{t_3}}=\text{constant}\ . \tag{8.5-12}$$

This can be written as:

$$m_c\frac{\sqrt{T_{t_2}}}{p_{t_2}}\frac{p_{t_2}}{p_{t_3}}\sqrt{\frac{T_{t_3}}{T_{t_2}}}=\text{constant}\ . \tag{8.5-13}$$

Making use of Equation (8.5-3), it follows that:

$$m_c\frac{\sqrt{T_{t_2}}}{p_{t_2}}=\text{constant}\times\left(\varepsilon_f\right)^{1-\frac{\gamma-1}{2\gamma\eta_{pol}}}\ . \tag{8.5-14}$$

With respect to the design point conditions we obtain:

$$\frac{m_{corr.f}}{m_{corr.f_d}}=\frac{\left(\varepsilon_f\right)^{1-\frac{\gamma-1}{2\gamma\eta_{pol}}}}{\left(\varepsilon_{f_d}\right)^{1-\frac{\gamma-1}{2\gamma\eta_{pol}}}}\ . \tag{8.5-15}$$

Finally, the air mass flow rate through the high-pressure compressor must be equal to the mass flow rate passing through the high-pressure turbine. This requirement can be written as:

$$m_h \frac{\sqrt{T_{t_2}}}{p_{t_2}} = m_h \frac{\sqrt{T_{t_6}}}{p_{t_6}} \frac{p_{t_6}}{p_{t_5}} \frac{p_{t_5}}{p_{t_4}} \frac{p_{t_4}}{p_{t_3}} \frac{p_{t_3}}{p_{t_2}} \sqrt{\frac{T_{t_5}}{T_{t_6}}} \sqrt{\frac{T_{t_2}}{T_{t_5}}} \ . \tag{8.5-16}$$

For a choked low-pressure turbine flow, the high-pressure turbine operating point is fixed (Figure 8.5-1). Hence, the corrected mass flow rate $m_h \sqrt{T_{t_6}} / p_{t_6}$ is constant, as well as the pressure ratio p_{t_5} / p_{t_6} and the temperature ratio T_{t_5} / T_{t_6}. Also, the combustion pressure ratio p_{t_5} / p_{t_4} is assumed to be constant.

Apparently, using the design point condition as a reference, the corrected mass flow rate through the high-pressure spool can be related to the fan and compressor pressure ratio and the quantity φ:

$$\frac{m_{corr.h}}{m_{corr.h_d}} = \frac{\varepsilon_c \ \varepsilon_f}{\varepsilon_{c_d} \ \varepsilon_{f_d}} \frac{1}{\sqrt{\varphi}} \ . \tag{8.5-17}$$

With the definition of the bypass ratio, $B = m_c / m_h$, follows:

$$\frac{B}{B_d} = \frac{m_c \ m_{h_d}}{m_{c_d} \ m_h} \ . \tag{8.5-18}$$

Since the corrected air mass flow rates in Equations (8.5-15) and (8.5-17) are related to the inlet conditions, substitution into Equation (8.5-18) yields:

$$\frac{B}{B_d} = \frac{m_{corr.c}}{m_{corr.c_d}} \frac{m_{corr.h_d}}{m_{corr.h}} = \frac{\left(\varepsilon_f\right)^{1 - \frac{\gamma - 1}{2 \gamma \eta_{pol}}}}{\left(\varepsilon_{f_d}\right)^{1 - \frac{\gamma - 1}{2 \gamma \eta_{pol}}}} \frac{\varepsilon_{c_d} \ \varepsilon_{f_d}}{\varepsilon_c \ \varepsilon_f} \sqrt{\varphi} \ . \tag{8.5-19}$$

This equation can be reduced to the form:

$$\frac{B}{B_d} = \left(\frac{\varepsilon_{f_d}}{\varepsilon_f} \right)^{\frac{\gamma - 1}{2 \gamma \eta_{pol}}} \frac{\varepsilon_{c_d}}{\varepsilon_f} \sqrt{\varphi} \ . \tag{8.5-20}$$

The procedure developed in the above consists of a set of three compatibility equations of flow and power, Equations (8.5-5), (8.5-11) and (8.5-20), which completely define the variation of fan pressure ratio, compressor pressure ratio and bypass ratio with engine temperature ratio.
It may be appreciated that the determination of the engine characteristics ε_f, ε_c and B, by solving the Equations (8.5-5), (8.5-11) and (8.5-20), require iterative computations.

8.6 GENERALIZED PERFORMANCE PRESENTATION

It may be clear from Sections 8.1 and 8.2 that for steady-state operating conditions of a jet engine, the performance parameters (1) gross thrust, T_G, (2) air mass flow rate, m, and (3) compressor rpm, n, are fully determined when the following five independent variables are specified:

- Total pressure at compressor entry, p_{t_2},
- Total temperature at compressor entry, T_{t_2},
- Fuel mass flow rate, m_f,
- Atmospheric pressure, p_0,
- Characteristic linear dimension of the engine, D.

In dry air, the solution of the problem may then be written in the form of three equations, in which a function of the variables is equated to zero. E.g., for the gross thrust, T_G:

$$f(T_G, p_{t_2}, T_{t_2}, n, p_0, D) = 0 \; . \qquad (8.6\text{-}1)$$

Although the only control to the engine is the variation in the fuel mass flow rate, m_f, practically, the compressor speed, n, is used to define the engine control setting. Therefore, in Equation (8.6-1) the independent variable m_f is replaced by n.

The variables T_G, p_{t_2}, n, p_0 and D have the dimensions mass [M], length [L] and time [T]. When considering the dimensions of the temperature T_{t_2}, it is convenient to associate it with the kinetic energy per unit mass given by $c_p\, T_t$ (see Equation (7.4-12)), which quantity has the dimensions $[L^2\, T^{-2}]$.

By the principle of dimensional analysis it is possible to reduce the function of six variables expressed by Equation (8.6-1) to a different function of three nondimensional groups formed from these six variables (Ref. 8):

$$f\left[\frac{T_G}{p_{t_2} D^2}, \frac{p_{t_2}}{p_0}, \frac{n\, D}{\sqrt{T_{t_2}}}\right] = 0 \; . \qquad (8.6\text{-}2)$$

Evidently, when in Equation (8.6-2) two parameters are known, also the third is fixed. This yield for the nondimensional gross thrust:

$$\frac{T_G}{p_{t_2} D^2} = f_1 \left[\frac{p_{t_2}}{p_0}, \frac{n D}{\sqrt{T_{t_2}}} \right]. \tag{8.6-3}$$

Similarly, we find the relations for nondimensional fuel and air mass flow rate as:

$$\frac{m_f}{p_{t_2} D^2 \sqrt{T_{t_2}}} = f_2 \left[\frac{p_{t_2}}{p_0}, \frac{n D}{\sqrt{T_{t_2}}} \right], \tag{8.6-4}$$

and

$$\frac{m \sqrt{T_{t_2}}}{p_{t_2} D^2} = f_3 \left[\frac{p_{t_2}}{p_0}, \frac{n D}{\sqrt{T_{t_2}}} \right]. \tag{8.6-5}$$

In principle, in Equation (8.6-1) also the internal friction between the particles of the gas flow, by means of the coefficient of dynamic viscosity (see Chapter 3), must be considered. Then another dimensionless group in the form of the Reynolds number will appear in the relations (8.6-2) to (8.6-5). However, it has been found that, for normal operation, the effect of change in Reynolds number is unimportant. Only at extreme altitudes, Reynolds number effects may be of significance to the engine efficiencies (Ref. 45). Moreover, at very high altitudes there may also occur a change for the worse in the efficiency of the combustion due to the low pressure levels in the combustion chamber.

Division of Equation (8.6-4) by Equation (8.6-5) furnishes:

$$\frac{m_f}{m T_{t_2}} = f_4 \left[\frac{p_{t_2}}{p_0}, \frac{n D}{\sqrt{T_{t_2}}} \right]. \tag{8.6-6}$$

When we are considering the performance of a given engine, the variable D can be left out from the equations.

Usually, the nondimensional groups for T_G , m_f, and m are given again the dimensions of thrust and mass flow rate by dividing the pressures p_{t_2} and p_0, and temperature T_{t_2} by sea-level atmospheric pressure $(p_0)_0$ and temperature $(T_0)_0$, respectively, where $(p_0)_0 = 101325$ N/m^2 and $(T_0)_0 = 288.15$ K (0 m ISA). By this operation we obtain the so-called *generalized performance quantities*:

$$\frac{T_G}{\delta_{t_2}} = f_5\left[\frac{p_{t_2}}{p_0}, \frac{n}{\sqrt{\theta_{t_2}}}\right] \tag{8.6-7}$$

$$\frac{m_f}{\delta_{t_2}\sqrt{\theta_{t_2}}} = f_6\left[\frac{p_{t_2}}{p_0}, \frac{n}{\sqrt{\theta_{t_2}}}\right] \tag{8.6-8}$$

$$\frac{m\sqrt{\theta_{t_2}}}{\delta_{t_2}} = f_7\left[\frac{p_{t_2}}{p_0}, \frac{n}{\sqrt{\theta_{t_2}}}\right]. \tag{8.6-9}$$

Note that in the Equations (8.6-7) to (8.6-9), the following pressure and temperature ratios are used:

$$\delta_{t_2} = \frac{p_{t_2}}{(p_0)_0} \quad \text{and} \quad \theta_{t_2} = \frac{T_{t_2}}{(T_0)_0}.$$

The significance of using the nondimensional quantities given by Equations (8.6-7) to (8.6-9) may be obvious, as it is only necessary to plot three sets of curves in order to document the performance of the engine. For nonstandard operating conditions, it is only needed to utilize the values of δ_{t_2} and θ_{t_2} to obtain the actual engine performance.

For a bypass engine, the typical form of generalized gross thrust curves are sketched in Figure 8.6-1.

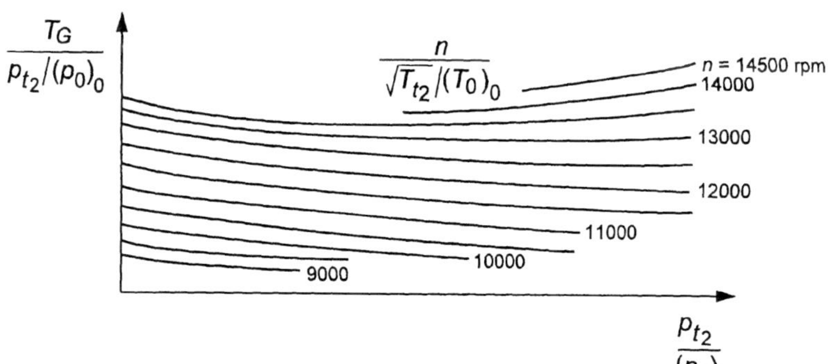

Figure 8.6-1. Typical generalized gross thrust data for bypass engine.

Note that when a figure is attached to the intake efficiency, also the airspeed is known from the previous Equations (7.4-12), (7.4-13), and (7.6-1), and by using the relation; $V_0 = M_0 \sqrt{\gamma R T_0}$. Ultimately, the matching net thrust follows from the relation: $T_N = T_G - mV_0$.

9 PROPELLER PROPULSION

9.1 PROPELLER PROPULSION SYSTEMS

The problem of the inefficiency of jet propulsion at low flight speeds can be removed by application of propeller propulsion. The essential difference between the two types of propulsion is the nature of the rearward mass flow. The propeller transports a large mass flow of cold air at a relatively low speed, whereas jet propulsion involves the acceleration of a much smaller amount of hot gas to a high velocity. As a consequence, the turbojet and the turbofan are only effective at supersonic and high subsonic flight speeds, respectively, whereas the propeller is a practical means of producing thrust at low subsonic airspeeds.
The propeller propulsion systems to be considered in this text are the turboprop engine (the turbo-engine and propeller combination) and the reciprocating piston engine with a propeller.

The turboprop and the piston engine are characterized by the fact that these systems convert the heat energy in the fuel into mechanical shaft power and that this power then drives the propeller. The latter device produces the thrust by accelerating atmospheric air backwards with respect to the airplane.
Thus, just like the operation of the turbofan engine, there are two different conversion processes involved; a heat energy conversion and a power conversion process.

The useful power available for propulsion is given by the product of propeller thrust T and airspeed V_0. As introduced in Chapter 7, Section 7.5, this quantity is called *power available* and denoted by the symbol P_a. The ratio of power available to shaft power, P_{br}, is the *propulsive efficiency* of the propeller:

$$\eta_j = \frac{T V_0}{P_{br}} = \frac{P_a}{P_{br}} . \tag{9.1-1}$$

Firstly, in Sections 9.1 to 9.5 of this chapter we will review the generation of shaft power by the turbo and piston engine, and after that the main characteristics of the propeller will be discussed in Sections 9.6 to 9.8.

In the gas generator of a single shaft turboprop engine, most of the turbine power is extracted from the hot gases by means of an enlarged turbine, which also drives the propeller (Figure 9.1-1).

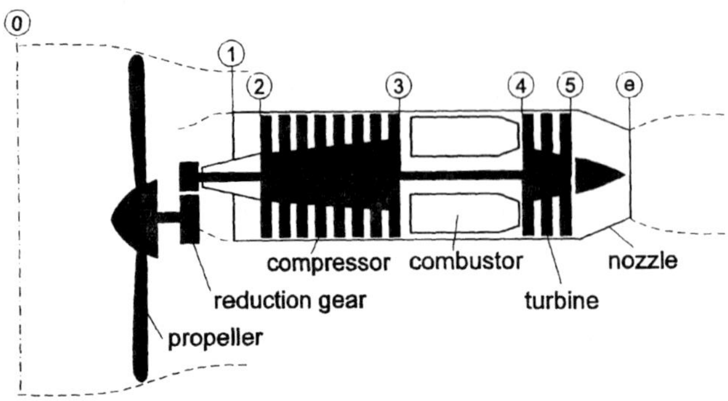

Figure 9.1-1. Layout of single shaft turboprop engine.

Figure 9.1-1 shows that compressor, turbine, and propeller are all mechanically connected. A reduction gear is present between the compressor and the propeller since the propeller must rotate at a much lower rotational speed than the gas generator in order to avoid extreme propeller blade tip speeds.
Normally, there is sufficient residual energy available from the turbine to release high velocity gases from the exhaust nozzle to produce some additional jet thrust.

Figure 9.1-2 shows a two-spool configuration of the turboprop engine. This arrangement has a separate high-pressure compressor-turbine system, which is mechanically independent of the low-pressure compressor-turbine system. The low-pressure turbine or power turbine extracts from the gas flow, the useful power to drive the propeller.

Of course, a turboprop uses a much larger air mass flow rate than other turbo-engine types and so obtains the highest propulsive efficiency. Nevertheless, as

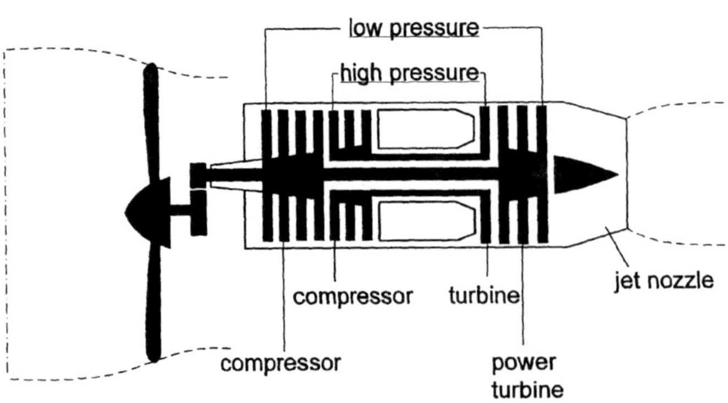

Figure 9.1-2. Layout of two-spool turboprop engine.

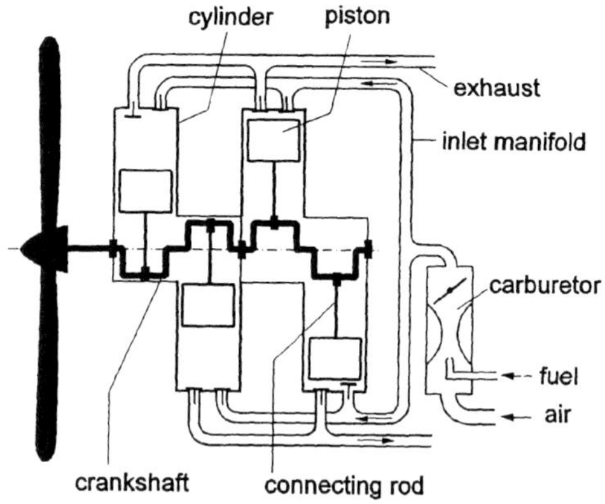

Figure 9.1-3. Four cylinder piston engine with propeller.

the flight speed increases up to, say, 500 km/h, the blade tips approach the speed of sound through which the propulsive efficiency is seriously impaired.

The piston engine plus propeller is used today for driving small general aviation airplanes only. These internal combustion engines have four or six cylinders, are air cooled, and operate on the four-stroke cycle principle. The cylinders are virtually always arranged in horizontally opposed pairs (Figure 9.1-3). Fuel may be supplied by individual injection to each cylinder or by a carburetor which mixes the fuel and air to the right proportions and leads this mixture to the cylinders via the inlet manifold. Each cylinder has an intake and an exhaust valve and inside the cylinder a piston is moving up and down by the high gas pressures obtained from burning the mixture of fuel and air. The pistons drive the crankshaft via connecting rods, which change the reciprocating motion of the pistons to the rotational motion of the crankshaft. The crankshaft, in turn, drives the propeller.

In contrast to gas turbine engines, the processes in the piston engine are not continuous. In combination with the presence of reciprocating parts, the piston engine runs less smooth and enables less heat energy to be released.

9.2. THE TURBOPROP ENGINE

We proceed with an examination of the thermodynamic processes in the gas generator of a single shaft turboprop engine. For simplicity, we will assume that the pressure of the gas falls completely to ambient pressure in going through the turbines.

Working through the cycle, the stations will be numbered as depicted in Figure 9.1-1. Thus, the total temperature at the compressor inlet comes from:

$$T_{t_2} = T_0 \left[1 + \frac{\gamma - 1}{2} M_0^2 \right] , \qquad (9.2\text{-}1)$$

where T_0 is the ambient temperature and M_0 the flight Mach number.
As used earlier for the turbojet and the turbofan, the efficiency of the intake may be described by the ratio of total pressure at the compressor inlet to the free stream total pressure (see also Equation (7.6-1)):

$$\eta_R = \frac{p_{t_2}}{p_{t_0}} , \qquad (9.2\text{-}2)$$

where η_R is the pressure recovery factor.

By combining the previous Equations (7.5-8) and (7.6-8), we obtain the turbine power required to drive the compressor per unit air weight flow rate as:

$$\frac{P_c}{m g} = \frac{c_p}{g} T_{t_2} \left[\varepsilon_c^{\frac{\gamma - 1}{\gamma \eta_{pol}}} - 1 \right] , \qquad (9.2\text{-}3)$$

where $\varepsilon_c = p_{t_3} / p_{t_2}$ is the compressor pressure ratio and η_{pol} the polytropic efficiency of the compressor.
Note that in deriving Equation (9.2-3) use is made of the assumption that the specific heat at constant pressure, c_p, is constant throughout the engine, and that the fuel mass flow rate is negligible. The quantity P_c / mg is called *specific compressor power*.
Likewise, we can use Equations (7.5-8) and (7.6-12) to solve for the total *specific turbine power*:

$$\frac{P_t}{m g} = \frac{c_P}{g} T_{t_4} \left[1 - \left(\frac{p_{t_5}}{p_{t_4}} \right)^{\frac{\eta_{pol}(\gamma - 1)}{\gamma}} \right] , \qquad (9.2\text{-}4)$$

where now η_{pol} is the polytropic efficiency of the turbine.
Using the simplification that in the turbine the flow expands to ambient pressure, the specific output of the engine, from Equations (9.2-3) and (9.2-4), then turns out to be:

$$\psi_P = \frac{P_{br}}{m\,g} = \frac{c_p}{g}\left[T_{t_4}\left[1-\left(\frac{p_{t_5}}{p_{t_4}}\right)^{\frac{\eta_{pol}(\gamma-1)}{\gamma}}\right] - T_{t_2}\left[\varepsilon_c^{\frac{\gamma-1}{\gamma\,\eta_{pol}}} - 1\right]\right]. \tag{9.2-5}$$

Contrary to jet engines, the specific fuel consumption of the turboprop is expressed as the fuel weight flow rate, F, divided by shaft power, P_{br}:

$$c_P = \frac{F}{P_{br}} = \frac{m_f\,g}{P_{br}}. \tag{9.2-6}$$

Using Equation (7.5-7) produces the following expression:

$$c_P = \frac{F}{P_{br}} = \frac{m_f\,g}{\psi_P\,m\,g} = \frac{c_P\,(T_{t_4} - T_{t_3})}{H\,\psi_P}, \tag{9.2-7}$$

where c_P has the dimensions N/Wh or N/kWh.

In Figure 9.2-1 are presented specific shaft power and specific fuel consumption of the gas generator for matched expansion in the power turbine $(p_{t_5} = p_e = p_0)$. These have been calculated for a range of turbine entry temperatures and compressor pressure ratios and apply to a flight Mach number of $M_0 = 0.4$ and an altitude of 4000 m (ISA). Used is the assumption that c_p and γ have everywhere in the flow through the engine the same values as in ambient air.
It is seen again that in a gas turbine engine an increase in turbine entry temperature results in a considerable increase in specific output. The curves in

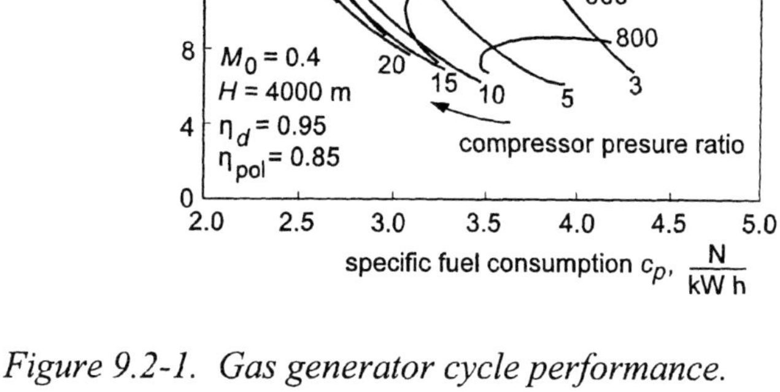

Figure 9.2-1. Gas generator cycle performance.

Figure 9.2-1 also show that there is for any given turbine entry temperature, an optimal compressor pressure ratio producing maximum specific power.
Increase in pressure ratio beyond this optimum reduces specific power, but provides an improvement of specific fuel consumption. The curves also make clear that for any given turbine entry temperature there exists an optimal value of compressor pressure ratio yielding the minimum specific fuel consumption. Comparison of the curves in Figure 9.2-1 and in Figure 7.8-5 shows a complete similarity, only that at a given pressure ratio, the specific fuel consumption in the turboprop tends to improve somewhat with increasing turbine entry temperature, whereas in the turbojet there is an attendant increase in specific fuel consumption.
The actual turboprop performance differs from the output in Figure 9.2-1 in that a part of the useful power becomes available as jet thrust. This is due to the fact that when designing a turboprop, the pressure of the gas at the turbine exit is always greater than ambient pressure in order to use the pressure drop after the turbines to eject the gases from a jet nozzle (Figures 9.1-1 and 9.1-2). In cruising flight the power produced from the exhaust thrust may amount to 10-20 percent of the useful power.

The contribution of the jet thrust, T_j, to the power output is usually expressed in terms of an apparent increase in shaft power.
The definition of the so-called *equivalent shaft power*, P_{eq}, follows from:

$$P_a = T V_0 = \eta_j P_{br} + T_j V_0 , \tag{9.2-8}$$

where P_a is the power available for propulsion, and η_j is the propulsive efficiency. Hence,

$$P_{eq} = \frac{P_a}{\eta_j} = P_{br} + \frac{T_j V_0}{\eta_j} . \tag{9.2-9}$$

When the airplane is not moving, it is assumed that the propeller generates a fixed amount of thrust per unit power output. Thus, the static equivalent shaft power can be expressed as:

$$P_{eq} = P_{br} + \frac{T_j}{K} . \tag{9.2-10}$$

For an average propeller, the factor K is approximately 15 N/kW.

Although the output of the turboprop often is specified in terms of equivalent shaft power, it might be worthwhile for any analysis to detail shaft power and jet thrust.

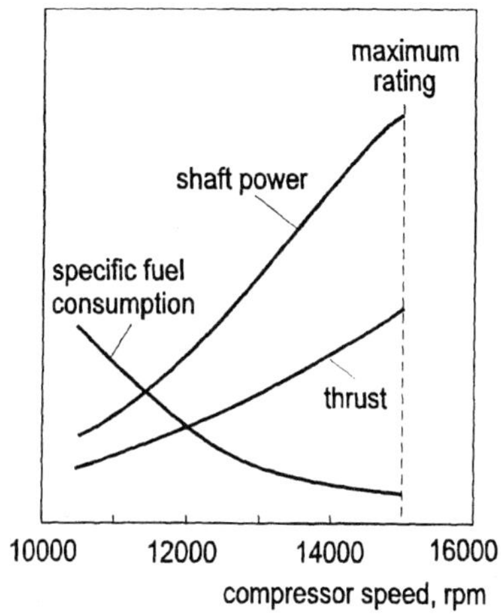

Figure 9.2-2. Static performance at sea level of turboprop engine.

From the above discussion it may be clear that the performance of a turboprop is dependent on the same parameters which also define the performance of jet engines (see Section 8.4).

When the restriction is made of operation in International Standard Atmosphere, we can write (cf. Equations (8.4-3) and (8.4-4)):

$$\left.\begin{aligned} P_{br} &= P_{br}(\Gamma, V_0, H) \\ T_j &= T_j(\Gamma, V_0, H) \\ c_p &= c_p(\Gamma, V_0, H) \end{aligned}\right\} \tag{9.2-11}$$

In Equation (9.2-11) the Greek capital letter "Γ" denotes engine control setting.

Figure 9.2-2 shows schematically the static performance of the turboprop at sea level as a function of engine control setting.

The typical variations of shaft power, jet thrust, and specific fuel consumption with flight Mach number are portrayed by the curves in Figure 9.2-3a. The shaft power and specific fuel consumption show some improvement with increasing forward speed, whereas the jet thrust strongly decreases as airspeed increases.

The qualitative effect of the altitude on turboprop performance is sketched in Figure 9.2-3b. It can be seen that there is a loss of power and thrust and an improvement in specific fuel consumption with increasing altitude.

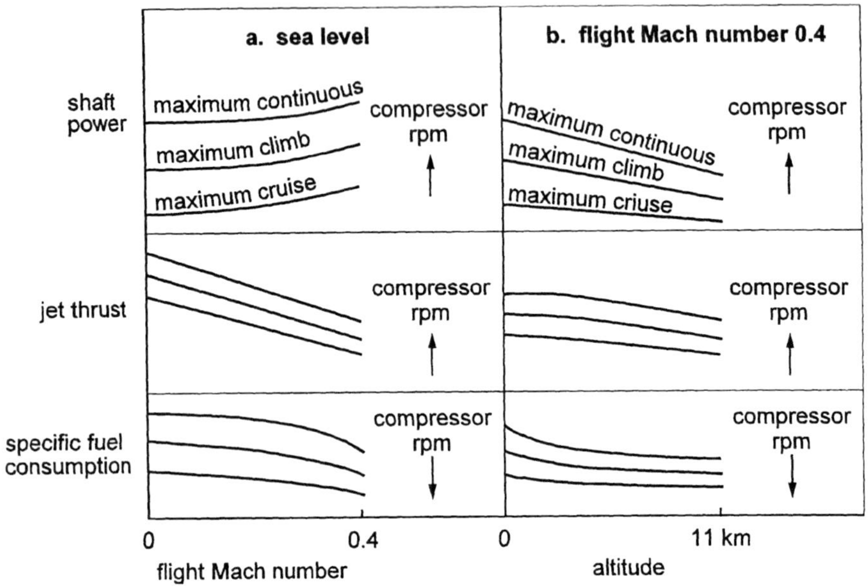

Figure 9.2-3. Typical turboprop performance.

Like the thrust of the turbojet engine, also the equivalent shaft power of the turboprop at any given altitude may be related to its sea-level value by the relationship:

$$\frac{P_{eq}}{(P_{eq})_0} = \left(\frac{\rho}{\rho_0}\right)^n , \qquad (9.2\text{-}12)$$

where in the troposphere, the exponent n has a value of approximately 0.75.

9.3. THE PISTON ENGINE

The working cycle of the four-stroke piston engine requires two revolutions of the crankshaft; two strokes down and two strokes up. The four events are illustrated in Figure 9.3-1 in the form of a plot of pressure inside the cylinder versus volume:

- The intake stroke (1). The piston travels from the cylinder head (top dead center) to the bottom of the cylinder (bottom dead center). An amount of fuel and air is sucked into the cylinder through the open intake valve.

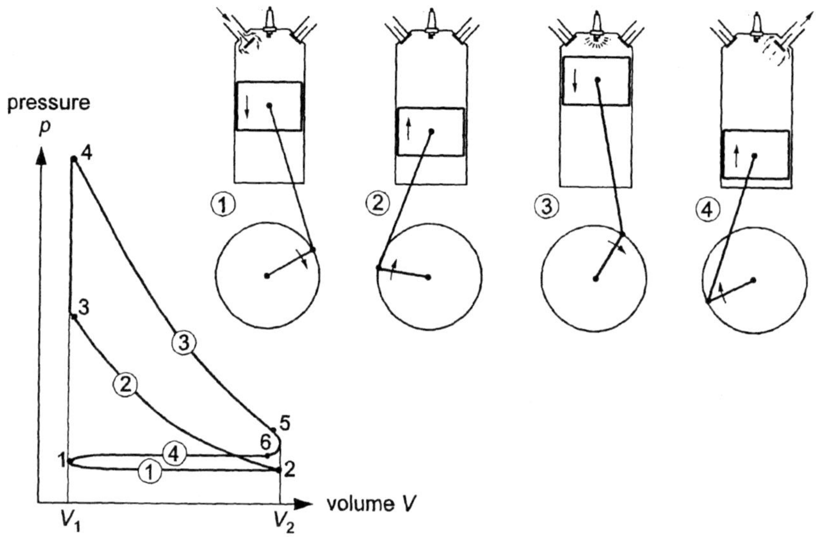

Figure 9.3-1. Working cycle of four-stroke piston engine.

- The compression stroke (2). The piston moves upward to the top of the cylinder, and the charge is almost isentropically compressed from the inlet-manifold pressure p_2 to a pressure p_3. When the piston approaches the top of the stroke, the mass of gas confined in the cylinder is ignited by the ignition system. Between points 3 and 4, combustion takes place at approximately constant volume and the pressure increases to p_4.

- The power stroke (3). The high pressure pushes the piston downwards and the burning gas expands more or less isentropically to the pressure p_5. Just before the piston reaches the bottom of the stroke the exhaust valve opens and the pressure inside the cylinder drops to the value p_6 ($p_6 \approx p_1$).

- The exhaust stroke (4). The piston again moves upward and the burned gases in the cylinder are forced out through the exhaust valve.

The compression ratio, ε, of the piston engine is the volume of space in the cylinder when the piston is at bottom dead center, V_1, divided by the volume when the piston is at top dead center, V_2 ($\varepsilon = V_2/V_1$). Therefore, the compression ratio of a piston engine is a constructional quantity.

The conversion of the heat energy released during combustion to mechanical power, usually, is analyzed by considering the *Otto-cycle*, named after Nikolaus August Otto (1832-1891), German engineer and inventor who built in 1876 the first four-stroke piston engine.

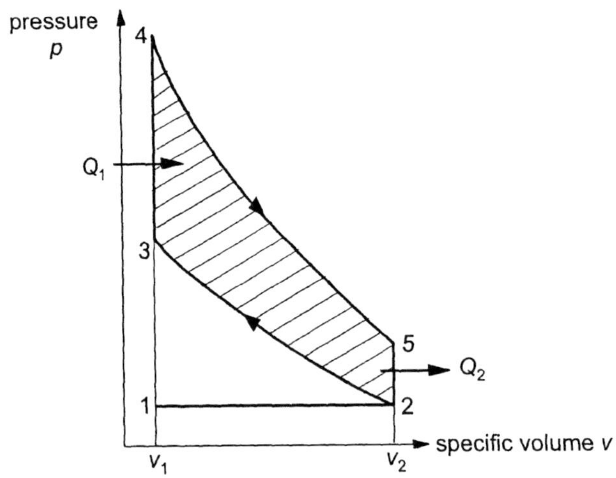

Figure 9.3-2. The Otto-cycle.

This cycle, shown in Figure 9.3-2 in terms of pressure versus specific volume $v = V/m$ in m^3/kg or in the reciprocal of the density ρ ($v = 1/\rho$), consists of four ideal processes. Point 2 in this figure stands for the state of the mixture of air and fuel in the cylinder at the inlet-manifold pressure that is compressed isentropically along the line 2-3. The vertical line between states 3 to 4 represents an *isochoric process*, where heat is added to the charge by burning the fuel at constant volume (constant density), thereby considerably increasing the pressure. Then, the gases are expanded isentropically along the line 4-5 to the pressure p_5, where it has the same volume as it had at the end of the inlet stroke. Finally, from 5 to 2, the cycle is completed by the discharge of the gases isochorically (at constant volume).

For the piston engine, the thermal efficiency of the cycle may be expressed as (cf. Equation (7.5-16)):

$$\eta_{th} = \frac{P_i}{m_{f_i} H} = \frac{Q_1 - Q_2}{Q_1} , \qquad (9.3\text{-}1)$$

where $P_i = (Q_1 - Q_2)$ is the power output per cylinder, $Q_1 = m_{f_i} H$ is the heat energy supplied to the air in the cylinder in unit time, and Q_2 is the heat energy discharged from the cylinder in unit time during the exhaust stroke.
Looking at the Otto-cycle in Figure 9.3-2 and assuming that the specific heat at constant volume, c_v, and the ratio of specific heats, γ, are constant throughout the cycle, we may write:

$$\eta_{th} = \frac{m c_v\left[(T_4 - T_3) - (T_5 - T_2)\right]}{c_v (T_4 - T_3)} . \qquad (9.3\text{-}2)$$

According to the Poisson relation, Equation (7.3-27), we have:

$$\frac{T}{\rho^{\gamma - 1}} = T v^{\gamma - 1} = \text{constant}. \qquad (9.3\text{-}3)$$

From Equation (9.3-3) we obtain:

$$\left(\frac{v_2}{v_1}\right)^{\gamma - 1} = \varepsilon^{\gamma - 1} = \frac{T_3}{T_2} = \frac{T_4}{T_5} . \qquad (9.3\text{-}4)$$

Hence, from Equations (9.3-2) and (9.3-4):

$$\eta_{th} = 1 - \frac{T_5 - T_2}{T_4 - T_3} = 1 - \frac{T_4 \dfrac{T_2}{T_3} - T_2}{T_4 - T_3} = 1 - \frac{T_2}{T_3} = 1 - \frac{1}{\varepsilon^{\gamma - 1}} . \qquad (9.3\text{-}5)$$

From Equations (9.3-5) and (7.5-26) we find that the ideal thermal efficiency of both the piston engine and the turbojet is only dependent on the compression ratio, and increases with increasing ε.
It should be noted that due to heat and friction losses, the actual value of the thermal efficiency is lower than that forecasted by Equation (9.3-5).

Figure 9.3-3 gives the representation of the Otto-cycle in the *h-s* diagram, where isochors are drawn between the state points 3 and 4, and 5 and 2.

The net work per cylinder done during one cycle, W_i, is represented by the hatched area in Figure 9.3-2 since this work is given by:

$$W_i = \oint p \, \mathrm{d}V . \qquad (9.3\text{-}6)$$

At an engine speed n (revolutions per second), the number of power strokes in one second is $n/2$. Hence, the power developed in one cylinder, the *indicated power* P_i, is:

$$P_i = W_i \frac{n}{2} . \qquad (9.3\text{-}7)$$

When we neglect the friction losses between the moving parts of the engine (η_m = 1.0), the power delivered to the shaft, P_{br}, will be equal to the total indicated power:

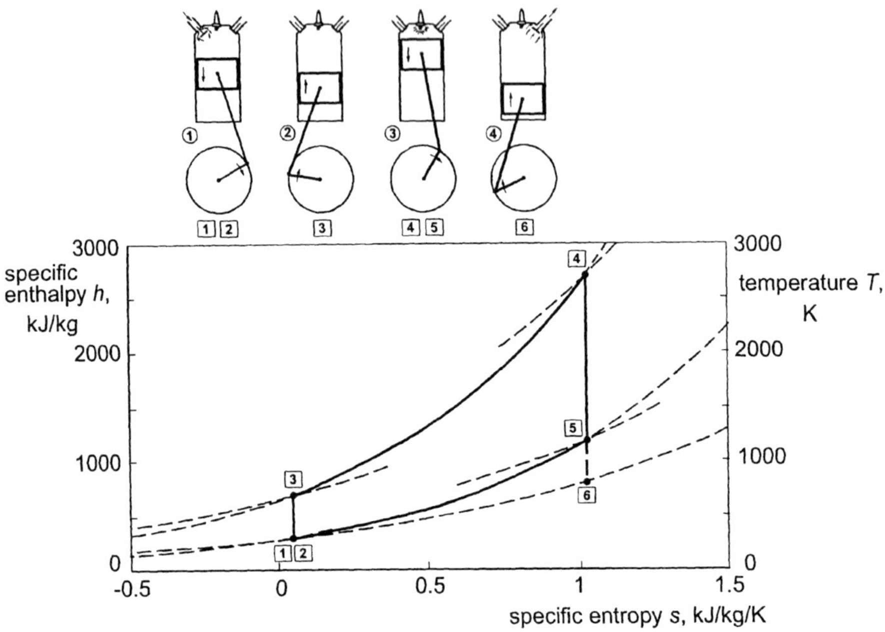

Figure 9.3-3. h-s diagram for Otto-cycle.

$$P_{br} = N P_i \, , \tag{9.3-8}$$

where N is the number of cylinders.
Combination of Equations (9.3-7) and (9.3-8) yields:

$$P_{br} = N W_i \frac{n}{2} \, . \tag{9.3-9}$$

The thermal efficiency of the engine, the ratio of the power developed in the engine and the thermal input per unit time, becomes:

$$\eta_{th} = \frac{N P_i}{m_f H} \, . \tag{9.3-10}$$

The heating value of the gasoline fuel used in airplane piston engines is, just like kerosine, about $H = 4.3 \times 10^7$ Joule/kg.

The working cycle of the piston engine gives a high thermal efficiency over a wide range of inlet manifold pressures, engine speeds, and flight altitudes. A cruising thermal efficiency of 35% (c_p = 2.4 N/kWh) is fairly representative of airplane piston engines.

The power P_i and the work W_i may be related to each other by means of the *mean effective pressure*, p_e, which quantity is defined by:

$$p_e = \frac{W_i}{\frac{\pi}{4} D^2 S} , \tag{9.3-11}$$

where D is the diameter of the piston (the bore) and S is the length of the displacement of the piston (the stroke).
Clearly, the pressure p_e in Equation (9.3-11) is a fictitious parameter, which is broadly used as an index in determining the shaft power. Insertion of Equation (9.3-11) into (9.3-9) yields:

$$P_{br} = p_e \frac{\pi}{4} D^2 S \frac{n}{2} N , \tag{9.3-12}$$

where the quantity $(\pi/4)\ D^2\ S\ N$ is the total volume displaced by the piston during one stroke (the total piston displacement of the engine).
Apparently, the shaft brake power delivered by a given engine is a function of mean effective pressure and engine speed:

$$P_{br} = f(p_e, n) . \tag{9.3-13}$$

Further thermodynamic analysis learns that the pressure p_e varies proportionally to the quotient of pressure and temperature of the fuel-air mixture at the beginning of the compression stroke (point 2 in Figures 9.3-2 and 9.3-3). This means that p_e depends on the density of the atmospheric air and on the throttle setting δ ($0 \leq \delta \leq 1.0$). Accordingly, we have:

$$P_{br} = f(H, \delta, n) = f(H, \Gamma) , \tag{9.3-14}$$

where the engine control setting Γ is a function of δ and n.

From Equation (9.3-14) it is apparent that for the piston engine there is no dependence of shaft power P_{br} on flight speed V_0.

9.4 MEASUREMENT OF SHAFT POWER

The power delivered to the crankshaft is called *shaft brake power*, which designation has as origin the measurement of the engine output by coupling the crankshaft to a brake system.
The shaft power may be determined by connecting the engine with a *dynamometer*, which is a power-absorbing apparatus that is able to correctly

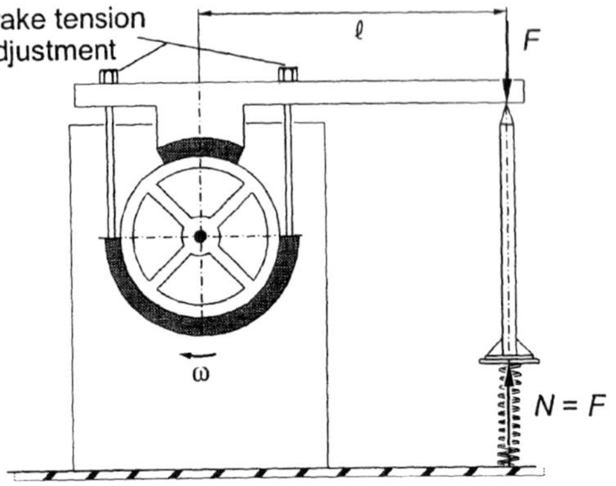

Figure 9.4-1. Dynamometer.

measure the generated power, when the engine is mounted on a ground-based test stand.

Sketched in Figure 9.4-1 is the Prony brake, which is a well-known example of a brake device to measure engine torque and shaft power.
In the equilibrium position at a given engine speed (revolutions per unit time), the power delivered by the engine equals the power dissipated by the brakes. Then the force F, exerted by the lever arm, balances the restoring force N produced by the vertically mounted spring. The defining values are now the force F, length l and engine speed n.
At this point the reader should remind from *dynamics* that the rotational motion of a rigid body about a fixed axis is given by the relationship:

$$P = M\omega \,, \tag{9.4-1}$$

where P is power, M is torque or moment of the force about the axis and ω is the angular velocity.
Hence, the power delivered by the engine is found from the following expression:

$$P_{br} = \frac{F\,l\,2\,\pi\,n}{60} \,, \tag{9.4-2}$$

where the engine speed n is in revolutions per minute (rpm).

Comparing Equation (9.4-2) with the previous Equation (9.3-12), shows that now also the mean effective pressure, p_e, is known.

Finally, we may mention that it is also possible to determine the shaft power in flight by means of a torque meter (Ref. 47).

9.5 PISTON ENGINE PERFORMANCE CHARTS

The charts as, generally, supplied by piston engine manufacturers are of the form sketched in Figure 9.5-1. In this so-called *standard power diagram* the shaft power in International Standard Atmosphere is given, assuming a mixture control which furnishes maximum power.

In the left-hand diagram the power at sea level is plotted against inlet manifold pressure for several values of engine speed. The power is limited by the maximum permissible engine speed and maximum inlet manifold pressure, which corresponds to the full-throttle condition (fully opened throttle valve).
The effect of the altitude on full-throttle power is shown in the right-hand diagram of Figure 9.5-1. The chart gives also the prevailing inlet manifold pressures. In the full-throttle condition and for a given engine speed the shaft brake power decreases with increasing altitude due to the decreasing air density. An empirical relationship often used to account for the effect of the altitude is:

$$\frac{P_{br}}{P_{br_0}} = 1.132 \frac{\rho}{\rho_0} - 0.132 \ , \qquad (9.5\text{-}1)$$

where the subscript "0" denotes sea-level conditions.

In order to determine the power for a part-throttle condition for given values of altitude, engine speed, inlet manifold pressure, and air temperature, the following procedure is employed;

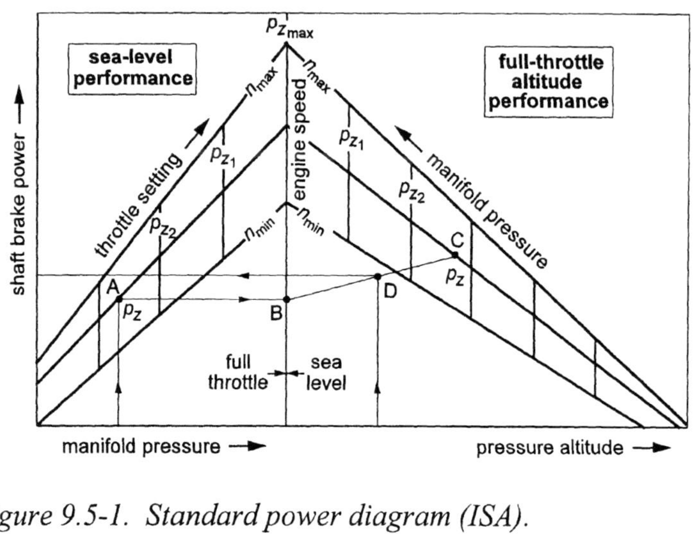

Figure 9.5-1. Standard power diagram (ISA).

1. Locate point A on sea-level curve for the known values of inlet manifold pressure and engine speed and transfer to B.

2. Locate point C on full-throttle altitude curve for the given inlet manifold pressure and engine speed.

3. Determine power output at the required height (point D) by linear interpolation between points B and C. Note that along the straight line B-C the power increases somewhat due to the decreasing air temperature with increasing altitude.

4. Modify the power P_{br} (ISA) at point D for the deviation of air temperature T from standard altitude temperature T_{ISA} by the following empirical formula:

$$P_{br} = P_{br_{\text{ISA}}} \sqrt{\frac{T_{\text{ISA}}}{T}} , \qquad (9.5\text{-}2)$$

where P_{br} is the actual power output and T and T_{ISA} are expressed in kelvin.

Equation (9.5-2) reveals, logically, that when the temperature of the air increases, the power delivered by the engine decreases.

In Figure 9.5-2 the ratio of shaft power in standard and off-standard atmosphere according to Equation (9.5-2), is plotted against air temperature. For comparison, the curves given in the previous Figure 8.4-6 are repeated in Figure 9.5-2, showing that due to its smaller air mass flow rate, the power of a piston engine is much less sensitive to air temperature variations than the thrust of jet propulsion

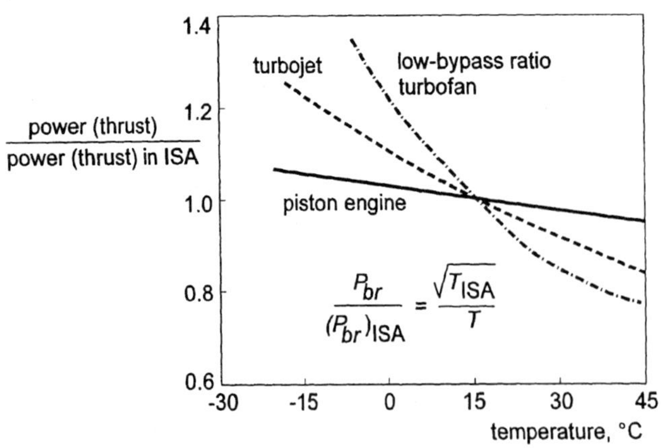

Figure 9.5-2. Effect of air temperature on shaft power.

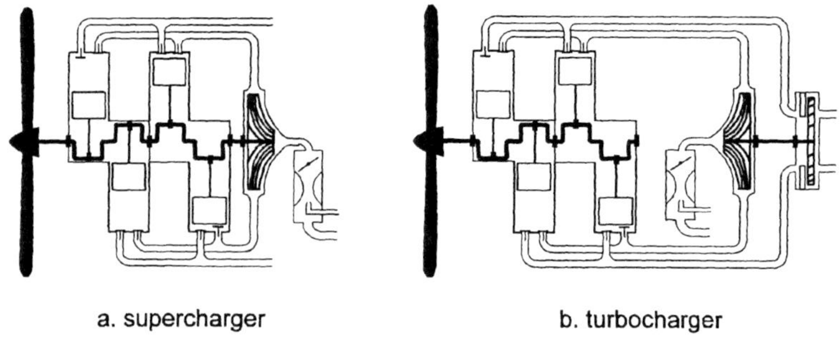

Figure 9.5-3. Supercharging.

devices.

To improve the performance of a normally aspirated engine a *supercharger* is used, which provides a greater pressure at the intake valve. The supercharger is usually positioned between the carburetor and inlet manifold. As sketched in Figure 9.5-3a, it is driven by the crankshaft. This arrangement may be called a *compound engine*. Alternatively, a *turbo-charger* may be employed, which is a supercharger impelled by a small turbine, whereby the latter device is driven by the exhaust gas stream (Figure 9.5-3b). The turbo-charger is more economic than a gear-driven supercharger since it utilizes the heat energy present in the exhaust gases.

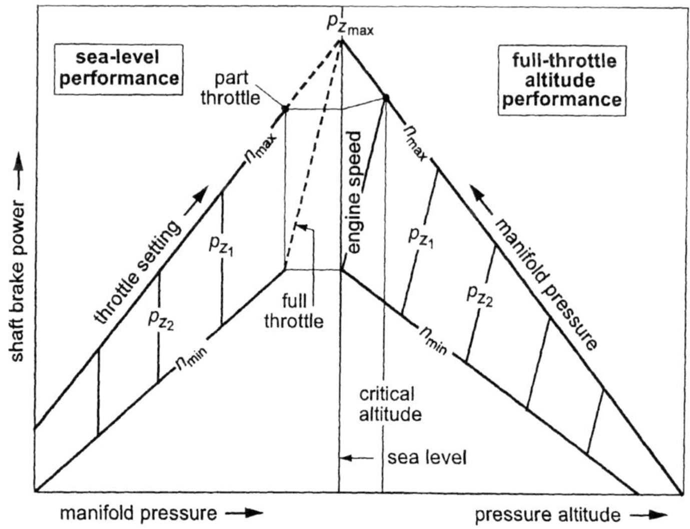

Figure 9.5-4. Standard power diagram (supercharged).

Figure 9.5-4 shows a picture of the standard power diagram of a supercharged engine. Depending on engine speed, the maximum inlet manifold pressure is increased at sea level and hence the altitude performance. In order to avoid unacceptable high engine loads, it is not possible to use the high inlet manifold pressures at lower altitudes. Therefore, limitations are put on the inlet manifold pressure, and only when altitude is gained it is licensed to open the throttle gradually in order to compensate for the loss in atmospheric pressure.
The altitude at which the throttle becomes fully open is termed the *critical altitude* of the engine for the given engine speed and inlet manifold pressure. Above the critical altitude, the power output decreases with increasing altitude, due to the reduction of the air density, conform to the property of an aspirated engine.

9.6. PROPELLER THRUST AND EFFICIENCY

As illustrated in Figure 9.6-1, an airplane propeller consists of two or more blades of which the blade cross sections are airfoil shaped. At given airspeed, the propeller blades convert the shaft power of the engine into a thrust by pushing air backwards, whereby the propeller thrust is given by the time rate of change of momentum of the air that passes through the propeller disk. The location of the control surface 0 in Figure 9.6-1 is far upstream of the propeller where air pressure and flow velocity have their free stream values p_0 and V_0. The aft plane j is sufficiently far downstream that the local pressure also equals the free stream pressure p_0. Hence,

$$T = m\,(w_j - V_0)\ , \tag{9.6-1}$$

where w_j is the final slipstream velocity.
Equation (9.6-1) defines the thrust which is obtained in the absence of engine and airframe. Therefore, this force may be called the *free air thrust*, contrary to the *effective thrust*, T_{eff}, which is the thrust corrected for all effects emanating

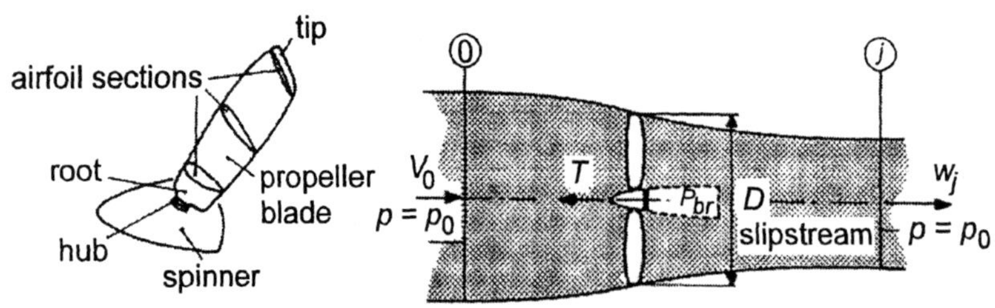

Figure 9.6-1. Shape and propulsive action of airplane propeller.

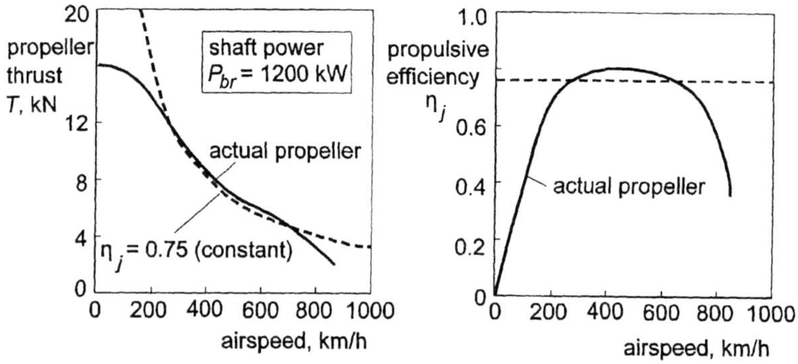

Figure 9.6-2. Typical propeller performance.

from the presence of airframe parts in the slipstream.

The work done by the thrust per unit time is the power available $P_a = T\,V_0$, and the ratio of P_a to shaft power P_{br} is the propulsive efficiency of the propeller (see also Equation (9.1-1)),

$$\eta_j = \frac{P_a}{P_{br}} = \frac{T\,V_0}{P_{br}} . \tag{9.6-2}$$

Consequently, the variation of thrust with airspeed is given by:

$$T = \frac{\eta_j\,P_{br}}{V_0} . \tag{9.6-3}$$

From our discussion in Section 9.3 it will be clear that at a given altitude and engine control setting, it is a good approximation to assume that the shaft power is independent of forward velocity. If we also assume that η_j has a constant value, the thrust predicted by Equation (9.6-3) will markedly diminish as airspeed increases. This strong dependence of thrust on airspeed holds also for an actual propeller, as sketched in Figure 9.6-2.

At zero velocity the actual propeller generates a finite thrust, which is called *static thrust*. Equation (9.6-2) manifests that at $V_0 = 0$ the propulsive efficiency is zero. In the first instance η_j increases considerably with airspeed, but beyond 200 km/h the propulsive efficiency remains more or less constant until, at about 700 km/h, the efficiency is seriously impaired as the blade tips approach the speed of sound.

Essentially, the speed of the propeller blade tip relative to the air is the vector sum of the forward velocity V_0 in the direction of the propeller axis and the rotational velocity ωR, directed perpendicular to the propeller axis (Figure 9.6-3). The magnitude of the propeller tip speed, V_t, is thus given by:

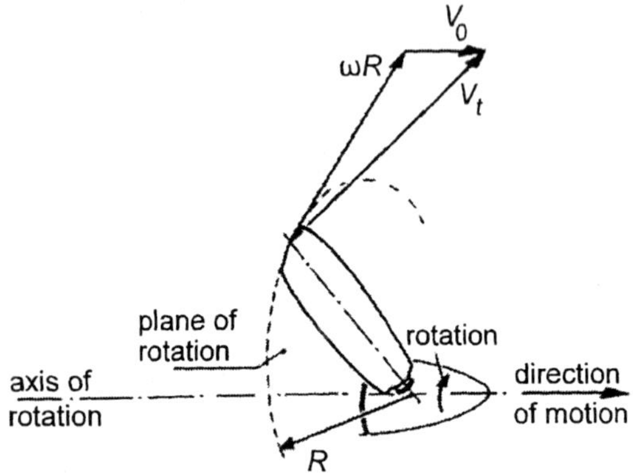

Figure 9.6-3. Propeller tip speed.

$$V_t = \sqrt{V_0^2 + (\omega R)^2} = \sqrt{V_0^2 + (\pi n_p D)^2} \,, \tag{9.6-4}$$

where R is blade radius, $D = 2R$ is propeller diameter, ω is angular velocity, and n_p is number of revolutions per second of the propeller ($\omega = 2\pi\, n_p$).

On the small airplanes with a piston engine, the propeller may be connected directly with the engine crankshaft. In this case the propeller speed n_p equals the engine speed n. On turboprop airplanes there is always a gearbox, which makes that the propeller turns much more slowly than the turbine shaft that drives it so that the ratio n_p / n is less than one.
When the tip speed V_t approaches the speed of sound, the compressibility effects will dramatically reduce the propulsive efficiency of the propeller. This factor limits the use of the current propellers to a maximum flight Mach number of about 0.6 (see Figure 9.6-2).

As measure for the propulsive efficiency, the *Froude efficiency* may be used, where the shaft power delivered to the propeller, P_{br}, is expressed as the increase of kinetic energy of the propulsive flow through the propeller (see also the derivations in Section 7.5):

$$P_{br} = \frac{1}{2} m \,(w_j^2 - V_0^2) \,, \tag{9.6-5}$$

yielding the familiar expression:

$$\eta_j = \frac{P_a}{P_{br}} = \frac{m\,(w_j - V_0)\,V_0}{\frac{1}{2} m\,(w_j^2 - V_0^2)} = \frac{2}{1 + \frac{w_j}{V_0}} \,. \tag{9.6-6}$$

Evidently, the propulsive efficiency according to Equation (9.6-6) can also be expressed in terms of the thrust and air mass flow rate:

$$\eta_j = \frac{2}{2 + \dfrac{T}{m\, V_0}} . \qquad (9.6\text{-}7)$$

Equations (9.6-6) and (9.6-7) show the general properties associated with propulsion. Clearly, η_j increases with increasing flight speed V_0 and decreasing slipstream velocity w_j. If the slipstream velocity is nearly equal to the flight speed, the kinetic energy of the jet is used very efficiently and the propulsive efficiency tends to unity. In order to achieve an improvement of η_j at constant thrust, the air mass flow rate passing through the propeller disk must be enlarged, which, in turn, requires a larger propeller diameter. On the other hand, from aerodynamic and acoustic considerations, it is known that there is a strong need to limit the tip speed of the blades.

Emphasis is made that Equations (9.6-6) and (9.6-7) represent the theoretical upper limit of the attainable propulsive efficiency since the Froude efficiency excludes rotational kinetic energy in the propeller slipstream and assumes that the axial velocity is uniform over the disk.
Clearly, the significance of the Froude efficiency lies especially in providing a qualitative appreciation of the way in which propellers are likely to act under different design and operating conditions.

A comparison of the relationship between propulsive efficiency and airspeed for the turbojet, the turbofan and propeller propulsion systems is indicated in Figure 9.6-4. Since turbojet thrust is produced at high jet velocities, the higher the flight speed, the higher is its propulsive efficiency. Due to the increased

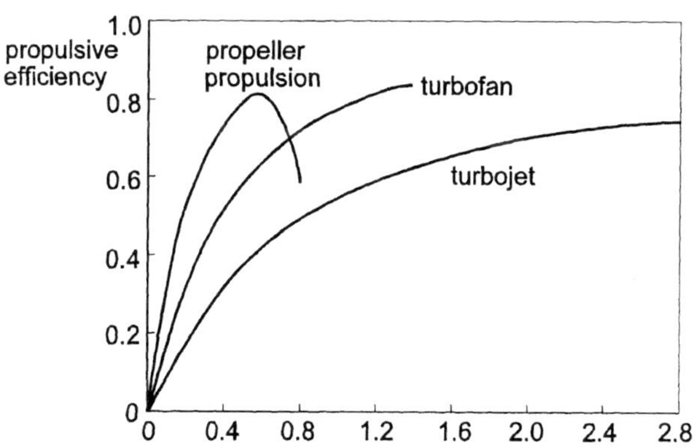

Figure 9.6-4. Typical propulsion efficiency trends.

mass flow rate and reduced mean exhaust velocity of the turbofan, this engine type is best in the high-subsonic speed range. At low-subsonic flight speeds, the propeller is much to be preferred to the other forms of propulsion.

As mentioned already in Section 9.2, the specific fuel consumption of propeller propulsion systems is defined as the fuel weight flow rate, F, divided by shaft power P_{br}:

$$c_P = \frac{F}{P_{br}} = \frac{m_f g}{P_{br}} , \qquad (9.6\text{-}8)$$

where c_P has the dimensions N/Wh or N/kWh.
A comparison of Equation (9.6-8) with Equations (9.3-8) and (9.3-10) reveals that the (power) specific fuel consumption is an inverse criterion of thermal efficiency:

$$c_P = \frac{g}{H \eta_{th}} . \qquad (9.6\text{-}9)$$

The overall efficiency of the complete propulsion process can be written as:

$$\eta_{tot} = \frac{T V_0}{m_f H} = \frac{T V_0}{P_{br}} \frac{p_{br}}{m_f H} = \eta_j \eta_{th} . \qquad (9.6\text{-}10)$$

This equation shows the established fact that the thermal efficiency is half the problem since the conversion of shaft power into power available is also coupled with losses, leading to $\eta_j < 1.0$.

9.7 PROPELLER GEOMETRY

Under normal flight conditions the direction of the propeller axis may be considered as coinciding with the direction of the flight velocity V_0 (Figure 9.7-1).

If motion induced in the flow through the action of the propeller is neglected, the velocity of an *airfoil section* relative to the air, V_r , is composed of a translation with airspeed V_0 and a simultaneous rotation about the propeller axis with rotational speed ωr:

$$V_r = \sqrt{V_0^2 + (\omega r)^2} = \sqrt{V_0^2 + (2\pi n_p r)^2} . \qquad (9.7\text{-}1)$$

Evidently, the component V_0 has the same value at each blade cross section, while the rotational speed is proportional to the distance r from the propeller

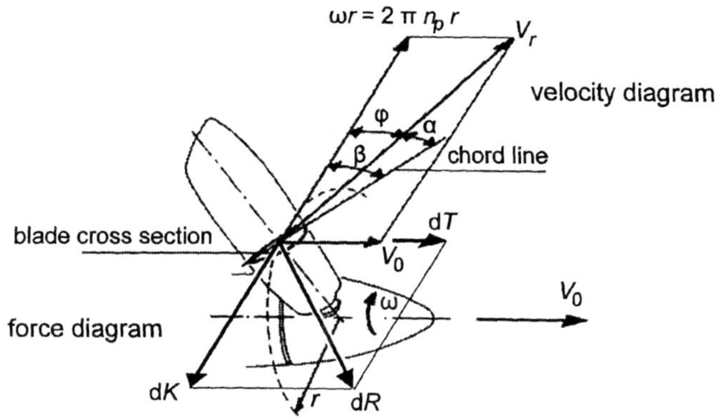

Figure 9.7-1. Propeller blade cross section in motion.

axis to the blade cross section concerned.

The angle between the relative velocity V_r and the plane of propeller rotation is called the *advance angle*, φ, and the angle from the plane of propeller rotation to the well-defined chord line of the airfoil or blade cross section is the *blade angle* β. This angle is also called the *pitch angle*. The blade angle β is the sum of the advance angle, φ, and the *angle of attack*, α,

$$\beta = \varphi + \alpha \, . \tag{9.7-2}$$

The chord line of the blade cross section is placed at the angle of attack to the relative velocity vector to obtain the *aerodynamic force* dR (see the text on the aerodynamics of flight in Ref. 61).

The component of dR parallel to the propeller axis, dT, furnishes the contribution of the blade cross section to the propeller thrust T:

$$T = B \int_0^R \mathrm{d}T \, , \tag{9.7-3}$$

where B is the number of blades.

The component dK in the direction of the plane of rotation is the force opposing the rotation of the blade cross section. The total power required to rotate the propeller, the *propeller power* P_p, is given by:

$$P_p = M\omega = 2\pi n_p B \int_0^R r \, \mathrm{d}K \, . \tag{9.7-4}$$

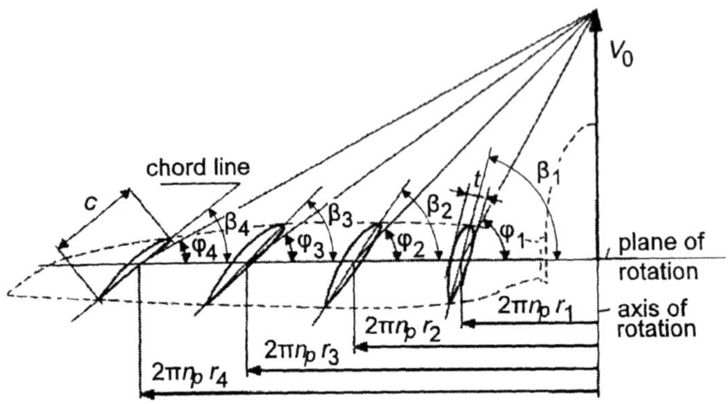

Figure 9.7-2. Variation of advance angle from hub to tip.

In equilibrium, the propeller rotates at a constant rpm determined by the balance between engine shaft power and propeller power:

$$P_{br} = P_p \ . \tag{9.7-5}$$

Returning now to consider the geometry of the propeller, we find from Figure 9.7-1 the advance angle as:

$$\tan\varphi = \frac{V_0}{\omega r} = \frac{V_0}{2\pi n_p r} \ . \tag{9.7-6}$$

For given values of airspeed V_0 and angular velocity ω, the advance angle decreases from the root to the tip of the blade, since the tip sections are moving faster than the root sections (Figure 9.7-2).
The advance angle of the tip section is given by:

$$\tan\varphi_t = \frac{V_0}{\omega R} = \frac{V_0}{\pi n_p D} \ . \tag{9.7-7}$$

The dimensionless quantity $V_0 / (n_p D)$ is called the *advance ratio* and given the symbol J:

$$J = \frac{V_0}{n_p D} = \pi \tan\varphi_t \ . \tag{9.7-8}$$

After one revolution of the propeller axis, the propeller has advanced a certain distance in the direction of flight. This length is called the *effective pitch*, p_e, of the helical motion performed by each point on the propeller (Figure 9.7-3a).

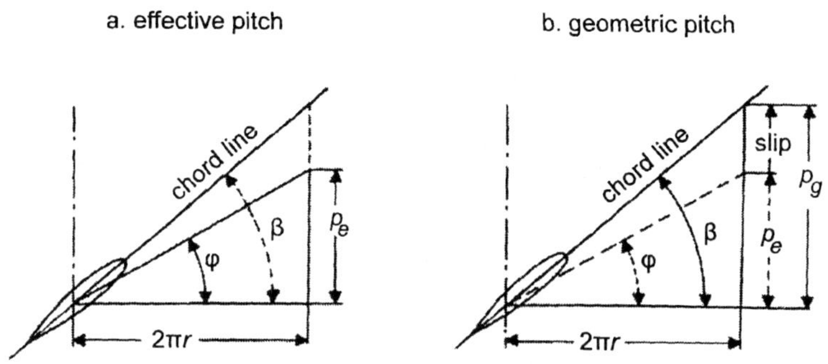

Figure 9.7-3. Effective and geometric propeller pitch.

The effective pitch of the propeller is related to the advance angle and propeller speed by:

$$p_e = 2\,\pi\, r \tan\varphi = 2\,\pi\, r \frac{V_0}{2\,\pi\, n_p\, r} = \frac{V_0}{n_p} \,. \tag{9.7-9}$$

It can be further seen that:

$$J = \frac{V_0}{n_p\, D} = \frac{p_e}{D} \,. \tag{9.7-10}$$

Hence the advance ratio is a measure of the effective pitch. A low value of J implies a fine pitch and a high J a course pitch.

In this respect we state that a low drag behavior of the propeller requires that the angle of attack is more or less constant along the blade radius (Ref. 8).
For that reason, the propeller blades always appear to be twisted with the smallest blade angle at the tip and the greatest at the hub, owing to the increase in the advance angle (see Equation (9.7-2) and Figure 9.7-2).
Figure 9.7-4 shows for a given propeller the variation of the blade angle along the blade radius, which, customarily, is designated by specifying β at 75% or 70% of the radial distance.
The higher the airspeed for which a propeller is designed, the greater the angle $\beta_{0.75}$ or $\beta_{0.70}$, and thus all blade angles at which the blade sections are set.

The blade angles are often given in terms of the *geometric pitch* p_g. This parameter is obtained from the geometry of Figure 9.7-3b. When expressed in terms of $\beta_{0.75}$ we get:

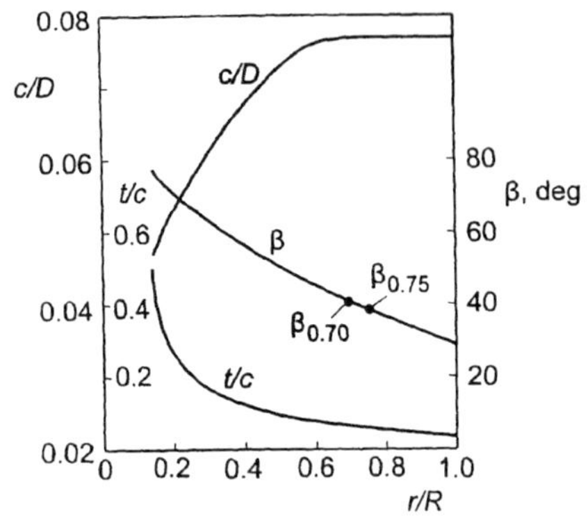

Figure 9.7-4. Geometric data of a propeller blade.

$$p_g = 2\pi r \tan\beta = 2\pi \frac{3}{4} R \tan\beta_{0.75} \tag{9.7-11}$$

or

$$\beta_{0.75} = \tan^{-1}\left[\frac{4}{3}\frac{p_g}{\pi D}\right] . \tag{9.7-12}$$

Geometric pitch is the distance that the propeller would travel per complete turning if it were rotated in a solid medium. The difference between the geometric pitch and the effective pitch is called the *propeller slip*.

Figure 9.7-4 also shows the typical distribution of maximum blade thickness-to-chord ratio, t/c, and relative chord length, c/D. Especially structural requirements make that near the root the thickness of the blade cross sections is relatively large. Also, for structural reasons the chord lengths over the outer part of the blade radius are kept constant.

9.8 PROPELLER CHARTS AND TYPES

Propeller thrust is always presented by using appropriate dimensionless groups of variables. Taking the thrust of a propeller with a given number of blades and given blade shape, as dependent variable, we may express it as a function of the following independent variables:

$$T = T(\rho, n_p, D, V_0, \beta_{0.75}, c, \mu) , \tag{9.8-1}$$

where ρ is the density of air, n_p is the (constant) propeller speed, D is the

propeller diameter, V_0 is the free stream velocity, $\beta_{0.75}$ is the blade setting, c is the speed of sound, and μ is the coefficient of dynamic viscosity of the air.
Utilizing the technique of dimensional analysis as described in Ref. 8, we can write the thrust in the form:

$$T = C_T \rho n_p^2 D^4 , \tag{9.8-2}$$

where C_T is the nondimensional thrust coefficient, of which the value may vary with blade angle $\beta_{0.75}$, advance ratio $J = V_0 / n_p D$, flight Mach number M_0, and Reynolds number $Re = \rho V_0 D / \mu$.
Carrying out dimensional analysis on propeller power, we find:

$$P_{br} = C_P \rho n_p^3 D^5 , \tag{9.8-3}$$

where C_P is the power coefficient.
Hence for a given propeller it follows that:

$$C_T = C_T(\beta_{0.75}, J, M_0, Re) \tag{9.8-4}$$

and

$$C_P = C_P(\beta_{0.75}, J, M_0, Re) . \tag{9.8-5}$$

Now, as shown in Figure 9.8-1, the propeller chart may give the values of C_T and C_P as functions of $\beta_{0.75}$ and J. In the graph, the points on the various curves C_P versus J for constant value of $\beta_{0.75}$ that correspond to the same value of C_T are connected.
The data in this graph represent experimental results obtained at fixed values of

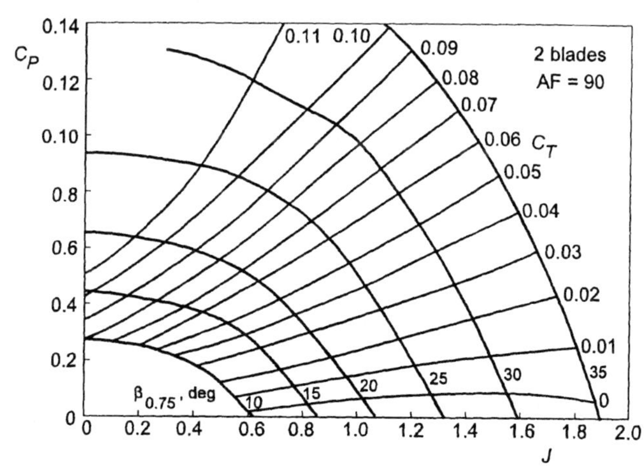

Figure 9.8-1. Propeller chart (NACA 16 blade section series).

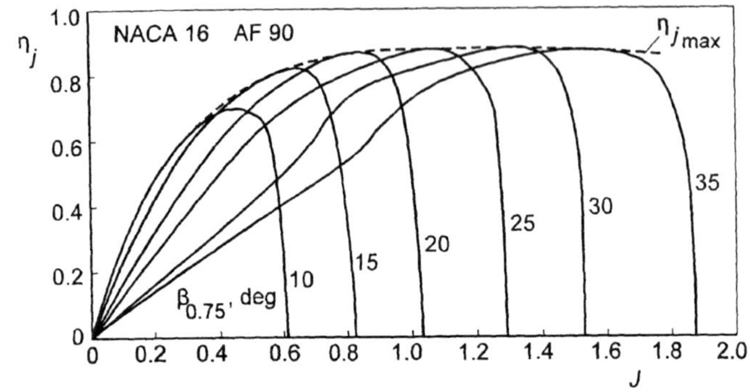

Figure 9.8-2. Propeller efficiency.

Mach number and Reynolds number.

By making use of Equations (9.8-2) and (9.8-3) we can express the propulsive efficiency in the form:

$$\eta_j = \frac{T V_0}{P_{br}} = \frac{C_T \rho n_p^2 D^4 V_0}{C_P \rho n_p^3 D^5} = \frac{C_T}{C_P} J \, . \qquad (9.8\text{-}6)$$

Figure 9.8-2 gives the propeller performance in terms of η_j, J and $\beta_{0.75}$. The optimum performance curve is indicated by the dashed line, joining the points of maximum propulsive efficiency. Notice that the graph refers to the same propeller as considered in Figure 9.8-1.

Inspection of Figure 9.8-2 shows that the blade setting has a marked influence on the propulsive efficiency. On a so-called *fixed-pitch propeller* the blade angle $\beta_{0.75}$ has one particular value so that optimum efficiency occurs for a given design condition of V_0 and n_p. It is also seen that for this type of propeller a high efficiency is gained in only a very narrow range of advance ratios. For example, low efficiencies are obtained at takeoff and climb if the propeller is designed for cruising conditions. On the other hand, the airplane may be equipped with a climb propeller, which has a smaller blade angle or lower pitch. This increases performance during takeoff and climb, but decreases performance during cruising flight.

Apparently, for optimum propulsive efficiency it would be of advantage to provide a pitch-change mechanism to change the blade setting as the advance ratio varies. This requirement has resulted in a solution which is known as the *variable-pitch propeller*. As can be seen from the curves in Figure 9.8-2, the variable-pitch propeller furnishes improved performance over a broad range of

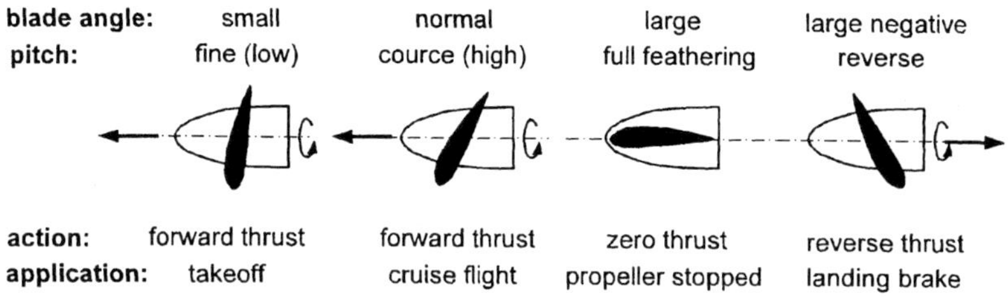

Figure 9.8-3. Application of pitch control.

operating conditions. This explains that nowadays (the cheaper) fixed-pitch propeller only is used on the small single-engine airplanes.

An airplane equipped with a variable-pitch propeller has an engine throttle control for the engine output and a pitch control to regulate the propeller speed. A so-called *constant speed propeller* is equipped with a governor, which changes the blade angles in response to any departure in propeller speed from the value selected by the pilot.

Constant speed propellers on most of the larger modern airplanes have the possibility for full feathering in flight, that is, the blades can be turned edge-on to the oncoming air to prevent the occurrence of high windmilling-drag in case of engine failure (Figure 9.8-3). Also, the pitch-change mechanism may be designed to permit the selection of a large negative blade angle after landing. Thus, creating *reverse thrust* for rapid deceleration of the airplane during ground run.

The governor adjusts the propeller speed by changing the blade setting to ensure that the power required by the propeller equals the shaft power, through which the selected engine speed is maintained.

For a constant speed propeller with given values for ρ, V_0, D, n_p and P_{br}, the power coefficient and the advance ratio follow from:

$$C_P = \frac{P_{br}}{\rho\, n_p^3 D^5} , \qquad (9.8\text{-}7)$$

$$J = \frac{V_0}{n_p D} . \qquad (9.8\text{-}8)$$

Now the blade angle $\beta_{0.75}$ and the thrust coefficient C_T are known from the propeller chart in Figure 9.8-1. The thrust, finally, is found from Equation (9.8-2).

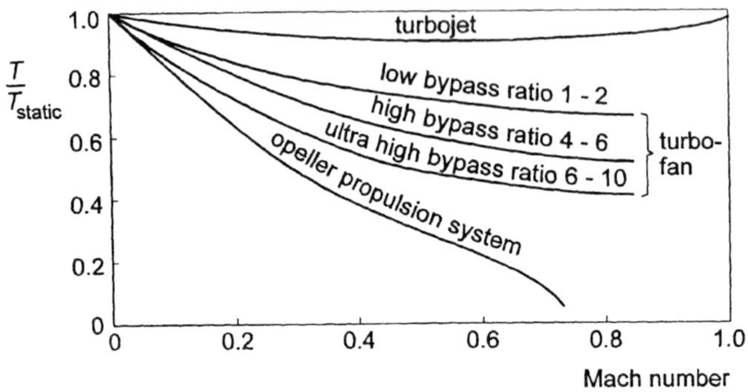

Figure 9.8-4. Typical variation of thrust with flight Mach number.

We go on to present Figure 9.8-4, showing the typical variation of thrust with flight Mach number for propeller propulsion systems, turbofans and turbojets. The data are expressed by the fraction of the thrust at a given Mach number to the static thrust ($M_0 = 0$). The curves show that throughout the subsonic flight regime propeller thrust declines rapidly with increasing Mach number. On the other hand, the thrust of a turbojet is reasonably constant with airspeed (see also the previous Figures 9.6-2 and 8.4-2 for propeller and the turbojet thrust, respectively).
For moderate bypass ratios, the turbofan thrust lies in the middle between that of propeller propulsion systems and turbojets.

At this point we remind the reader that the propeller characteristics shown in Figures 9.8-1 and 9.8-2 represent experimental results from wind tunnel tests executed at fixed values of flight Mach number and Reynolds number.
On the other hand, theoretical methods have been developed for the prediction of the performance of propellers of known design. Three types of theories are available to describe the performance of uninstalled propellers, namely, the momentum theory, the blade element theory, and the vortex theory. However, an analysis of these theories is beyond the scope of this introductory treatment, so that for a discussion of the topics the interested reader must turn to specialized texts on propeller propulsion (see, e.g., Ref. 61).

The calculation methods mentioned above require the input of airfoil aerodynamic data. In this light, it is of importance to mention that during the last twenty-five years the airfoil sections available for propellers have been improved considerably.

We end this chapter by noting that the theoretical methods on propeller performance prediction, commonly, provide the free air thrust (see also the comment made in Section 9.6). Obviously, the installation of propeller and engine in the airplane will change the flow through the propeller and over the engine and airframe, thereby affecting the propulsive efficiency. Further details of these interference effects can be found in Ref. 8.

10 POLLUTANT EMISSIONS OF AERO-ENGINES

10.1 THE COMBUSTOR

The combustor is an essential turbo engine component, in which the chemical energy of the fuel is converted into thermal energy. Despite its great importance to the generation of thrust, the dimensions of the combustor are relatively small. This may be illustrated by the sketches of the turbofan engine in the previous Figures 7.1-3 and 8.1-1. Clearly, only a limited space is required to store the combustion system.

In Figure 10.1-1 a conventional combustion chamber is outlined. It illustrates the division of the inner space of the chamber into a *primary zone,* an *intermediate zone,* and a *dilution zone.* The three zones are characterized by the occurrence of different temperature levels (Ref. 48). With regard to the function of these zones, in the following a few remarks will be made.

1. The function of the *primary zone* is to stabilize the flame and to provide sufficient time, temperature, and turbulence to achieve essentially complete combustion of the fuel. Typically, current airplane power plants are equipped with a *non premixed* combustion system. This means that the fuel is sprayed into the primary zone. Thus, the fuel consists of droplets, which have to evaporate, mix with air and subsequently oxidize. In the non premixed (or diffusion) flame, fuel and air are initially separated, so that the combustion only

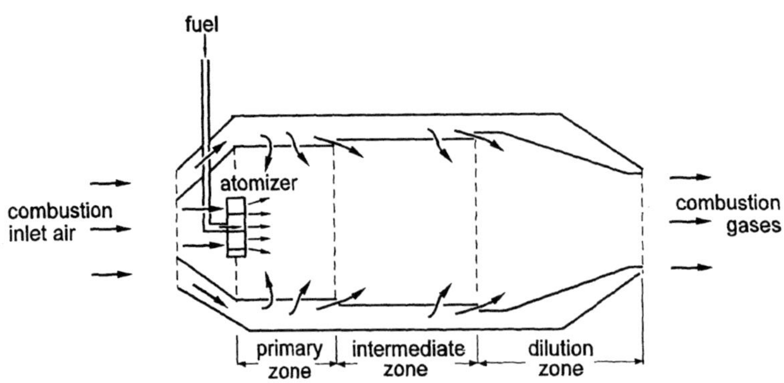

Figure 10.1-1. Main zones in an airplane gas turbine combustor.

occurs at the interface between fuel and oxidizer. This is accomplished by mixing of a portion of the compressed air with the fuel. Characteristic of a non premixed combustion is the occurrence of high primary-zone temperatures. The determining quantity is the *adiabatic flame temperature*, which, for given reactants is the maximum temperature that can be obtain. This temperature occurs when the reaction takes place adiabatically and goes to completion. A first approach to its magnitude follows from the flow energy equation in the form (see Chapter 7):

$$m_f H = m\, c_p \Delta T_t \; . \tag{10.1-1}$$

The concept of the adiabatic flame temperature is simple. However, the numerical evaluation of this quantity may be difficult, because of the dependence of c_p on temperature through dissociation of the combustion products, which can lead to the formation of many unknown species.
Anyhow, the adiabatic flame temperature is often higher than 2000 K. At this temperature level, dissociation reactions take place, which result in the appearance of significant concentrations of CO and atomic oxygen O in the efflux gases. Should these gases pass directly to the dilution zone and be rapidly cooled by the addition of massive amounts of air, the gas composition would be "frozen", and unburned CO would be discharged from the combustor. Dropping the temperature to an intermediate level by the injection of small amounts of air in the intermediate zone encourages the burnout of soot and allows the combustion of CO and other unburned hydrocarbons to proceed to completion.

2. In the *intermediate zone,* in which small amounts of air are added to the gas flow through the combustion chamber, two functions, depending on operating conditions, can be distinguished.
   Firstly, at low altitudes the intermediate zone serves as a region in which incomplete combustion products are burned, and recovery of dissociation losses takes place. These losses result from chemical instability of the combustion products. By dropping the temperature to an intermediate level through the addition of air, CO and any partly unburned fuel are allowed to oxidate finally. Secondly, at high altitudes the intermediate zone serves as an extension of the primary zone. Due to the low ambient pressure, the reaction rate is slower, and the combustion is far from complete at the exit of the primary zone. The intermediate zone provides an increased residence time at high temperature, resulting in an improved combustion of the fuel.

3. In the *dilution zone* additional compressed air enters the combustion chamber from the surrounding outer annular air flow passage. In principle, the dilution

zone acts as an extension of the intermediate zone as it effectuates an almost complete combustion of the fuel.

In the intermediate and dilution zone air is introduced through holes in the liner walls. The extra air cools the wall of the combustion chamber and provides a hot gas stream with a proper temperature and temperature distribution at the exit area of the burner suitable for entry into the turbine. It is advisable to create a low temperature at the turbine blade roots, where the stresses are greatest. Characteristic is an increasing temperature from the blade roots through a maximum and then a decreasing temperature toward the blade tips (Ref. 38). Optimizing the temperature distribution avoids the occurrence of hot spots throughout the cross section at the turbine entry that can damage the turbine blades and allows the maximum average turbine inlet temperature, resulting in a high engine thermal efficiency.

10.2 COMBUSTION OF HYDROCARBON FUELS

Before considering the combustion in turbo engines, some attention will be given to the chemistry of hydrocarbon compounds.

The hydrocarbon types involved comprise *chain* and *ring* structures, of which the difference is illustrated in Figure 10.2-1. Each line represents a pair of electrons. The figure also illustrates the definition of saturated and unsaturated hydrocarbons. On the basis of its valence, a carbon atom is able to share with four other atoms to form a molecule. In a *saturated* hydrocarbon all the carbon atoms are joined by a single bond, whereas an *unsaturated* hydrocarbon has two or more adjacent carbon

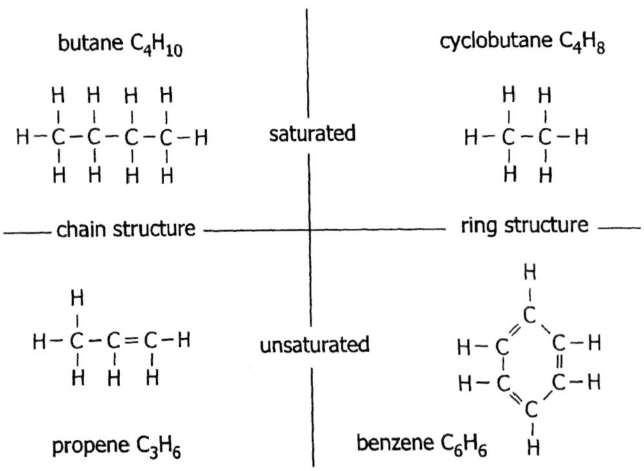

Figure 10.2-1. Molecular structure of a few hydrocarbons.

atoms joined by a double or triple bond.
It is also important to mention that two hydrocarbons having the same number of carbon and hydrocarbon atoms but different chemical structures are called *isomers*. An example is a liquid octane, C_8H_{18}, with its 18 isomers, of which each has a different structure.

Fossil fuels are best divided into chain structured *aliphatic hydrocarbons* and closed ring-structured *alicyclic hydrocarbons*.
Aliphatic hydrocarbons are further subdivided into alkanes, alkenes, and alkynes. The basic series of the aliphatics, also called paraffins, is formed by the alkanes. They are presented by the formula C_nH_{2n+2}, and consist of open chains of saturated hydrocarbons (see Figure 10.2-1). Alkenes are unsaturated hydrocarbons, and are represented by the formula C_nH_{2n}. Alkynes are also unsaturated hydrocarbons, with the difference that they have triple bonds. They are represented by the formula C_nH_{2n-2}.
The series of ring-structured hydrocarbon fuels, especially, consists of aromatic hydrocarbon compounds. The aromatic family includes the naphthalene series, C_nH_{2n-12}, and the unsaturated benzene series, C_nH_{2n-6}.

Most liquid and gaseous hydrocarbon fuels are a mixture of many different hydrocarbons that are derived from crude oil through distillation and cracking processes. Distillation is obtained by slowly heating of the fuel so that it vaporizes, while cracking includes the break down of heavy hydrocarbon molecules into lighter molecules by means of high temperatures or catalysis.
The above remarks are only a summary review of the topic. Of course, the reader who wishes to know more about organic chemistry must turn to an advanced textbook.

In dealing with the combustion of liquid hydrocarbon fuels it is convenient to express the composition in terms of a single hydrocarbon. Therefore, we continue with a treatment of combustion reactions with considering *complete combustion* of a hydrocarbon fuel given by C_xH_y. Complete combustion means no dissociation, so that all fuel carbon C is found in CO_2 and all fuel hydrogen H is found in H_2O. The necessary condition for obtaining complete combustion is that the air-fuel ratio is *stoichiometric*, i.e., the quantity of oxidizer air is just the amount required to completely burn a quantity of fuel C_xH_y.
During combustion the mass of each element remains the same. Thus, the stoichiometric relation for complete combustion of a hydrocarbon fuel is described by the following equilibrium reaction of the fuel with oxygen (Ref. 31):

$$C_xH_y + a\,(O_2 + b\,N_2) \rightarrow x\,CO_2 + \tfrac{1}{2}\,y\,H_2O + a\,b\,N_2\,, \qquad (10.2\text{-}1)$$

where a, being the stoichiometric coefficient of oxidizer, is obtained from conservation of oxygen atoms, yielding:

$$a = (x + y/4) . \qquad (10.2\text{-}2)$$

It is noted that all the initial substances are termed reactants, and the substances that result from the combustion process are termed *product species*.
When we use the simplification that the composition of air is 21% O_2 and 79% N_2 by volume (see Table 1.2-1), we find that for each mole of O_2 in the air, there is 79/21 = 3.762 mole of N_2. Hence, the factor b in Equation (10.2-1) is equal to 3.762. Inserting this result into Equation (10.2-1) gives:

$$C_xH_y + a\,(O_2 + 3.762\ N_2) \rightarrow x\ CO_2 + ½\,y\ H_2O + 3.762\ a\ N_2 \qquad (10.2\text{-}3)$$

The ratio of carbon to hydrogen depends on the type of fuel. For $x = 0$, the oxidation reaction reduces to complete combustion of hydrogen H_2. With an engine operating with hydrogen as fuel, the principal product is water, according to the exothermic (heat-producing) reaction:

$$2\ H_2 + O_2 \rightarrow 2\ H2O. \qquad (10.2\text{-}4)$$

For complete combustion of methane CH_4, for example, we have:

$$CH_4 + 2\ O_2 + 7.524\ N_2 \rightarrow CO_2 + 2\ H_2O + 7.524\ N_2 . \qquad (10.2\text{-}5)$$

From Equations (10.2-2) and (10.2-3), we get the stoichiometric fuel-air ratio as:

$$\left(\frac{m_f}{m_h}\right)_{\text{stoichiometric}} = \frac{M_f}{4.762\,(x + y/4)\,M} , \qquad (10.2\text{-}6)$$

where m_f is the fuel mass flow rate, m_h the mass flow rate of the hot air stream through the engine core, M_f the molecular mass of the fuel, and M is the molecular mass of air.
The emission of CO_2 and H_2O can be deduced from the Equations (10.2-2) and (10.2-3) by substitution of the appropriate atomic masses (carbon C = 12.011, hydrogen H = 1.00794, oxygen O = 15.9994).
Rounding off the atomic mass for carbon to C = 12, for hydrogen to H = 1, and for oxygen to O = 16, we find from Equation (10.2-5) that 1 kmol of the CH_4 fuel (=16 kg) with 4 kmol of oxygen (= 64 kg) generates 1 kmol CO_2 (= 44 kg) and 2 kmol H_2O (= 36 kg). Hence, for combustion of 1 kg methane about 4 kg of oxygen is required, producing 2.75 kg carbon dioxide CO_2 and 2.25 kg water vapor H_2O.

The value of the stoichiometric fuel-air ratio for methane is obtained from Equation (10.2-6) by inserting M_f = 16 kg/kmol and M = 28.96442 kg/kmol, yielding $(m_f / m_h)_{stoichiometric} = 0.058$.

The equivalent composition of *kerosene* fuel may be assumed to consist of $C_{12}H_{23}$ (Ref. 49). Then, for complete combustion we have:

$$C_{12}H_{23} + 17.75\ O_2 + 66.7755\ N2 \rightarrow 12\ CO_2 + 11.5\ H_2O + 66.7755\ N_2\ .(10.2\text{-}7)$$

From Equation (10.2-7) we find that 1 kmol of $C_{12}H_{23}$ fuel (=167 kg) with 17.75 kmol of oxygen (= 568 kg) provides 12 kmol CO_2 (= 528 kg) and 11.5 kmol of H_2O (= 207 kg). Apparently, for complete combustion of 1 kg kerosene about 3.40 kg oxygen is necessary, producing 3.16 kg CO_2 and 1.24 kg H_2O.
From Equation (10.2-6), using M_f= 167 kg/kmol and M= 28.96442 kg/kmol, we directly find that the stoichiometric fuel-air ratio, $(m_f/m)_{stoichiometric}$, for kerosene is about 0.068.
If more than the stoichiometric quantity of air is supplied to the fuel, the mixture is said to be *lean*. We speak of a *rich mixture* when less than the stoichiometric quantity is supplied.

Instead of the fuel-air ratio, normally, the *equivalence ratio*, Φ, is used to indicate quantitatively whether a fuel-air mixture is lean, stoichiometric, or rich.
The equivalence ratio is defined as the fuel-air ratio of a mixture relative to the fuel-air ratio for stoichiometric combustion:

$$\Phi = \frac{(m_f/m)}{(m_f/m)_{stoichiometric}} \ . \qquad (10.2\text{-}8)$$

Clearly, for a lean mixture $\Phi < 1$, for a stoichiometric mixture $\Phi = 1$, and for a rich

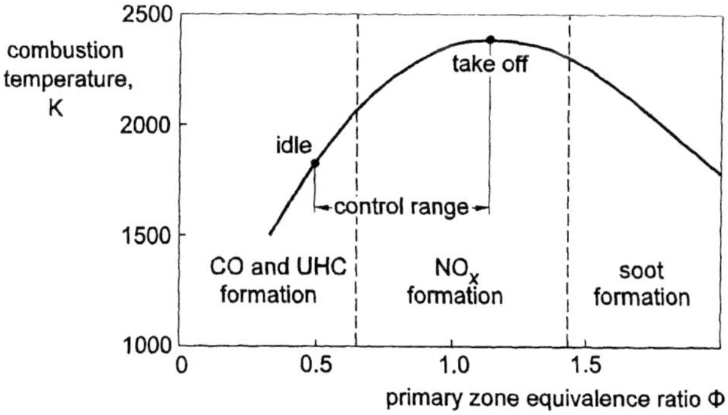

Figure 10.2-2. Combustion temperature in the primary zone (Ref. 32).

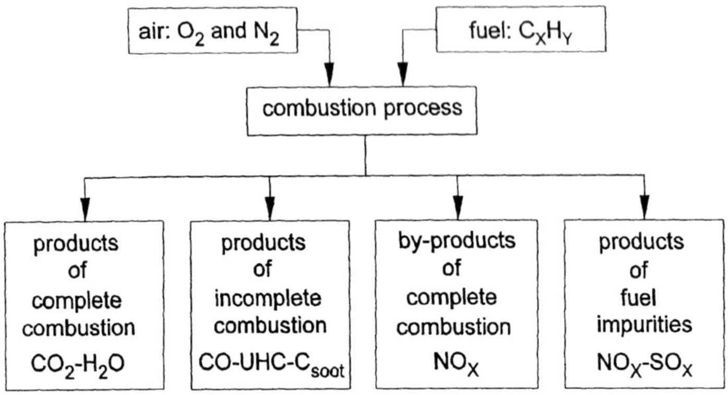

Figure 10.2-3. Combustion products.

mixture $\Phi > 1$. At takeoff thrust the primary zone in the combustion chamber is operating virtually stoichiometrically ($\Phi = 1.0$), and at idle thrust the equivalence ratio amounts to about $\Phi = 0.5$ (Figure 10.2-2).

From Equation (10.2-1) we know that the products of combustion from the burner of an air breathing aircraft engine using kerosene fuel, would ideally be composed of carbon dioxide CO_2 and water vapor H_2O. However, referring back to Figure 1.6-1, it can be seen that due to imperfect combustion, high combustion temperatures, and fuel impurities, there are also trace species to be found in the exhaust gases of aero-engines. As illustrated in Figure 10.2-3, the various emission products can be divided into four groups, namely:

- Products of complete combustion: CO_2 and H_2O.
- Products of incomplete combustion: UHC, CO and C_{soot}.
- By-products of complete combustion: NO_x.
- Products of fuel impurities: NO_x and SO_x.

Apparently, the constituents that give cause for concern, primarily, are unburned hydrocarbons UHC, carbon monoxide CO, smoke or soot particles C_{soot}, nitrogen oxides NO_x (nitric oxide NO and nitrogen dioxide NO_2), and sulfur oxides (sulfur dioxide SO_2 and sulfur trioxide SO_3). By far the biggest part of the nitrogen oxides NO_x is formed during combustion as a result of the required high turbine inlet temperatures.

Sulfur oxides SO_x arise from sulfur compounds that are present as impurities in the fuel. However, in the kerosine-type fuels currently used for aero-engines there is very little sulfur, of which the emission is related to the fuel consumption of the engine.

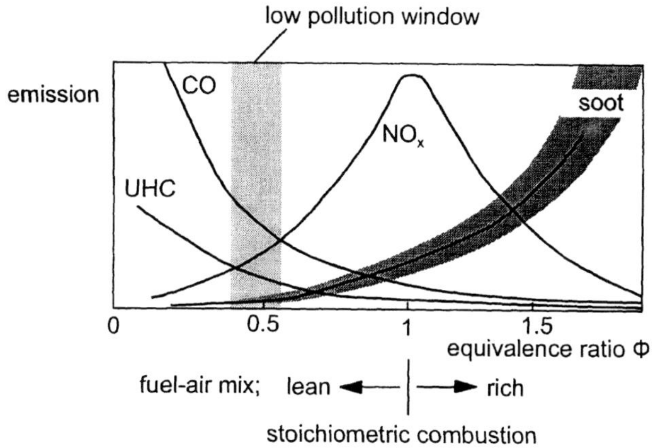

Figure 10.2-4. Formation of pollutants.

Figure 10.2-4 shows qualitatively the emission levels of the four major pollutants as a function of the equivalence ratio Φ, which quantity directly defines the flame temperature (cf. Figure 1.6-2). Apparently, at values of Φ near stoichiometric combustion ($\Phi = 1$), the NO_x production is at maximum, while comparatively speaking, CO and UHC emissions are low in this area. On the contrary, at values of $\Phi < 0.5$ and $\Phi > 1.5$ the emissions of NO_x are low, while below $\Phi = 0.5$ the emissions of CO and UHC rapidly increase. Soot formation, especially, occurs at values of $\Phi > 1.0$. Therefore, as pictured in Figure 10.2-4, a low pollution window can be distinguished in the fuel lean mixture area ($0.4 < \Phi < 0.6$). In this area, NO_x emissions are considerably reduced, while other pollutants are present at possibly more acceptable levels.

In agreement with the predictions made earlier, Table 10.2-1 indicates that irrespectively of thrust setting, at stoichiometric combustion each kg of kerosene yields circa 3.15 kg of carbon dioxide CO_2 and about 1.25 kg of water vapor H_2O. Compared to these values, at the various operating regimes only small and very small amounts of carbon monoxide CO, unburned hydrocarbons UHC, nitrogen oxides NO_x, and soot particles are formed (Ref. 1). Table 10.2-1 also shows that the emission levels of sulfur oxides SO_x indeed are so low that they may be neglected.

As is usual, in the table the emission levels of combustion products are expressed in terms of the *emission index*, *EI(i)*, which is the ratio of the mass flow rate of the emitted substance m_i to the mass flow rate m_f of the fuel burned by the combustion process:

$$EI(i) = \frac{m_i}{m_f} . \tag{10.2-9}$$

Therefore, Table 10.2-1 also presents the values of the emission index $EI(i)$ for the

species arising from combustion of kerosine. *EI(i)* is defined as the amount of a pollutant substance *(i)*in gram per kg of fuel consumed.
Clearly, the emission index can be seen as a measure of the efficiency of a particular combustion process in generating a particular pollutant.

The emissions from aero-engines may also be expressed as *mass specific emissions* $MSE(i)$. For jet engines this measure is defined as the mass flow rate of a pollutant (i) divided by the engine thrust, having the unit g / N.s.

Mass specific emissions are conveniently related to the emission index as:

$$MSE(i) = \frac{m_f \, EI(i)}{T} , \qquad (10.2\text{-}10)$$

where m_f is the fuel mass flow rate and T is the nett thrust produced.
Another specific emission measure is the mass of pollutant emitted per amount of heat energy supplied, which is expressed by the ratio: $EI(i) \,/\, H$, where H is the heating value.

In the dynamometer testing of piston engines, the measure $MSE(i)$ is defined as the mass flow rate of a pollutant *(i)* divided by engine power, where the unit, e.g., is g /kW.h. These mass specific emissions are related to the emission index as:

$$MSE(i) = \frac{m_f \, EI(i)}{P_{br}} , \qquad (10.2\text{-}11)$$

where P_{br} is the engine power produced.

substance	mass of combustion product per unit mass of fuel, kg	emission index, g/kg fuel	emission depends on:
CO_2	3.15	3150	fuel consumption
H_2O	1.25	1250	
CO	0.0004 - 0.065	0.4 - 65	thrust setting; max. production at idle
UHC	0.0002 - 0.012	0.2 - 12	
Soot	± 0.000015	± 0.015	thrust setting; max. production at full thrust
NO_X	0.004 - 0.030	4 - 30	
SO_X	0.00002 - 0.006	0.02 - 6	

Table 10.2-1. Typical kerosine emission levels.

Basically, the emission quantities of carbon dioxide CO_2 and water vapor H_2O are proportional to the fuel consumed. However, the amount of unburned hydrocarbons UHC and the formation of nitrogen oxides NO_x, carbon monoxide CO and soot particles depend on engine type, thrust setting, flight altitude (air pressure and temperature), and airspeed.

The products resulting from ideal conbustion of kerosene (CO_2 and H_2O) may be considered as "normal" or "regular" combustion products, notwithstanding the fact that both gases are enhancing the greenhouse effect. In this respect, the role of water vapor is even twofold, as it is not only an important greenhouse gas, but also it may cause contrail formation due to the injection of water into the very cold atmospheric air near the tropopause. The occurrence of contrails will be discussed in Section 10.8.

Aside from CO_2 and H_2O, combustion products such as CO, UHC, NO_x, SO_x, and soot may be denoted as "undesired" or "unwanted". As stated earlier, generally, at low engine control settings the formations of CO and UHC predominate, whereas at high control settings NO_x and soot arise (see Figures 1.6-2 and 10.2-4, and Table 10.2-1).

As will be demonstrated in Chapter 12, nitrogen oxides NO_x are the most significant undesired emission species from aero-engines, typically comprising more than 90 percent of NO and 10 percent of NO_2, of which the latter are formed in a later stage when NO is oxidized further into NO_2 (Ref. 34).
There are various mechanisms through which NO_x may be formed, but the most important for aero-engines is the so-called *thermal-NO_x* formation process (see Section 10.4). During this process nitrogen N_2, comprising 78.084 percent by volume in clean air (see Table 1.2-1), is oxidized by oxygen radicals O, which are formed by dissociation of oxygen molecules O_2 at high temperatures.

Soot emissions arise at high engine control settings, i.e., at high fuel-air ratios, which normally occur during aircraft takeoff. Soot consists almost entirely of carbon particles, which tend to color the exhaust gas flow black. The formation of soot is a relatively poorly understood process. Unlike CO, UHC and NO_x, it is not an equilibrium product of combustion. The rate of soot formation is more governed by physical processes of fuel atomization and fuel-air mixing than by chemical kinetics (Ref. 48).

The presence of sulfur oxides SO_x from aviation depends only on the quality of the kerosene fuel. Refining processes are available to remove these impurities, so that, as mentioned earlier, kerosene hardly contains sulfur.

Any possible emission of SO_2 may produce sulfuric acid (H_2SO_4) droplets according to the reactions given in Section 6.2 of Chapter 6.
Then, rain water acidifies by forming the sulfate ion SO_4^{2-}, according to the reation (6.5-8) repeated here:

$$H_2SO_4 + 2\,H_2O \rightarrow 2\,H_3O^+ + SO_4^{2-}\,. \qquad (6.5\text{-}8)$$

Apparently, when oxidized to sulfuric acid salt, the airborne sulfate particles (aerosols) will act as condensation nuclei and in this way encourage cloud formation, thereby shading the surface of the Earth and helping to offset greenhouse warming.

The magnitude of the engine thrust is controlled by varying the turbine entry temperature, which is achieved by adjusting the fuel-air ratio in the combustion chamber (see Figure 10.2-2).
At full-thrust setting the equivalence ratio of the primary zone is nearly stoichiometric, while at the combustor exit the equivalence ratio amounts to about $\Phi = 0.3$. Consequently, in the primary zone under full-thrust operating conditions, NO_x is produced in large doses. It is plain that at maximum engine control setting there is also a tendency toward the formation of soot, especially at local fuel-rich mixtures, where $\Phi > 1$.

At idle thrust, the equivalence ratio in the primary zone decreases to a value of about $\Phi = 0.5$, leading to a lower NO_x formation (Figures 10.2-2 and 10.2-4). Also the residence time in the burner is smaller than at full thrust (Ref. 50). This leads to incomplete combustion and so to the formation of UHC and CO. Furthermore, small fuel flow rates through the injection nozzles tend to lead to poor atomization, decreasing the combustion efficiency.

In addition to above mentioned undesired substances from incomplete combustion of fossil fuels, into the atmosphere a variety of hydrocarbons may be released, which are not totally oxidized during combustion. For this reason, the following substances are also of significance for the quality of the ambient air (see also Chapters 1 and 6):

- Polycyclic aromatic hydrocarbons (PAH), originating from incomplete combustion at relatively low temperatures, and processing of tarry substances.

- Methane (CH_4), landing up in the atmosphere through anaerobic micro-biological processes, for instance through ruminants, bio-gas from domestic refuse, anaerobic water purification.

Emission of hydrocarbons into the atmosphere also occurs by evaporation of volatile organic compounds (VOC). These substances are released for instance during storage and transshipment of fuels, when using as a solvent or during incomplete combustion of fossil fuels.

10.3 HEATING VALUE

It is appropriate to pay some attention to the *heating value* of a fuel. This quantity is the amount of heat energy produced when one mole or one kilogram of a fuel is completely burned in oxygen. Its unit may be expressed in J/kmol or J/kg.
In explanation of the heating value we consider a flow through a channel (Figure 10.3-1). At cross-section 1, a stoichiometric mixture of air and hydrocarbon fuel (the reactants) enters the control volume (the reactor or calorimeter) where combustion takes place. At cross-section 2, the combustion products leave the control volume. Both, the reactants and the combustion products are at the same temperature T and pressure p. The combustion process is assumed to be complete, i.e., all of the fuel carbon is converted to CO_2 and all of the fuel hydrogen is converted to H_2O.
If the changes in kinetic and potential energies are negligible, the steady flow energy equation per unit mass of fuel (see the previous Equation (7.4-8)) reads:

$$h_r + q_c = h_p \,, \tag{10.3-1}$$

where h_r is the absolute enthalpy of the reactants at temperature T, h_p the absolute enthalpy of the products at temperature T, and q_c is the heat transfer.
Equation (10.3-1) can be written as:

$$q_c = h_p - h_r = h_{rp} \,. \tag{10.3-2}$$

The quantity h_{rp} is called the *enthalpy of combustion* and is the change in absolute enthalpy during the combustion process. We note from Equation (10.3-2) that it is

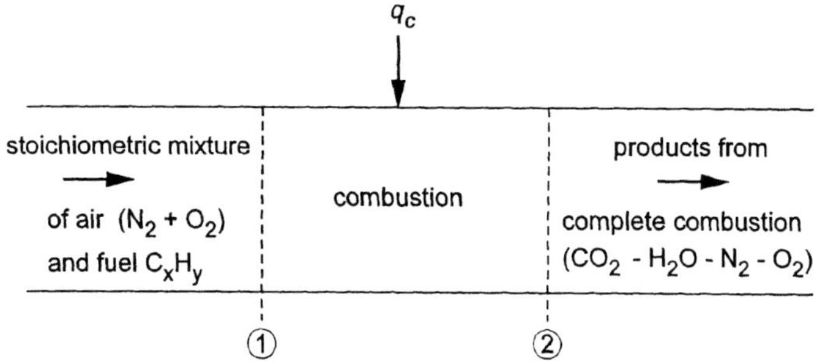

Figure 10.3-1. Determination of heating value.

equal to the heat transfer per unit mass of fuel. If we require that at the exit the combustion products have the same flow condition as the mixture at entrance, heat energy must be removed from the flow, so that the enthalpy of combustion is negative.

The quantity *heating value* or *heat of combustion* in this text, denoted by the capital letter H, is defined as the heat evolved from the products per unit mass of fuel. Thus, the heating value and enthalpy of combustion have the same magnitude but an opposite sign:

$$H = -h_{rp} \ . \tag{10.3-3}$$

Obviously, the heating value is always positive. Since the heat transfer occurs in a constant-pressure process at a given temperature, heating values associated with enthalpy of combustion are usually called *constant-pressure heating values*.

The above approach offers a *lower* and a *higher heating value*. The lower heating value, LHV, gives the heat of combustion obtained when all of the water in the combustion products remains in the vapor state. The higher heating value, HHV, corresponds to the case where all of the water has condensed to liquid.

In determining the heating value for various fuels, we can make use of tables of thermodynamic properties for various substances. In these tables the enthalpy is given relative to an arbitrarily assumed standard reference state of $T_0 = 25\,°C$ and $p_0 = 101325$ N/m^2.
The above technique assumes that at this base the enthalpy of all the elements is zero. Then, the absolute enthalpy at temperature T, is known as the sum of an enthalpy that takes into account the energy associated with chemical bonds, the so-called *enthalpy of formation* at T_0 and p_0, and an enthalpy that is associated only with temperature, the so-called *sensible enthalpy*, being the enthalpy change in going from T_0 to T.
A complete description of this approach can be found in Ref. 31, where also numerical values of enthalpy of formation at the reference state are listed for a number of species, as well as the accompanying sensible enthalpy as a function of temperature. Moreover, standard-state heating values for a variety of hydrocarbon fuels are given.
Lower heating values for a number of fuels were earlier cited in Table 1.5-1. Doubtless, some of these data were measured using a *bomb calorimeter* in which the reaction is carried out in an enclosed, pressurized chamber immersed in a known mass of water.

10.4 FORMATION OF UNDESIRED COMBUSTION PRODUCTS

In this section some attention will be given to the process of pollutant formation in a conventional combustor chamber.

Figure 10.4-1, illustrates the progress of a fuel-air mixture, as it passes through a combustor. The temperature increases very quickly as the fuel burns up to more than 2000 K, forming CO, CO_2, H_2O, and NO. Apparently, the formation of NO is somewhat delayed. If the gases would be held at the high peak temperature, NO would build up to an equilibrium value for that temperature. In the intermediate and dilution zone, CO is oxidating into CO_2, due to the injection of excess air. The excess air is responsible for the temperature drop behind the primary zone down to a combustion chamber exit temperature of about 1800 K.

Obviously, the formations of unburned hydrocarbons UHC, carbon monoxide CO and smoke are caused by incomplete combustion of the fuel, which is due to shortage of oxygen O_2 and/or low temperatures in certain regions of the combustion chamber. Unburned hydrocarbons UHC are formed as an intermediate product at an early stage of the oxidation process. Subsequently, they oxidize to form carbon monoxide CO. If the residence time at high temperature is long enough and sufficient oxygen O_2 is provided, CO oxidizes to CO_2. Hence, formation of UHC is more or less proportional to the formation of CO, however at a lower level.

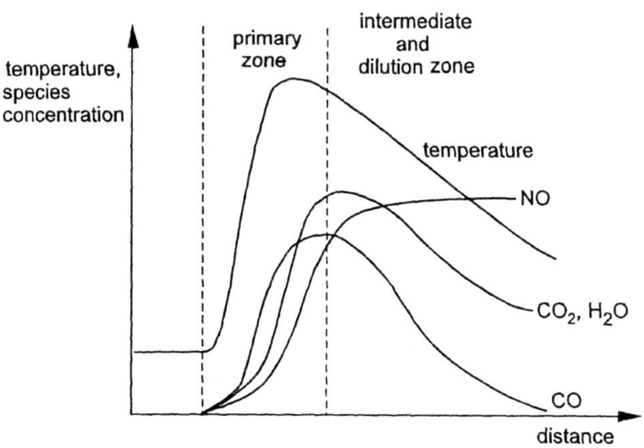

Figure 10.4-1. Variation of temperature and species concentration in an air-fuel mixture passing through a combustor (Ref. 50).

At low values of the equivalence ratio Φ, which is at low fuel-air ratios, the primary zone temperature is lower, leading to a smaller reaction rate. Accordingly, the oxidation process cannot be completed successfully, giving rise to the formation of UHC and CO. Low fuel-air ratios occur at low thrust settings of the engine. They are accompanied by small mass flow rates of fuel through the fuel injection nozzles, which tend to lead to poor atomization. This also enhances UHC and CO formation. At very high fuel-air ratios, CO formation occurs due to lack of oxygen, because no complete combustion of the fuel can be achieved.

In the high temperature zones of the combustor, nitrogen oxides NO_x, being the sum of NO and NO_2, are formed through the reaction of atmospheric nitrogen N_2 with oxygen or through the combustion of fuel-bound nitrogen. In combustion processes, NO is far more important than NO_2.

There are three mechanisms of NO formation, depending on particular conditions of temperature, concentration, residence time and type of fuel, namely: *thermal-NO, prompt-NO,* and *fuel-NO formation*.
The mechanism of the *thermal-NO formation* occurs according to the following reactions (Ref 50):

$$N_2 + O \leftrightarrow NO + N \tag{10.4-1}$$

$$N + O_2 \leftrightarrow NO + O \tag{10.4-2}$$

$$N + OH \leftrightarrow NO + H \tag{10.4-3}$$

The first of these reactions is endothermal, which means that an energy input is required, and that the change of the reaction rate is very rapid. The rate at which NO is formed, therefore, is largely determined by the rate of the reaction (10.4-1), as this reaction is considered to proceed at a much greater rate than the other two reactions. Consequently, the reaction (10.4-1), usually, is the only reaction considered in the formation of thermal NO (Ref. 51).
The concentrations controlling the reactions (10.4-1) and (10.4-3) are determined by thermal dissociation of O_2 and H_2O, which also occurs at high temperatures. This means that the rate of the NO formation is strongly affected by the magnitude of the flame temperature (Ref. 50).
The fuel-air ratio plays an important role in this process. While excess air, on the one hand, means that there are more oxygen radicals O available for the NO formation, it lowers, on the other hand, the temperature by cooling the gas stream. However, the overall effect is a reduction in NO_x formation.

As noted before, if the gases were held at a high temperature for sufficient time, the NO concentration would build up. For a combustion chamber this means that the residence time, i.e., the time that the mixture is held in the high temperature reaction zone, is an important parameter in the NO_x-formation process.

Under fuel-rich conditions, i.e., under lean-premixed conditions which occur early in the flame region, NO can also be produced by the reaction of fuel radicals (e.g., CH) with nitrogen from the air. This mechanism is called *prompt-NO formation*, where the following reactions are presumed to govern the process (Ref. 5):

$$CH + N_2 \leftrightarrow HCN + N \tag{10.4-4}$$

$$C + N_2 \leftrightarrow CN + N \tag{10.4-5}$$

$$CN + H_2 \leftrightarrow HCN + H \tag{10.4-6}$$

$$CN + H_2O \leftrightarrow HCN + OH \tag{10.4-7}$$

$$HCN, CN + O \leftrightarrow NO + R \tag{10.4-8}$$

In the reaction (10.4-8), R denotes an organic residue.

According to Ref. 5, prompt NO formation starts at higher temperatures than thermal-NO_x formation. Therefore, it may be expected that prompt-NO formation in technical flames is of secondary significance. However, in Ref. 51 it is stated that at lower combustor inlet temperatures, the prompt-NO_x formation predominates the thermal-NO_x formation. This is however not considered a problem since lower inlet temperatures correspond to engine-idle thrust setting, where NO_x formation is negligible.

Fuel-NO formation, finally, takes place through oxidation of nitrogenous compounds, which mainly are found in fuel oils and coal (Ref. 5). As far as aero-engines are concerned, kerosene contains little or no so-called *fuel-bound nitrogen*, so that fuel NO may be considered to be of no importance to aircraft pollution (Ref. 48).

In addition to the above remarks about the formation of NO_x, it can be said that nitrogen dioxide NO_2 is only encountered in low-temperature regions of the combustion chamber. There, the concentrations of the hydroperoxy radical HO_2 can be more than enough to form NO_2 from NO (Ref. 49):

$$NO + HO_2 \leftrightarrow NO_2 + OH \tag{10.4-9}$$

However, this formation of NO_2 is quickly removed by the following reactions:

$$NO_2 + H \leftrightarrow NO + OH \tag{10.4-10}$$

$$NO_2 + O \leftrightarrow NO + O_2 \tag{10.4-11}$$

Significant amounts of NO_2 can result if the reactions (10.4-10) and (10.4-11) quench because of decreasing radical concentrations due to mixing with cold fluid elements. Moreover, NO_2 may be formed in the dilution zone, through the reaction:

$$2\,NO + O_2 \leftrightarrow 2\,NO_2 \tag{10.4-12}$$

In conclusion, the temperature and the residence time in a combustion chamber determine the amount of NO_x being formed.

To conclude, we mention the formation process of soot or smoke, which occurs at high fuel-air ratios, in too fuel-rich areas of the flame. Insufficient vaporization leads to large fuel droplets, which, together with the lack of oxygen, cause soot formation when the combustion process is completed before the droplets are vaporized and mixed with air.

10.5 EMISSION CERTIFICATION

In order to improve the air quality around airports being affected by engine pollutant emissions, the International Civil Aviation Organization (ICAO) has set emission standards on a global basis (Ref. 52). These regulations impose restraints on the emissions of environmentally harmful species by turbojet and turbofan engines, designed for subsonic and supersonic aircraft propulsion. Since 1981 the regulations are included in the engine certification programs.
The following species have to comply with the standards:

- smoke (soot)
- carbon monoxide, CO
- unburned hydrocarbons, UHC
- nitrogen oxides, NO_x.

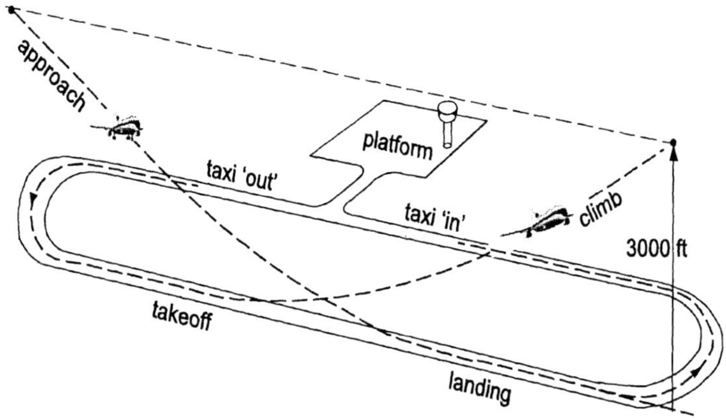

Figure 10.5-1. ICAO reference emissions landing and takeoff cycle.

The species have to be measured in static tests for a number of engine control settings. As we already know, smoke or soot consists of carbon-rich particles, causing a black exhaust gas flow color. To quantify the smoke formation use is made of the *smoke number*, *SN*, which quantity indicates to what degree darkening occurs ($0 < SN < 100$). The properties of the fuel to apply during the tests are accurately specified in the regulations (Table 10.5-1). The measured emission levels during static tests are converted to standard sea-level conditions (0 m ISA) and a relative humidity $H = 60\%$.

The regulations are based on the emission level during the so-called *reference emissions landing and takeoff cycle* (LTO cycle), portrayed in Figure 10.5-1. This flight-cycle consists of an approach flight and a landing from an altitude of 3000

property	allowable range of values
Density kg/m^3 at 15°C	780 - 820
Distillation temperature, °C	
10% boiling point	155 - 201
Final boiling point	235 - 285
Net heat of combustion, MJ/kg	42.86 - 43.50
Aromatics, volume %	15 - 23
Naphthalenes, volume %	1.0 - 3.5
Smoke point, mm	20 - 28
Hydrogen, mass %	13.4 - 14.1
Sulfur, mass %	less than 0.3%
Kinematic viscosity at -20°C, mm^2/s	2.5 - 6.5

The smoke point of a fuel is the height in millimeters of the flame in a wick lamp at which the flame starts to smoke. The lower this height, the higher is the tendency of the fuel to emit soot.

Table 10.5-1. Specification for fuel to be used in aircraft turbine engine emission testing.

phase	engine control setting		duration, minutes	
	subsonic	supersonic	subsonic	supersonic
idle/taxi-out	0.07 T_{max}	0.058 T_{max}^*	19	19
take-off	T_{max}	T_{max}^*	0.7	1.2
climb	0.85 T_{max}	0.65 T_{max}^*	2.2	2.0
approach		0.15 T_{max}^*		1.2
landing	0.30 T_{max}	0.34 T_{max}^*	4.0	2.3
taxi-in/idle	0.07 T_{max}	0.058 T_{max}^*	7.0	7.0

Table 10.5-2. Engine performance in LTO cycle.

flight-cycle consists of an approach flight and a landing from an altitude of 3000 ft (975 m), taxiing, and a takeoff and climb to 3000 ft. The emission levels are determined at a dictated engine control setting, during a specified time span for each flight phase, as indicated in Table 10.5-2.

Here T_{max} refers to the maximum takeoff thrust expressed in newton at 0 m ISA, and at a relative humidity of 60%. T_{max}^* refers to the maximum thrust of all types of jet-engines used for propulsion at supersonic speeds.
For engines designed for subsonic aircraft, the maximum permissible amount of smoke follows from:

$$SN \leq 83.6\,(T_{max})^{-0.274}\,, \tag{10.5-1}$$

or a value of $SN = 50$, whichever is lower.

The maximum allowable masses of gaseous emissions (D_P) for a subsonic turbofan or turbojet engine, with a maximum thrust greater than 26.7 kN, are determined from the following formulae, where the emissions are quantified as the amount of pollutant in grams per kN rated thrust:

for CO: $\frac{D_p}{T_{max}} \leq 118$ g/kN (date of production after 12-31-1985) (10.5-2)

for UHC: $\frac{D_p}{T_{max}} \leq 19.6$ g/kN (date of production after 12-31-1985) (10.5-3)

for NO_x: $\frac{D_p}{T_{max}} \leq 40 + 2\,\varepsilon_c$ g/kN (date of production between 12-31-1985 and 12-31-1995) (10.5-4)

Here ε_c denotes the compressor pressure ratio, i.e., the ratio of the total pressure at the compressor exit plane and the total pressure at the compressor entry plane, for takeoff conditions at 0 m ISA.

For jet engines with a production date after 12-31-1996 the NO_X emissions will have to comply with:

$$\frac{D_p}{T_{max}} \leq 32 + 1.6\,\varepsilon_c \;\; g/kN \;. \qquad (10.5\text{-}5)$$

Future engines are likely to have to comply with even more stringent regulations (Ref. 29):

$$\frac{D_p}{T_{max}} \leq 16 + 0.8\,\varepsilon_c \;\; g/kN \;. \qquad (10.5\text{-}6)$$

Apparently, the permissible emission levels of NO_X are functions of the compressor pressure ratio.

For engines designed for the propulsion of supersonic aircraft, with a production date after 02-17-1982, the following constraints are applicable:

$$\text{for soot:} \quad SN \leq 83.6\,(T_{max})^{-0.274} \;. \qquad (10.5\text{-}7)$$

with a maximum smoke number of $SN = 50$.

$$\text{for CO:} \quad \frac{D_p}{T_{max}*} \leq 4550\,\varepsilon_c^{-1.03} \qquad (10.5\text{-}8)$$

$$\text{for UHC:} \quad \frac{D_p}{T_{max}*} \leq 144\,(0.92)\,\varepsilon_c \qquad (10.5\text{-}9)$$

$$\text{for } NO_X\text{:} \quad \frac{D_p}{T_{max}*} \leq 36 + 2.42\,\varepsilon_c \;. \qquad (10.5\text{-}10)$$

The expressions (10.5-4) to (10.5-6) and (10.5-10) are shown graphically in Figure 10.5-2.

The dependence of the admissible NO_X levels on the compressor pressure ratio is consistent with the fact that the actual NO_X emission increases with increasing ε_c (see Section 10.6).

Emission figures for the LTO-cycle are available from the ICAO exhaust emission data bank in Ref. 53.

For the airplane in the cruise, no regulations have been formulated yet. However,

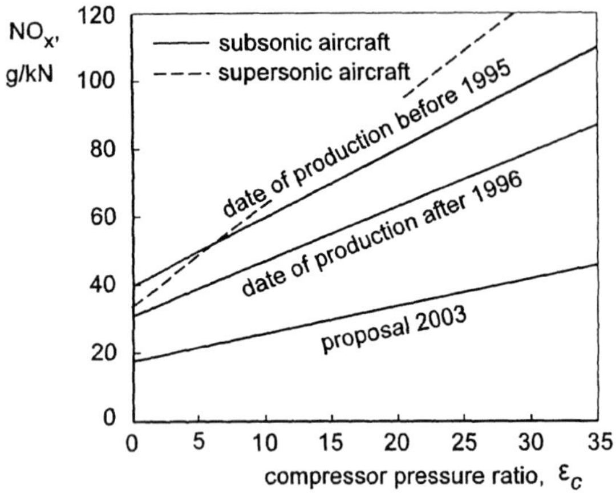

Figure 10.5-2. NO_x emission requirements.

since at present the attention, especially, is focused on the effects of NO_x emissions from aircraft engines upon the upper layer of the troposphere and the lower part of the stratosphere, additional regulations are likely to follow. Certainly, under in-flight circumstances, the values of mass specific emission, $MSE(i)$, will be difficult to measure, since for a given aircraft they are dependent on geopotential pressure altitude H_p, air temperature T, thrust setting Γ, and airspeed V.

10.6 RELATIONS FOR THE EMISSION OF NITROGEN OXIDES

Generation of nitrogen oxides, NO_x, occurs during combustion of hydrocarbon fuel when atmospheric air is used as an oxidizer. Atmospheric air consists for approximately 78% of nitrogen N_2, which oxidizes to NO_x at high combustion temperatures. The main features of the NO_x-formation processes have been described in Section 10.4.

The existing NO_x-prediction models vary from simple empirical relations, via semi-empirical models, to very complex mathematical models, incorporating the complicated flow behavior inside the combustor and chemical kinetics.

In principle, the rate of formation of NO_x of a gas turbine engine varies exponentially with the stoichiometric flame temperature and may be described by the following empirical relation (Ref. 38):

$$\mathrm{NO_x} = A\, e^{\frac{T_{st}}{B}} , \qquad (10.6\text{-}1)$$

where T_{st} denotes the stoichiometric flame temperature, and A and B are constants. Such empirical NO_X prediction models, mostly, are based on the so-called *Lipfert plot*. The scientist F.W. Lipfert showed in 1972 for a large number of engines types, the existence of an excellent correlation between combustion chamber inlet temperature, T_{t_c}, and NO_X-emission index (Ref. 54).

In Figure 10.6-1 the Lipfert plot is reproduced, where the data are valid for standard day atmospheric conditions (0 m ISA). The plot manifests that a major parameter is the combustor inlet temperature. This result leads to the conclusion that the use of a high engine pressure ratio has a detrimental effect on the NO_X emission. However, meanwhile one clearly recognizes that not only the pressure ratio is important, but also the actual primary zone temperature, which need not to have a direct relation to the stoichiometric flame temperature. Consequently, the combustor inlet temperature is important in so far that it influences the temperature level in the combustion chamber.

In Ref. 51 an empirical relation is suggested, which is a function of the combustor inlet temperature and the ambient pressure:

$$EI_{\mathrm{NO_x}} = 10^{[1 + 0.0032\,(T_{t_c} - 581.25)]} \sqrt{\frac{p}{p_0}} , \qquad (10.6\text{-}2)$$

where T_{t_c} is the combustor inlet total temperature (= compressor discharge total

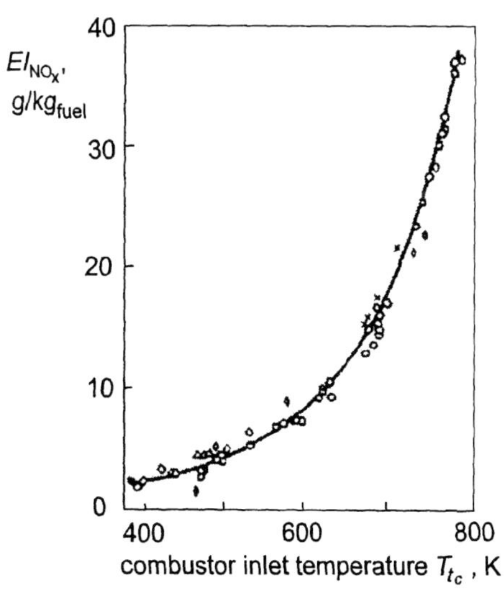

Figure 10.6-1. Lipfert plot.

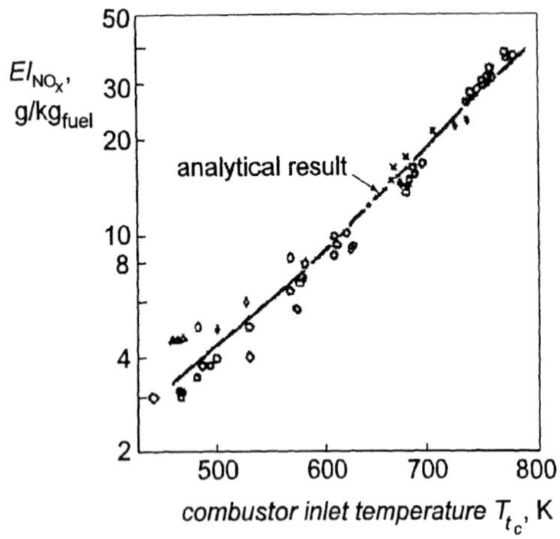

Figure 10.6-2. Correlation between model equation and Lipfert plot (Ref. 51).

temperature), p is atmospheric pressure and p_0 is the sea-level atmospheric pressure in ISA.

Figure 10.6-2 shows the correlation of the relation (10.6-2) with the data from the Lipfert plot.

Another example of an empirical NO_x-prediction model is (Ref. 55):

$$EI_{NO_x} = 0.00036 \sqrt{p_{t_c} \, e^{\frac{T_{t_c}}{200}}} \,, \tag{10.6-3}$$

where p_{t_c} is the total pressure at the entry of the combustion chamber. The constant before the square root sign was derived from published data of the Olympus straight jet engines of the Concorde aircraft in the cruise.

A semi-empirical relation for the quantification of the NO_x-emission index at the combustor exit has been established in References 48 and 56 to predict emission levels of high bypass ratio turbofan engines at takeoff and climb-out conditions defined by the ICAO regulations described in Section 10.5.
The correlation in question embodies the main variables of combustor size and inlet operating conditions:

$$EI_{NO_x} = \frac{A \, p_{t_c}^{1.25} \, V_c \, e^{0.01 \, T_{st}}}{m \, T_{pz}} \,, \tag{10.6-4}$$

where $A = 2\times10^{-8}$ and:

V_c is combustion chamber volume,

m is air mass flow per unit time into the combustion chamber,

T_{pz} is primary zone temperature.

Using Equation (10.1-1), the relationship between T_{t_c} and T_{st} can be written as:

$$T_{st} = T_{t_c} + \Delta T = T_{t_c} + \frac{m_f H}{m c_p} . \qquad (10.6\text{-}5)$$

Substituting in Equation (10.6-2) the value $c_p = 1005$ m^2/s^{2}K (air), $H = 4.3\times10^7$ J/kg (kerosene), and $m_f / m = 0.068$ (stoichiometric fuel-air ratio), we get $\Delta T_t \approx$ 2900 K.

Owing to real-gas effects, actually, the temperature rise ΔT_t will be substantially lower. This is shown in Figure 10.6-3, giving the relation between ΔT_t and T_{t_c} for kerosene at constant pressure as deduced from data in Ref.48.

As can be seen from the graph, the increase in flame temperature is only about one-half of the increase in inlet temperature. This is due to dissociation of combustion products at high temperatures. At temperatures above 1800 K, the products of complete combustion CO_2 and H_2O become unstable, leading to dissociation of these species into simpler molecular and atomic forms, such as CO, H_2, H, O, and OH. It is important to remember that dissociation is a highly endothermal process, absorbing a substantial amount of heat, through which the stoichiometric flame temperature drops.

In Figure 10.6-4 the variation of the stoichiometric flame temperature with gas pressure is indicated. Only at lower pressures some effect is noticeable. Given is

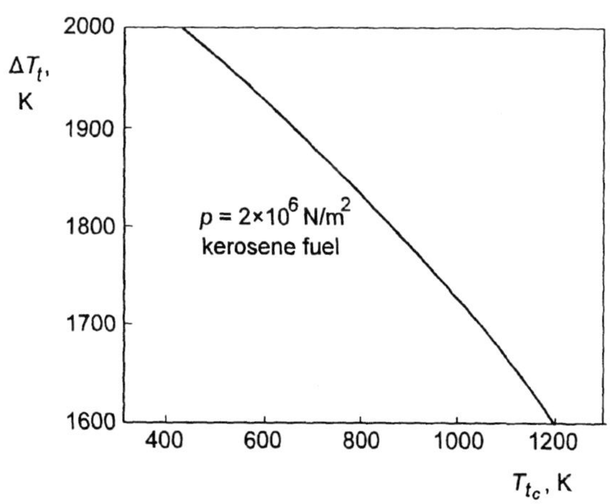

Figure 10.6-3. Effect of combustor inlet temperature on stoichiometric flame temperature.

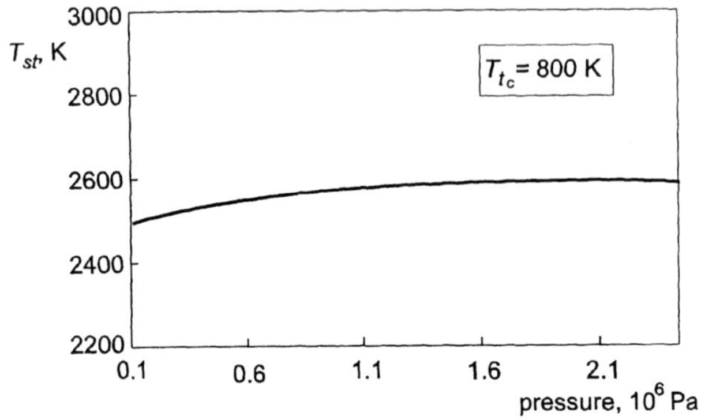

Figure 10.6-4. Effect of pressure on stoichiometric temperature (Ref. 48).

that the engine design pressure ratio at the combustor inlet is 30 or more. This means that, e.g., at an altitude of 11 km ISA, the combustor inlet pressure amounts to about $30 \times 22632 = 678960$ N/m^2 (numerical values of ambient temperature and pressure are tabulated as a function of the altitude in Ref. 8).
In view of the curve in Figure 10.6-4, it is reasonable to ignore the influence of pressure on the flame temperature of a stoichiometric mixture.

Next we consider the *primary zone temperature* T_{pz} in Equation (10.6-4). Due to practical difficulties involving temperature measurements in the primary zone, the temperature in that region, usually, is deduced from the *adiabatic flame temperature*. This temperature is defined as the temperature that a mixture would attain, if no heat would be lost by convection and radiation. The energy released

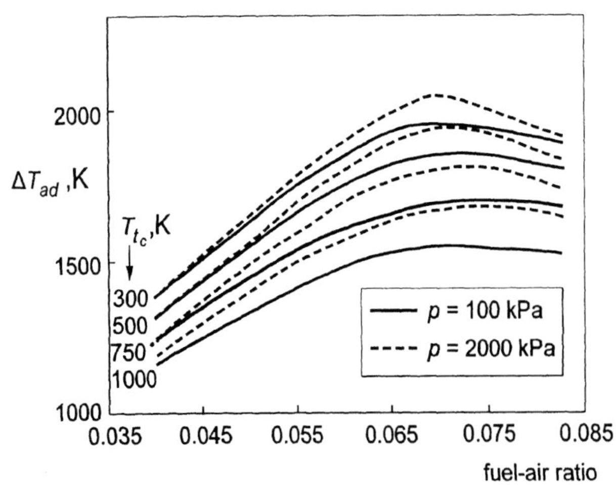

Figure 10.6-5. Adiabatic temperature-rise curves (Ref. 48).

is then fully utilized in heating the gas flow. If the fuel-air ratio in the primary zone is known, the adiabatic temperature rise ΔT_{ad} follows for a given combustor inlet temperature T_{t_c} and pressure p from the curves in Figure 10.6-5. The curves are applicable to fuels with a (lower) heating value of 4.308×10^7 J/ kg. To account for the heat losses due to transfer from the flame to the channel walls and heat radiation, the primary zone temperature is written as:

$$T_{pz} = T_{t_c} + \eta_{pz} \Delta T_{ad} , \qquad (10.6\text{-}6)$$

where η_{pz} is the primary zone combustion efficiency.

Figure 10.6-6, also extracted from Ref. 48, shows η_{pz} as a function of the combustion chamber loading, defined as $m_f / V_c p^2$, where m_f is the fuel mass flow rate.

The variation of the primary zone combustion efficiency with the combustion chamber loading in Figure 10.6-6 is derived from analysis of several conventional aero-engine combustors, and can be expressed adequately by an equation of the form (Ref. 48):

$$\eta_{pz} = -38.14 \log \frac{m_f}{V_c p^2} - 343.18 . \qquad (10.6\text{-}7)$$

The above expression enables the determination of the primary zone temperature.

In the past, during the development of the jet engine, the combustion temperatures have been constantly increased, to accomplish higher thermal efficiencies. However, these high temperatures are favorable to thermal-NO_x formation. Measures to reduce formation of nitrogen oxides, therefore, should be focused on avoiding stoichiometric flame temperatures. As discussed earlier, at stoichiometric combustion, the fuel-air mixture contains just enough oxygen for complete

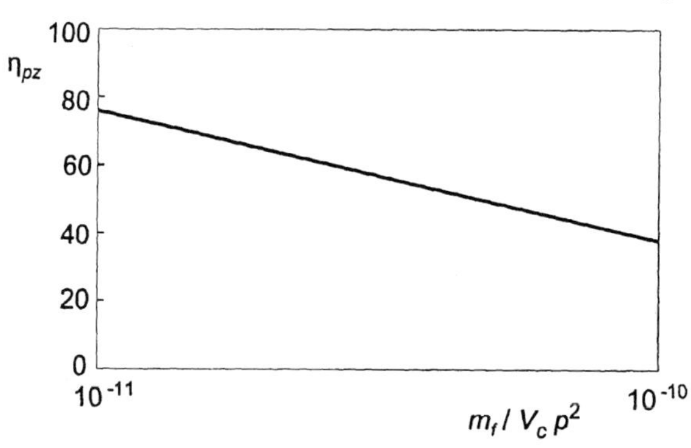

Figure 10.6-6. Primary zone combustion efficiency.

combustion of the fuel, forming CO_2 and H_2O.

As already said, the stoichiometric fuel-air ratio of kerosene is approximately 0.068. At this value of m_f/m, the rise of temperature in the combustion chamber amounts to more than 2000 K. Referring to Figure 10.2-4, it can be seen that two pollution windows occur, where the NO_x formation is significantly lower. One is in the lean region, at low fuel-air ratios, the other in the rich region, thus at high fuel-air ratios. The formation of carbon monoxide and unburned hydrocarbons are still acceptable in the lean region; this also applies to soot formation in the rich region.

10.7 NO_x REDUCTION IN AERO-ENGINES

Contrary to the emissions of UHC, CO and soot, which have been considerably reduced in modern aero-engines, the emission of NO_x has increased due to the high gas temperatures in the newer engines. The most important source of NO_x is thermal NO, of which the formation is described by the Equations (10.4-1) to (10.4-3). The initiating reaction in this mechanism is strongly dependent on temperature and the concentration of oxygen radicals O. Therefore, lowering the highest temperatures in the combustion chamber is an important measure to reduce NO_x emissions.

A reduction of NO_x formation may be achieved by injection of liquid water or steam into the combustion chamber. Because the latent heat of vaporization of water is relatively high, a lower combustion temperature and so a reduced amount of thermal NO is achieved. This method is often applied in industrial gas turbines. Unfortunately, the applicability for aero-engines is very limited, because of the large amounts of water which would have to be carried by the airplane. Combustion chamber designs without any application of water injection are called *dry systems*.

The only practical way to lower the combustion temperature seems to modify the design of the combustion chamber.

As discussed before, in a conventional combustion chamber the primary zone is supplied with only a portion of the air from the compressor, so that at full thrust virtually stoichiometric combustion in the primary zone occurs, leading to a high NO_x production.

In order to improve this condition, the engine operating point can be shifted into the lean combustion region by supplying the primary combustion zone with an increased amount of compressed air (Figure 10.7-1). To effectuate a lower

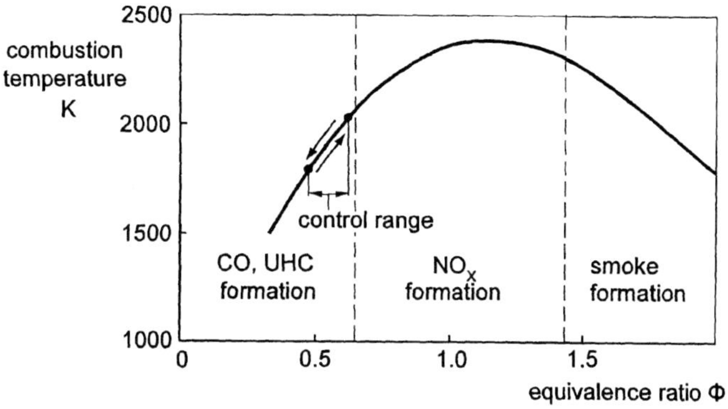

Figure 10.7-1. Engine control in a lean premixed prevaporized combustor (Ref. 18).

combustion temperature, also the condition must be satisfied that the fuel-air mixture remains homogeneous. Otherwise, local peak temperatures in the primary zone will occur and NO_x may be formed.
In any case, it is obvious from the previous Figure 10.2-2, that controlling the equivalence ratio Φ, is an effective means to minimize the emission of pollutants, especially when combined with a shortening of the residence time of the gases at high temperatures. These measures form the basis of new combustor concepts for low NO_x formation. They are referred to as *dry low* NO_x *combustors.*
The dry low NO_x combustor types considered in the following are: (1) the lean premixed pre vaporized (LPP) combustor, (2) the staged combustor type, and (3) the rich burn quick quench lean (RQL) combustor.

The primary zone combustion in the *lean premixed pre vaporized* (LPP) *combustor* takes place in the lean region ($\Phi < 1$), where low pollutant formation occurs. To achieve this by feeding the whole of the compressor air to the primary combustion zone, the process of atomizing, vaporizing and mixing of the fuel should be completed before starting the combustion. This can be accomplished inside the combustion chamber, or within an external premixing section. As the primary zone temperature will be lower than the stoichiometric flame temperature, LPP combustors are considered to be an effective means to decrease NO_x emissions. Figure 10.7-1 shows that the engine control range for this combustor type is quite narrow, which may lead to the problem of maintaining stable combustion at low thrust conditions (a lean blowout). Other problems associated with this type of combustor is danger for a flashback, i.e., the flame entering the premixing zone, and lack of cooling air.

The *staged combustor* type can be based on either fuel staging or air staging,

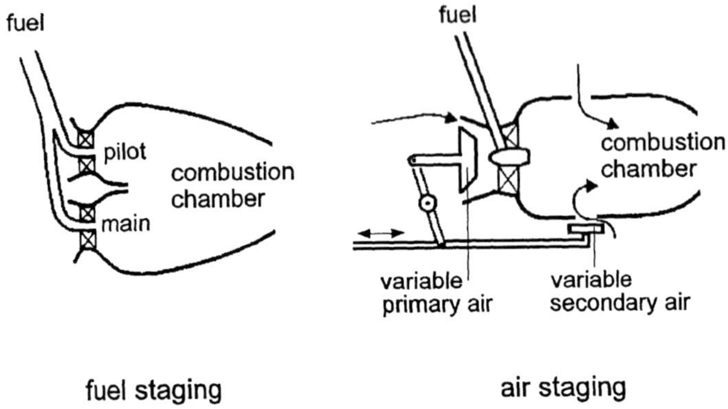

Figure 10.7-2. Staged combustion (Ref. 18).

operating lean or rich, respectively. As depicted in Figure 10.7-2, this means that the control of the equivalence ratio is achieved in two different ways; (a) by adjusting the amount of fuel, called *fuel staging*, or (b) by adjusting the amount of air, called *air staging*.

When fuel staging is applied, the combustion chamber consists of at least two separate combustion zones, named *the pilot stage* and *main stage*, arranged either radially or axially. In the pilot stage, a small portion of the fuel is burnt at a high equivalence ratio, whilst the bulk of the fuel is burnt in the main stage, at a low equivalence ratio. As a result, lower levels of NO_x can be achieved at high thrust settings. At low thrust settings, only the pilot stage is used to burn fuel. By dividing the fuel among two spray nozzles, a better atomization and vaporization may be realized, also leading to lower UHC and CO emissions. Unsteady control

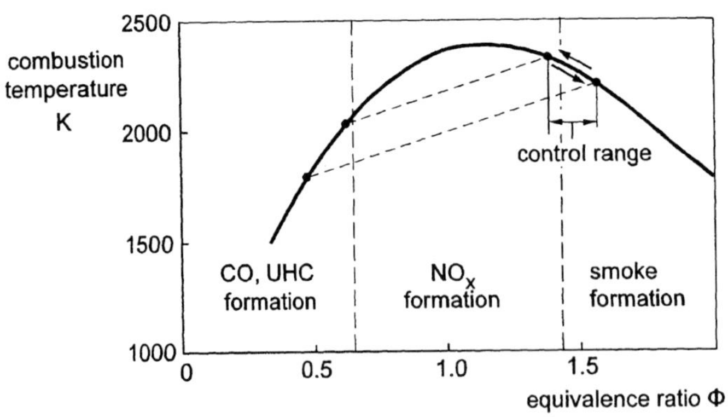

Figure 10.7-3. Control range for the rich burn quick quench lean combustor (Ref. 18).

in engine load may occur, especially when shifting from the pilot stage to the main stage.
An optimum fuel-air ratio in the primary zone can also be achieved by application of air staging, i.e., by varying the proportion of air entering the combustion chamber. This requires a control mechanism capable of varying the cross section determining the air supply. However, accommodating moving parts in the vicinity of the high combustion temperatures may lead to great difficulties.

The *rich burn quick quench lean* (RQL) *combustor* uses the two low NO_x-emission windows in the rich and the lean region (Figures 10.7-3 and 10.2-4). The primary zone operates in the rich region, where low emissions occur. It is followed by a secondary zone, operating in the lean region where also low emissions are obtained. Transition from the rich to the lean zone is achieved by rapid addition of air, avoiding near-stoichiometric fuel-air ratios.
In a general sense, lean combustion introduces problems with respect to the amount of air available for cooling of the combustor walls, an increased complexity of the engine control system, and the risk of unacceptable increases in UHC and CO emissions.

10.8 AIRCRAFT CONTRAILS

A clearly visible effect of air transport on the atmosphere is the formation of line-shaped contrails by high-flying aircraft. The exhaust gas flows behind the engines release into the atmosphere, in addition to a large quantity of heat, a considerable amount of water vapor. As a result, aircraft contrails arise from the mixing of the hot and humid engine exhaust gases with ambient cold and dry air. The condensation processes are strongly enhanced by the massive presence of condensation nuclei in the engine exhaust flows. Due to the very low temperatures of the outer air, a contrail starting as a water cloud in the warm air mixture, is soon transformed into an ice cloud.
Referring to the previous Figure 4.2-1, it is apparent that a heat release into the atmosphere causes a local decline in relative humidity, while the emitted water vapor causes an increase in relative humidity. In Figure 10.8-1 the saturation vapor pressure of water is presented again, for convenience for water vapor only.
Points P_1 and P_2 on the plot indicate two initial atmospheric conditions, both with a relative humidity of, for example, 50 percent. Point Q indicates the atmospheric condition behind the engine exhaust. After some time, when the aircraft has moved away from point Q, the atmosphere returns to its initial condition due to mixing of the exhaust gases with the ambient air.
The change in temperature and vapor pressure of the mixture may be represented

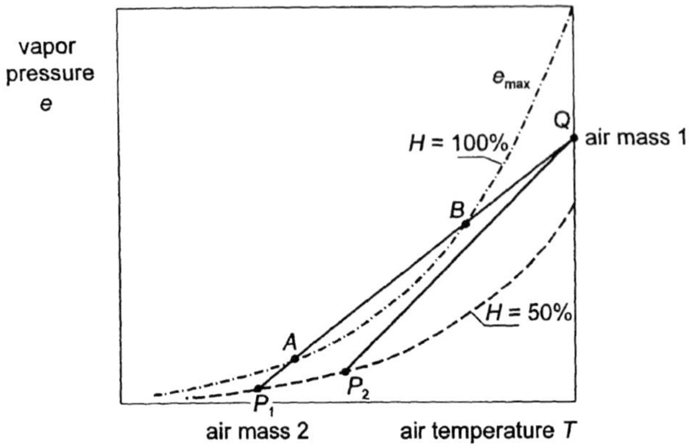

Figure 10.8-1. The occurrence of jet airplane contrails.

by the straight lines, QP_1 and QP_2. The slope of these lines is fully determined by the location of point Q, hence by the fuel properties and the engine control setting, which both define the heat release and the water vapor emission rate. Line QP_2 does not intersect with the curve for maximum saturation vapor pressure, unlike line QP_1. This is caused by the fact that the curve for maximum saturation vapor pressure at low temperatures has a less steep slope than at high temperatures. Figure 10.8-1 shows that if the line QP exceeds the e_{max}-T curve, condensation occurs with the subsequent formation of a contrail.

The phenomenon of aircraft contrails mainly occurs at low temperatures ($T < -40°$ C), which are generally found at high altitudes. The contrail moves along with the aircraft, because it arises at a fixed distance from the engine exhaust plane, determined by point B in Figure 10.8-1. The length of the contrail is fully defined by the length of the line BA. If the air in the initial condition is already saturated, then P and Q are located on the curve for e_{max}, and hence the contrail will not dissolve any more. However, the length of the contrail is usually limited by the time required to transform the air from condition B to A, by means of mixing. However, this length does not emerge from Figure 10.8-1.

Also, the formation of aircraft contrails has led to an increase in environmental concern. Dependent on air temperature, two types of contrails behind a jet-driven airplane can be distinguished:

1. *Short length contrails*, which are manifested by sharp white lines, having constant lengths of about 1000 m, and thus disappear as fast as the velocity of the airplane.
   They occur when the ambient air is very dry but not extremely cold, through which the ice particles forming the contrail can evaporate very quickly.
2. *Persistent contrails*, which are observed when the ambient air is extremely cold, such as occurs at flight altitudes greater than 10 km ($T \approx -60°$C). Then, the ice

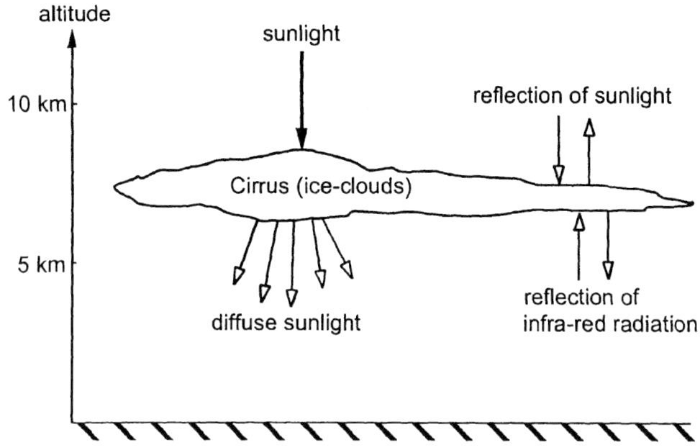

Figure 10.8-2. Effects from cirrus clouds.

particles evaporate so slowly, that the contrails may have lengths of dozens of kilometers.

It may be observed that during clear weather with sufficient atmospheric humidity, the persistent contrails initiate the formation of extensive and long-lasting cirrus cloud fields.

There still is considerable discussion of the possible effects of persistent condensation trails produced by high-flying airplanes. A first effect of such man-made clouds could be an increase in the global albedo, which in the first instance is associated with cooling of the Earth, while a second effect could be a reinforcement of the greenhouse effect by reflecting the heat from the Earth back toward the ground (Figure 10.8-2).

A third effect could be the dropping of ice crystals from these high clouds into lower cloud layers, where they might induce precipitation.

Therefore, also contrail-induced cloudiness may noticeably influence life on Earth (see also the previous Sections 4.4 and 5.4).

About 30% of the Earth is covered with cirrus clouds, occurring mainly in the upper troposphere (see the previous Figure 4.4-1). An estimate for airplane-induced cirrus cloud cover for the late 90's ranges from 0 to 0.2% of the surface of the Earth.

For the reference scenario for the expected development of future global air traffic used in Ref. 1(see Section 10.9), around the year 2050 cirrus cloud cover may increase by a factor 4 (0 to 0.8%).

10.9 CLIMATE EFFECT ESTIMATES

The amount of pollutant emitted by airplanes mainly depends on distance flown and level of technology. Growth in air transport will lead to an increase in global fuel consumption, while technology is focused on minimizing specific fuel consumption and emission indices.
In order to provide an insight into the potential effects that aviation has had since 1992 and may have in the future on both stratospheric ozone depletion and global climate change, a range of future unconstrained emission scenarios has been examined in Ref. 1. These scenarios are based on different scenarios for the global growth of air traffic, NO_x-reduction technology, and fuel-efficiency technology (Table 10.9-1).

The scenario (a) is utilized to illustrate the plausible effects of aviation and is named the reference scenario, while the other scenarios represent a range of possible growth of air travel.
In Figure 10.9-1 the development of CO_2 emissions is given according to the various scenarios. From the graph it can be seen that emissions of carbon dioxide by aviation were 0.51×10^{12} kg of CO_2 per year in 1992. This is about 2% of total anthropogenic carbon dioxide emission in the same year or about 13% of carbon dioxide emissions from all transportation sources.

The range of scenarios considered indicates that aviation emissions of CO_2 will continue to grow, and by 2050 will be 0.84×10^{12} to 5.32×10^{12} kg of CO_2 per year.

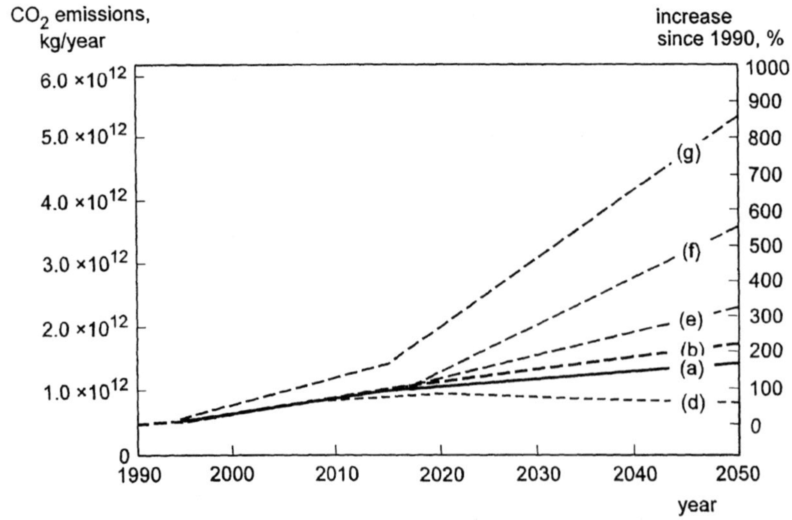

Figure 10.9-1. Total carbon dioxide emissions by airplanes.

For the reference scenario this emission increases by 2050 to 1.47×10^{12} kg per year. In Figure 10.9-2 are shown the estimated values of globally and annually averaged radiative forcing (RF) from subsonic airplane emissions for the period 1992 to 2050. As discussed in Chapter 5, the concept of radiative forcing can be seen as the contribution of aviation to the global warming effect.
The best estimate of the radiative forcing in 1992 by airplanes is 0.05 W/m^2, or about 3.5 % of the total radiative forcing by all anthropogenic activities. For the reference scenario, the radiative forcing by airplanes in 2050 is 0.19 W/m^2. According to the range of scenarios in Figure 10.9-2, the forcing is projected to

scenario	avg. traffic growth per year (1990-2050)[1]	avg. annual growth rate of fuel burn (1990-2050)[2]	avg. annual economic growth rate	avg. annual population growth rate	ratio of traffic (2050/1990)	ratio of fuel burn (2050/1990)	notes
(a)	3.1%	1.7%	2.9% 1990-2025 2.3% 1990-2100	1.4% 1990-2025 0.7% 1990-2100	6.4	2.7	reference scenario developed by ICAO Forecasting and Economic Support Group (FESG); mid-range economic growth from IPCC (1992); technology for both improved fuel efficiency and NO_x reduction
(b)	3.1%	2.0%	2.9% 1990-2025 2.3% 1990-2100	1.4% 1990-2025 0.7% 1990-2100	6.4	3.3	(a) traffic and technology scenario with a fleet of supersonic aircraft replacing some of the subsonic fleet
(c)	3.1%	1.7%	2.9% 1990-2025 2.3% 1990-2100	1.4% 1990-2025 0.7% 1990-2100	6.4	2.7	(a) traffic scenario; technology with greater emphasis on NO_x reduction, but slightly smaller fuel efficiency improvement
(d)	2.2%	0.8%	2.0% 1990-2025 1.2% 1990-2100	1.1% 1990-2025 0.2% 1990-2100	3.6	1.6	FESG low-growth scenario; technology as for (a) scenario
(e)	3.9%	2.5%	3.5% 1990-2025 3.0% 1990-2100	1.4% 1990-2025 0.7% 1990-2100	10.1	4.4	FESG high-growth scenario; technology as for (a) scenario
(f)	4.0%	3.2%			10.7	6.6	traffic-growth scenario based on IS92a developed by Environmental Defence Fund (EDF); technology for very low NO_x assumed
(g)	4.7%	3.8%			15.5	9.4	high traffic-growth EDF scenario; technology for very low NO_x assumed

[1] Traffic measured in terms of revenue passenger.km.
[2] All aviation (passenger, freight, and military).

Table 10.9-1. Summary of future global aircraft scenarios used in Ref.1.

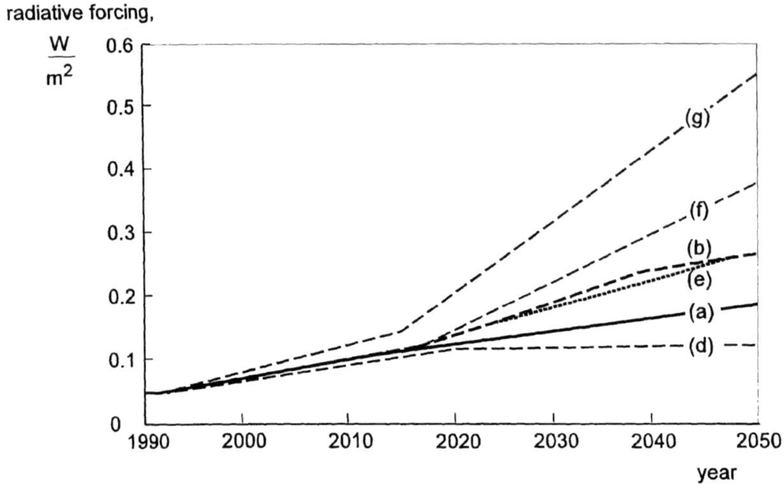

Figure 10.9-2. Total radiative forcing from subsonic airplane emissions.

grow to 0.13 to 0.56 W/m^2 in 2050, which is a factor of 2.6 to 11.0 times the value in 1992. The estimate of RF in Figure 10.9-2 combines the effects from changes in concentration of CO_2, O_3, CH_4, H_2O, line-shaped contrails, and aerosols, but excludes possible changes in cirrus clouds.
For the reference scenario, globally averaged values of the radiative forcing from different components in 1992 and 2050 are shown in Figure 10.9-3.

The graphs indicate that the most important contributions come from CO_2 and NO_X emissions, and the occurrence of contrails. Apparently, the effect of NO_X emissions

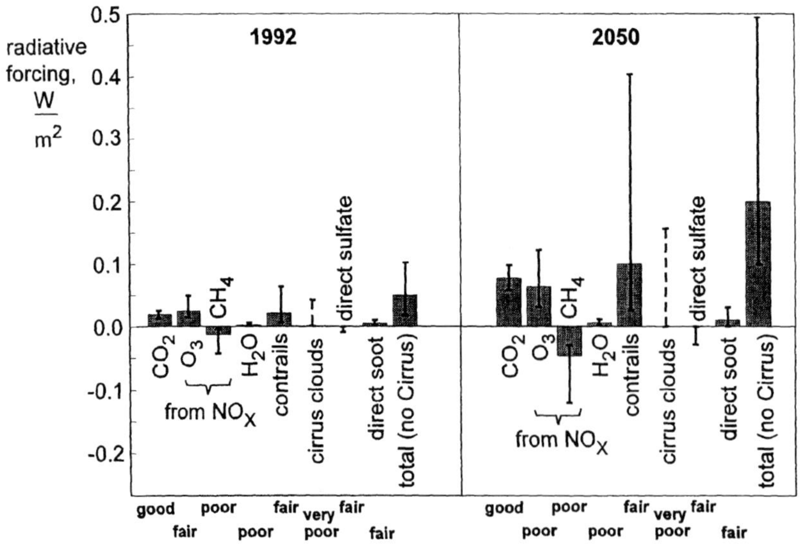

Figure 10.9-3. Radiative forcing from different components.

from subsonic airplanes in the upper troposphere is twofold. For one thing, they are estimated to have increased the ozone concentrations, for another they are expected to have decreased the methane concentrations. Similarly, increases in soot tend to warm the Earth's surface, whereas increases in sulfate tend to cool the Earth.

The bars in Figure 10.9-3 define the best estimates of radiative forcing, while the line associated with each bar is a two-thirds uncertainty range obtained using the best knowledge and tools available. In other words, there is a 67% probability that the true value falls within this range. Since the available information on cirrus clouds was insufficient, the dashed lines only indicate ranges of possible best estimates. The qualifications below the graphs reflect the level of scientific understanding of the results.

11 FLIGHT PERFORMANCE

11.1 INTRODUCTION

During the last decades air traffic has increased enormously due to the increasing demand for air transport. At the same time, people have become more aware of the fact that airplanes may also affect the environment by releasing harmful substances.

The main problems during taxiing are the emissions of carbon monoxide and unburned hydrocarbons. At the high engine thrust settings which appear during takeoff, climb, and cruise, the primary concern is the production of soot and nitrogen oxides. Especially, the formation rates of NO_X during the cruise may contribute to global warming and ozone depletion. At their cruise altitudes near the tropopause, the NO_X is injected directly into the free atmosphere, where NO_X concentrations normally are very low and practically no other anthropogenic sources are found.

Therefore, in the following the attention is centered on flight performance and the gases ejected at the greater flight altitudes used in climb, cruise, and descent. For long-range flights, naturally, by far the biggest part of the NO_X will be emitted during the cruise part of the flight.

In a general way, *flight performance* is the study of the translational motion of aircraft. We will assume that the readers of this book have a sufficient knowledge of this branch of science, so that it is not necessary to trouble them with a time-consuming treatment of the subject. For comprehensive descriptions of airplane performance the reader may turn to References 8, and 58 to 61. Therefore, this chapter only briefly reviews the performance in the various flight phases encountered by an airplane during a trip over a given travel distance. However, in view of the remarks made above, cruise performance will be treated in more detail.

Figure 11.1-1 illustrates the phases in a regular long-distance flight, without a diversion to an alternate airport. Successively, we distinguish:

- start engines and taxi to the runway,
- takeoff and initial climb, e.g., to 1500 ft (460 m),
- continued climb to the initial cruise altitude,
- cruise at a selected flight technique,
- descend to the altitude for start of approach flight,
- approach flight and landing,
- taxi from runway to the terminal.

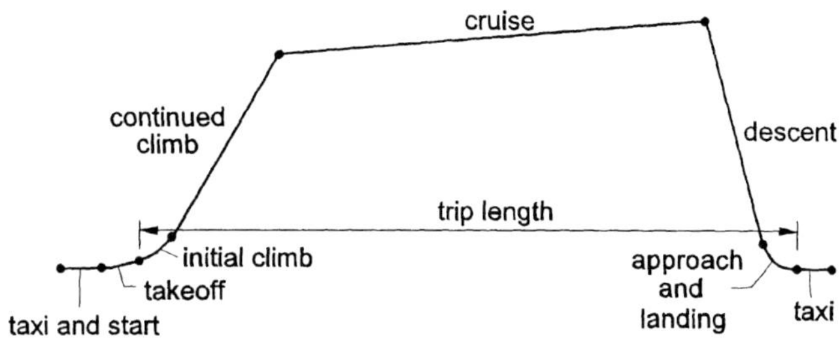

Figure 11.1-1. Phases of flight.

Although ground operations only may take about 25 minutes (see the previous Table 10.5-2), nowadays, taxying with one or more inoperative engines is a practice applied by many operators. This is a consequence of the fact that also during ground operations, large amounts of fuel are consumed, through which these maneuvers are of significance to air pollution near the ground. Also, it should be appreciated that operations such as takeoff, climb, descent, holding and landing, typically, are performed in confined and densely populated areas near the airports.

The diagram in Figure 11.1-2 illustrates the division of the trip fuel among the various flight phases as a function of flight distance. Sketched is the fuel used in terms of a percentage of the total fuel load at takeoff. The diagram shows the importance of the cruise on fuel consumption and so on high-altitude gas emissions for long-haul flights (see also Figure 11.3-1).

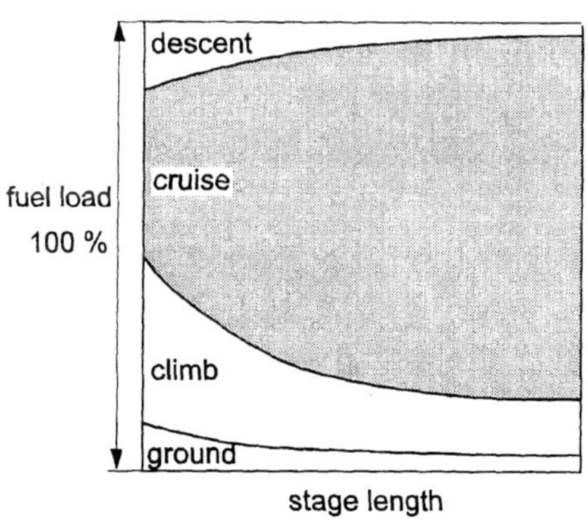

Figure 11.1-2. Typical division of fuel load versus phase of flight and stage length.

11.2 PATH AND POINT PERFORMANCE IN SYMMETRIC FLIGHT

Distinction can be made between airplane performance characteristics describing the instantaneous or *point performance* capabilities and those representing the *path performance* capabilities. The latter performance elements are related to the course of the flight and are established by integration, either analytically or numerically, of the point performance capabilities.

When flying from A to B, for most of the time, the airplane executes a so-called *symmetric flight,* in which both the velocity vector and the weight vector are parallel to the plane of symmetry of the airplane. Consequently, all four basic forces, lift L, drag D, airplane weight W, and thrust T, lie in this plane (Figure 11.2-1). In view of the importance of this type of flight for transport airplanes, in this section attention is given to the variables governing the control of the airplane in straight, symmetric flight. To this end, we will use the air-path axis (X_a, Y_a, Z_a) system for describing airplane motion. The origin of the system is at the center of gravity of the airplane. The X_a- axis lies along the velocity vector V. The Z_a-axis is taken in the plane of symmetry of the airplane, and is positive downwards for a normal airplane attitude. The Y_a-axis is positive to starboard.

Also the force system ($\boldsymbol{T}$, $\boldsymbol{R}$, $\boldsymbol{W}$) is shown in Figure 11.2-1, assuming that thrust T points into the direction of airspeed V. Then, from Newton's law of motion along the X_a- axis we obtain:

$$\frac{W}{g}\frac{\mathrm{d}V}{\mathrm{d}t} = T - D - W\sin\gamma \;, \tag{11.2-1}$$

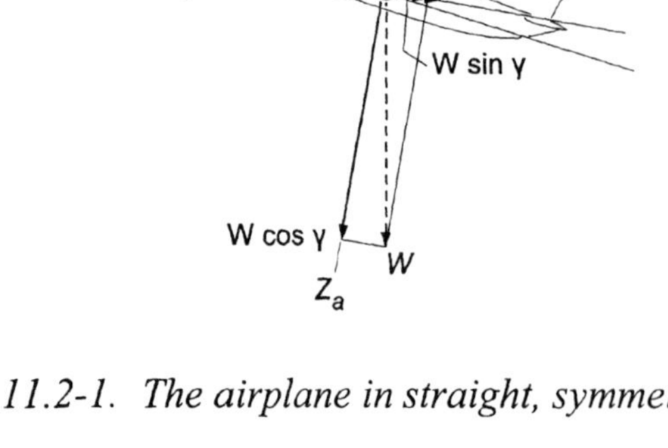

Figure 11.2-1. The airplane in straight, symmetric flight.

where dV/dt is the acceleration tangent to the flight path and γ is the flight-path angle. This angle is positive if the airplane climbs, and negative if the airplane descends.

Along the Z_a-axis we have in (quasi) rectilinear flight:

$$0 = W\cos\gamma - L . \qquad (11.2\text{-}2)$$

The lift and drag can be written as: $L = C_L \frac{1}{2} \rho V^2 S$ and $D = C_D \frac{1}{2} \rho V^2 S$, where C_L and C_D are the lift and drag coefficient, respectively, ρ is air density, V is true airspeed, and S is wing planform area.

The equations of motion can thus be written in the form:

$$\frac{W}{g}\frac{dV}{dt} = T - C_D \frac{1}{2} \rho V^2 S - W\sin\gamma , \qquad (11.2\text{-}3)$$

$$0 = W\cos\gamma - C_L \frac{1}{2} \rho V^2 S . \qquad (11.2\text{-}4)$$

Since the weight of the airplane decreases continuously with time due to the consumption of fuel by the engines, we have the relationship:

$$F = \frac{dW_f}{dt} = -\frac{dW}{dt} , \qquad (11.2\text{-}5)$$

where W_f denotes fuel weight and F fuel consumption per unit time (fuel weight flow rate).

Assuming the absence of wind, the variation of the altitude per unit time is the rate of climb, RC, of the airplane, which is equal to the vertical component of airspeed V:

$$RC = \frac{dh}{dt} = V\sin\gamma . \qquad (11.2\text{-}6)$$

When we consider flight in the International Standard Atmosphere (ISA), the atmospheric conditions are completely determined by the geopotential altitude H (see Chapter 3). Using the approximation that $h \approx H$, Equation (11.2-6) becomes:

$$RC = \frac{dH}{dt} = V\sin\gamma . \qquad (11.2\text{-}7)$$

From aerodynamics it is known that for an airplane with a given configuration (i.e., given flap setting, undercarriage position etc.), the coefficients C_L and C_D are functions of Reynolds number Re, flight Mach number M, and angle of attack α (Ref. 8):

$$C_L = C_L(\alpha, M, Re) \text{ and } C_D = C_D(\alpha, M, Re).$$

The Reynolds number, named after its discoverer the Irish scientist Osborne Reynolds (1842-1912), is defined by:

$$Re = \frac{\rho V_0 l}{\mu} , \qquad (11.2\text{-}8)$$

where μ is the coefficient of dynamic viscosity of the air, and l is a characteristic length. For airplanes *Re,* is based on the mean aerodynamic chord of the wing (see Chapter 12, Section 12.4).
The Reynolds number is a nondimensional parameter, representing the ratio of the pressure forces (kinetic or inertia forces) to the shearing stresses in the flow. In a low Reynolds number flow the viscous forces are important compared to the pressure forces, whereas in a high Reynolds flow the viscous forces are small relative to the inertia forces. The Reynolds number gives a quantitative description of the transition between laminar and turbulent flow. At low Reynolds numbers laminar flow occurs. In a high Reynolds number flow, such as typically experienced in the flight of airplanes, both a laminar and a turbulent state of the boundary layer flow are found. In the latter case the boundary layer thickens, which leads to an increased airplane drag (see also Chapter 12, Section 12.4).
The Mach number, $M_0 = V_0/c$, defines to what extent the compressibility of the air is of significance to the occurring aerodynamic force $\boldsymbol{R} = \boldsymbol{L} + \boldsymbol{D}$. For example, at flight Mach numbers $M_0 < 0.5$, the density of the air flowing over the airplane may be assumed constant, whereas at higher Mach numbers the air must be treated as compressible.

As depicted in Figure 11.2-1, the angle of attack is the angle at which the airplane meets the oncoming air, i.e., the angle between the X_a- axis and the X-axis of body axis system which is fixed to the airplane.
Since in International Standard Atmosphere $M = M(V, H)$ and $Re = Re(V, H)$, it follows that: $C_L = C_L(\alpha, V, H)$ and $C_D = C_D(\alpha, V, H)$. Furthermore, from Chapter 7, we know that thrust T and fuel flow rate F are functions of thrust setting Γ, airspeed V, and altitude H: $T = T(\Gamma, V, H)$ and $F = F(\Gamma, V, H)$.
Evidently, Equations (11.2-3) and (11.2-4) contain the following seven variables:

$$t, W, \alpha, V, H, \gamma, \Gamma . \qquad (11.2\text{-}9)$$

If we take time t as independent variable, the other six are dependent variables, which define the flight condition at each point of time.
To find these six unknowns, four equations (Equations (11.2-3) to (11.2-5) and (11.2-7)) are available, so that the course of the flight is fully defined if two dependent variables are prescribed as functions of time.
The two variables, which can be chosen freely, are named *control variables*. The remaining four variables are *state variables*.

The functions which describe the time history of the control variables are called *control laws*, e.g., $\alpha = \alpha\,(t)$ and $\Gamma = \Gamma\,(t)$.
With these two control laws we have a system of six equations to solve the six unknowns at each point of time.
The two basic controls on the airplane about which the pilot has the disposal in symmetric flight, namely, the elevator to adjust the angle of attack (pitch control) and the engine control lever to select the thrust setting (engine control), allow the actualization of the prescribed control laws. The name given to any combination of two control laws is *flight technique.*

Equations (11.2-3) to (11.2-5) and (11.2-7) completed with the two control laws describe the linear motion of the airplane, and define the *path performance values.* Examples of path performance values are time to climb to a certain altitude, range and endurance in cruise, and takeoff and landing distances.
As mentioned before, we also distinguish the *point performance* of an airplane. This name indicates the performance values at a given point of time, that is, at a given point on the flight path. This means that point performance values refer to instantaneous quantities, such as maximum speed in level flight and maximum rate of climb.

11.3 FUEL LOAD

The total weight of an airplane W_{to}, may be written as the sum of the airframe weight W_c, the weight of the propulsion system W_e, the weight of the operational items W_i, the payload W_p, the fuel weight W_f, and the weight of the reserve fuel W_{fr},

$$W_{to} = W_c + W_e + W_i + W_p + W_f + W_{fr}\ . \tag{11.3-1}$$

The airframe weight, the weight of the propulsion system, and the operational items weight may be combined into the *operational empty weight (OEW)*, $W_b = W_c + W_e + W_i$ which is the weight of the airplane prepared for service. Likewise, the sum of payload and trip fuel forms the *useful load*, $W_u = W_p + W_f$ (operational characteristic).
Dividing the constituent parts by the total weight gives the weight breakdown in terms of weight fractions,

$$1 = \frac{W_b}{W_{to}} + \frac{W_p}{W_{to}} + \frac{W_f}{W_{to}} + \frac{W_{fr}}{W_{to}}\ . \tag{11.3-2}$$

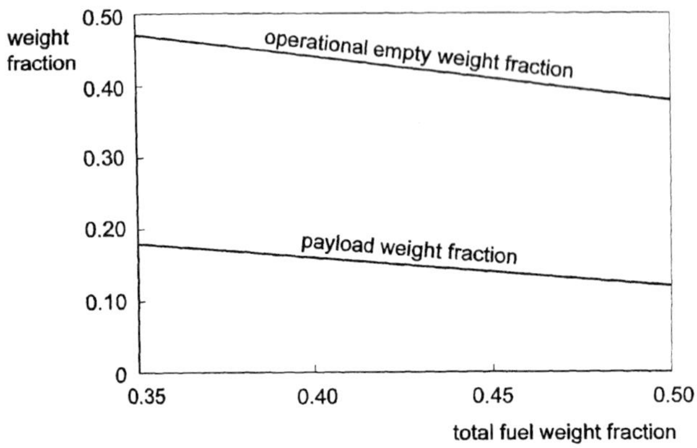

Figure 11.3-1. Typical weight fractions for long-haul aircraft.

The total weight is the weight at takeoff (TOW), which for a given airplane depends on the payload, the amount of fuel required for a particular flight, and the reserve fuel. The maximum allowable takeoff weight may be determined by structural or performance limitations.

In Figure 11.3-1 are shown the typical weight fractions for long-haul aircraft.

It is important to realize that the total fuel for a particular flight is noticeably affected by the reserve fuel, which is required to allow a diversion to an alternate airport if the airplane cannot land at its original destination. The reserve fuel policy for international routes may typically assume a continuation of the cruise for a time equal to 10% of the basic flight time at the normal cruise altitude, followed by the execution of a missed approach, and a cruise and descent to the alternate airport at 200 nautical mile (370 km) distant. In addition, it is assumed that the airplane will be required to hold in a stack for 30 minutes at a height of 1500 m in the vicinity of the alternate airport, awaiting permission to land (Figure 11.3-2). Obviously, also at the end of the basic trip the airplane may be brought into a "hold", until it can be cleared to start the approach to landing.

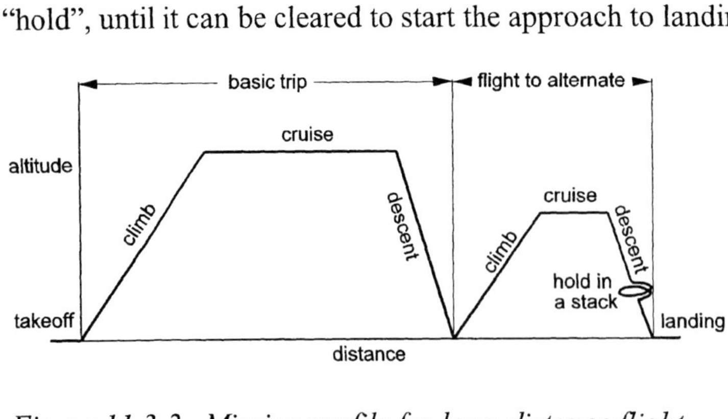

Figure 11.3-2. Mission profile for long-distance flight.

11.4 TAKEOFF AND LANDING

Figure 11.4-1 illustrates the takeoff maneuver of an airplane, involving the motion of the airplane from a standstill on the runway to a height of about 1500 ft, at which point the airplane starts the continued climb to the initial cruise altitude.

The takeoff path can be divided into two sections:

- The takeoff distance, which comprises an acceleration of the airplane along the runway to the *liftoff speed* V_{LOF}, followed by an accelerated climb until the airplane reaches the screen height specified in the airworthiness requirements (for airplanes certified in the transport category 35 ft (10.7 m)). The so-called *takeoff climb speed* V_2 achieved at the screen location must be greater than 1.2 V_S, where V_S is *the minimum stalling speed* for the airplane in the takeoff configuration (see Section 11.5).
  Always use is made of wing flaps because the extension of these devices has a large influence on maximum lift coefficient. The lower flight speeds achieved with the (partly) deflected flaps lead to reduced takeoff distances.
  During ground run, flap deflection remains constant and full thrust setting is maintained. However, to improve climb performance, the landing gear is retracted soon after the airplane has become airborne.
  The takeoff ground run includes the pre rotation phase and the rotation phase, where the airplane successively accelerates to the *rotation speed* V_R and from V_R to the liftoff speed V_{LOF}.
  At the speed V_R , the pilot initiates upward rotation of the airplane by using the elevator control. This speed is somewhat higher than the *decision speed* V_1, which is selected such that when at this speed an engine failure is recognized, the pilot is able to make a full stop on the runway or to continue the takeoff to the

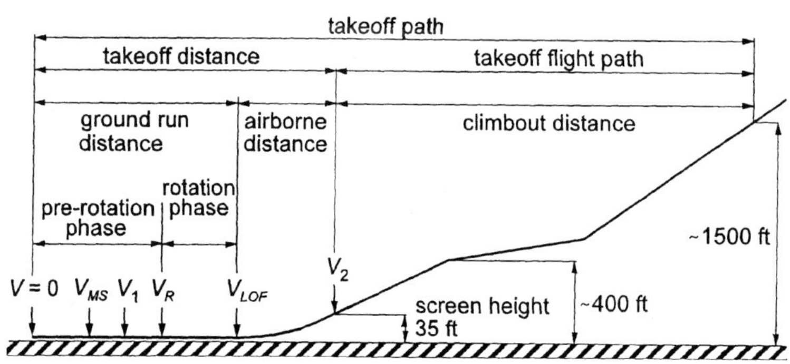

Figure 11.4-1. Takeoff maneuver.

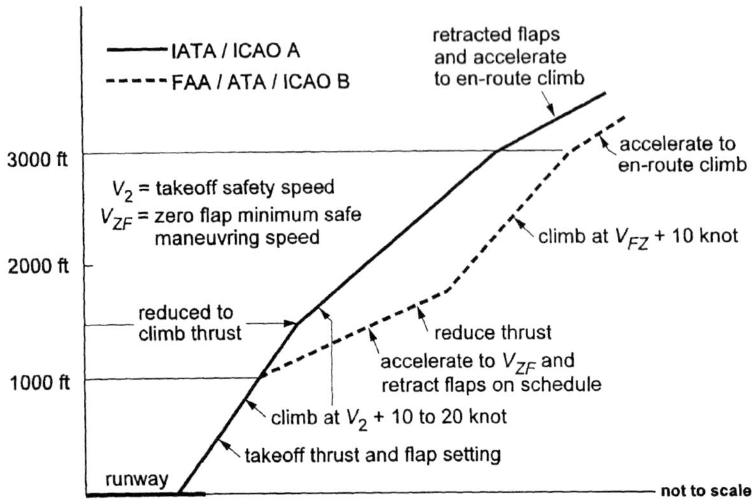

Figure 11.4-2. ICAO noise abatement takeoff and climb profiles.

screen height with one engine out, in the same distance, the so-called *balanced field length*. This procedure yields the required target speed $V_2 \geq 1.2\ V_S$ at the screen height, which guarantees a safe climb-out in the event of an engine failure at V_1.

During the pre rotation phase, the airplane incidence remains fairly constant. As rotation starts the angle of attack and thus the lift coefficient increase, so that soon after reaching V_R the nose wheel leaves the ground. At V_{LOF} the lift becomes equal to the weight of the airplane, and directly beyond the liftoff speed the lift exceeds the weight. Then also the main wheels leave the runway, and the airplane is airborne.

The airborne section, usually, is divided into the transition to climbing flight and the rectilinear climb to the screen height. In the transition phase is flown with an incremental lift coefficient in order to provide sufficient lift to accomplish an adequate curvature of the flight path, through which the flight-path angle is increased from zero at V_{LOF} to a finite value above the screen.

- The takeoff flight path traveled after passing the screen, where the airplane initially climbs to a height of about 400 ft (120 m) before the wing flaps are retracted and thrust setting is reduced to climb thrust. Finally, the airplane reaches a safe flight condition at an altitude of about 1500 ft, where the continued climb to the cruise altitude begins.

At this point it is worth mentioning that ICAO has published a standardization in operating procedures for takeoff and climb, to be used for noise abatement purposes (References 22 and 62).

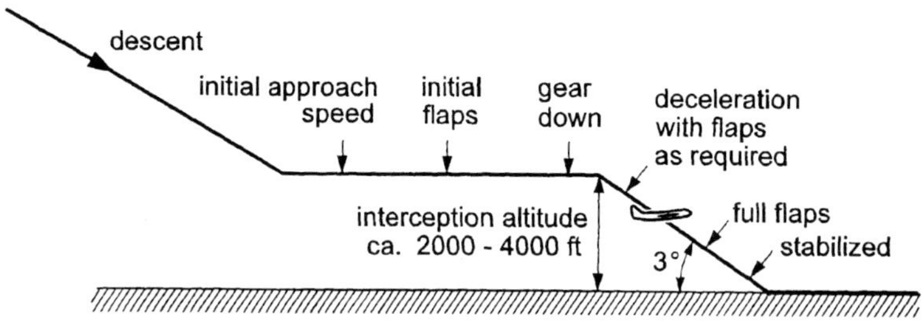

Figure 11.4-3. Approach flight.

The two departure techniques, which are depicted in Figure 11.4-2, have been laid down to ensure that the necessary safety of flight operations is maintained.
Turning our attention to the landing maneuver, Figure 11.4-3 shows that at the end of the descent, the airplane executes a level flight to intercept the ILS (Instrument Landing System) glide path to a particular runway. The ILS consists of a localizer beacon whose beam, radiating along the straight line of approach to the runway, guides the airplane onto the runway centerline (horizontal guidance), and a glide path transmitter positioned aside of the runway sending beams above and below the descent-path centerline for vertical guidance. During approach flight, where normally is flown at 3 ° to the horizontal, landing gear and wing flaps are extended.
The actual landing, finally, is the maneuver by which the airplane is brought from a steady approach speed V_A over a 50 ft (15 m) obstacle at the runway threshold (screen location) to a standstill on the runway (Figure 11.4-4).
Airworthiness requirements stipulate that V_A will not be less than 1.3 V_S, where now V_S is the minimum stalling speed for the airplane in the landing configuration (full flaps and undercarriage down).

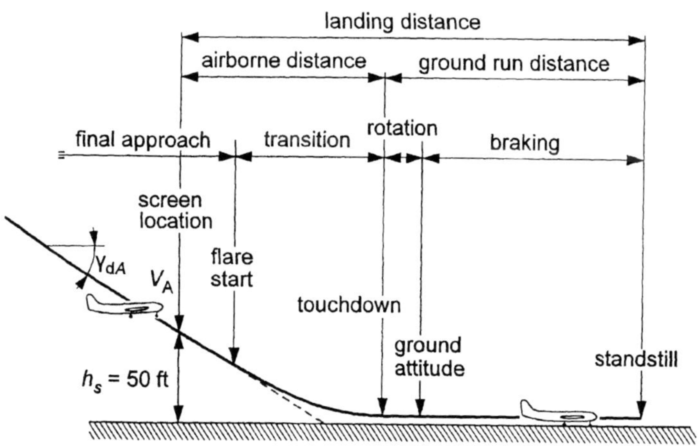

Figure 11.4-4. Landing maneuver.

The landing distance can be divided into the following two sections:

- The airborne distance, comprising a part of the final approach in the direction of the extended runway center line, followed by the transition flare, where the airplane is brought to horizontal motion by the time that the main wheels touch the runway. During the flare the engine thrust setting is throttled back to flight idle so that at touchdown the thrust is zero and the speed is below the approach speed V_A.

- The ground run distance, which consists of a free roll where the airplane is rotated down to the ground attitude, and the major ground run where the retarding force is increased by application of the wheel brakes, spoilers and thrust reversers.

We remark that for takeoff and landing, airplanes normally are operated in the direction in which the component of the wind along the runway is a headwind. The presence of a headwind component has a favorable effect in shortening the takeoff and landing distances. Of great importance to safe operation is the possibility to avoid lateral drift of the airplane relative to the runway due to the component of the wind normal to the runway. If the runway is dry, maximum allowable crosswind component velocities on takeoff may vary from about 7.5 m/s for light airplanes up to 20 m/s for large transport airplanes (References 8 and 67).
Finally, we note that also from an air pollution point of view optimized takeoff and landing techniques may be of importance.

11.5 THE PERFORMANCE DIAGRAM

In the previous Figure 11.2-1, we considered the force system for the airplane in straight, symmetric flight. Summing these forces parallel and normal to the flight path in steady flight, and assuming that the thrust line lies along the flight path, furnishes the following two equations:

$$L = W\cos\gamma \ , \tag{11.5-1}$$

$$T = D + W\sin\gamma \ . \tag{11.5-2}$$

Using the assumption that the flight-path angle γ is sufficiently small to replace its cosine by unity, Equation (11.5-1) reduces to:

$$L \cong W \ . \tag{11.5-3}$$

From Equation (11.5-2) and we find:

$$\sin\gamma = \frac{T - D}{W} , \tag{11.5-4}$$

where $(T - D)$ is called the *excess thrust.*

When Equation (11.5-2) is multiplied with airspeed V, we get:

$$TV = DV + WV\sin\gamma . \tag{11.5-5}$$

The rate of climb, RC, is the vertical component of the flight speed:

$$RC = V\sin\gamma . \tag{11.5-6}$$

Insertion of Equation (11.5-6) into Equation (11.5-5) gives:

$$TV = DV + WRC . \tag{11.5-7}$$

The left-hand side of Equation (11.5-7) is the power delivered by the power plant at the airspeed V, and is called *power available* P_a (see Chapter 7):

$$P_a = TV . \tag{11.5-8}$$

The term DV on the right-hand side of Equation (11.5-7) represents the power demanded for flight at airspeed V, and is called *power required* P_r:

$$P_r = DV . \tag{11.5-9}$$

The difference between P_a and P_r is the *excess power* P_c:

$$P_c = P_a - P_r . \tag{11.5-10}$$

Thus, the excess power refers to the power, which in steady flight is used to climb:

$$P_c = WRC . \tag{11.5-11}$$

The rate of a climb follows from Equations (11.5-10) and (11.5-11) as:

$$RC = \frac{P_a - P_r}{W} . \tag{11.5-12}$$

From Section 11.2 we know that at a given point of time (given W and H), the flight condition depends on the four variables: α, V, γ, and Γ.
Using the lift and drag coefficient in the above equations, we recall that these quantities are dependent on the airplane's angle of attack α, Reynolds number Re, free stream Mach number M_0, and the airplane's configuration.

For a low-subsonic airplane, the typical variations of lift and drag coefficient as a function of the angle of attack are sketched in Figure 11.5-1a. The curves show the angle of attack α_{crit} at which separation of the flow from the wing begins, leading

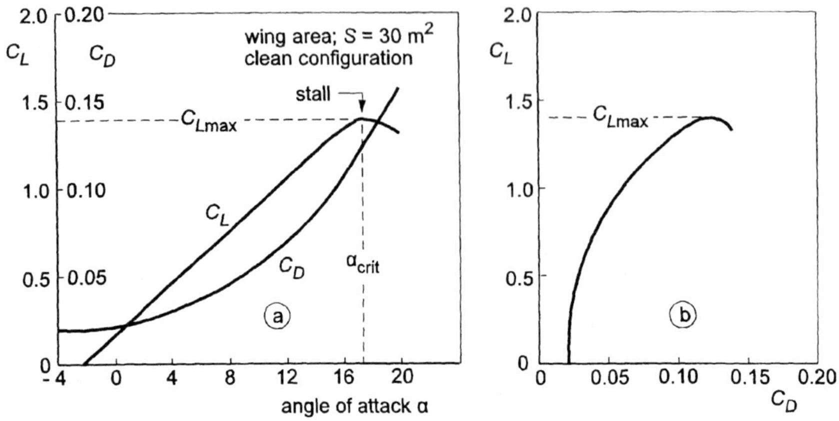

Figure 11.5-1. Aerodynamic characteristics of a propeller-driven airplane.

to a loss of lift and a substantial increase in airplane drag. Therefore, the airplane can only maintain a steady flight condition at an angle of attack below α_{crit} ($C_L < C_{Lmax}$).

Elimination of α from the relations $C_L = f(\alpha)$ and $C_D = f(\alpha)$ in Figure 11.5-1a yields the so-called *lift-drag polar*, the direct relationship between C_L and C_D (Figure 11.5-1b).

It will be evident that any presentation of a lift-drag polar should be accompanied by a record of the particular conditions which apply to the curve. In this respect it is of importance that for the cruise conditions, the effects of Reynolds number changes are small and usually may be neglected. On the other hand, the effect of flight Mach number on the aerodynamic coefficients is obvious.

Figure 11.5-2 shows the estimated lift-drag polars for a typical high-subsonic two-engine turbofan airplane. The airplane is in its so-called *clean configuration*, being the normal en route condition, in which landing gear and flaps, are retracted.

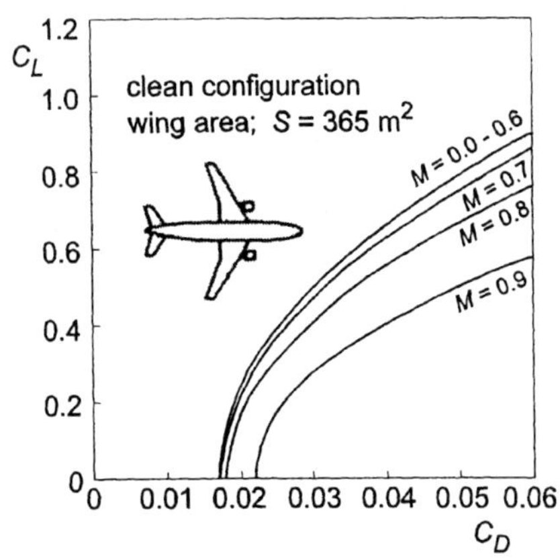

Figure 11.5-2: Lift-drag polars of high-subsonic transport airplane.

Airplane weight Wing area Flight altitude			W = 2500 kN S = 365 m^2 H = 0 m ISA			
C_L	V m/s	M	C_D	C_L/C_D	D kN	P_r kW
1.50	86.3	0.254	0.1548	9.69	258	22265
1.40	89.4	0.263	0.1193	11.74	213	19042
1.20	96.5	0.284	0.0917	13.09	191	18432
1.00	105.7	0.311	0.0684	14.62	171	18075
0.80	118.2	0.347	0.0499	16.03	156	18439
0.70	126.4	0.371	0.0423	16.56	151	19086
0.60	136.5	0.401	0.0355	16.89	148	20202
0.50	149.5	0.439	0.0298	16.78	149	22276
0.40	167.2	0.491	0.0248	16.13	155	25916
0.30	193.1	0.567	0.0211	14.21	176	33986
0.20	236.5	0.695	0.0186	10.73	233	55105
0.15	273.0	0.802	0.0190	7.91	316	86269

Table 11.5-1. Drag and power required for a high-subsonic transport airplane.

To solve Equations (11.5-2) to (11.5-5), the *performance diagram* may be used, where for a given airplane, configuration, altitude, airplane weight, and engine thrust setting, the drag or power required and the thrust or power available are plotted against airspeed.

Figures 11.5-3 *a* and *b* show for the airplane of Figure 11.5-2, the performance diagram successively in terms of force and power. The thrust curve is sketched according to the typical variation of high-bypass ratio turbofan thrust at constant engine thrust setting (see Chapter 8).
Putting together the two curves of thrust and drag is useful when analyzing climb gradients, whereas combining the curves of power required and power available directly delivers the relation between climb speed and airspeed.

The calculation of the drag and power required curves shown in Figure 11.5-3 is carried out in Table 11.5-1 for the aerodynamic data given in Figure 11.5-2, assuming an airplane weight of 2500 kN, a wing area of 365 m^2, and a flight altitude of 9000 m ISA, indicating an air temperature and density of T = 229.65K and ρ = 0.4663 kg/m^3, respectively (see Chapter 3).
From Equations (11.5-3) and (11.5-9), and the relationships that $L = C_L \tfrac{1}{2} \rho V^2 S$ and $D = C_D \tfrac{1}{2} \rho V^2 S$ we find that:

$$V = \sqrt{\frac{W}{S}\frac{2}{\rho}\frac{1}{C_L}} , \qquad (11.5\text{-}13)$$

$$D = \frac{C_D}{C_L} W , \tag{11.5-14}$$

$$P_r = W \sqrt{\frac{W}{S} \frac{2}{\rho} \frac{C_D^2}{C_L^3}} . \tag{11.5-15}$$

For the calculation of the performance diagram, we obtain from Equation (11.5-13) the airspeed at each value of C_L. The Mach number at given airspeed follows from:

$$M = \frac{V}{c} = \frac{V}{\sqrt{\gamma R T}} . \tag{11.5-16}$$

For each pair of C_L and M values, the drag coefficient C_D is found from Figure 11.5-2. Using these data in Equations (11.5-14) and (11.5-15) furnishes drag and power required.
From the curves in Figure 11.5-3 a number of special flight speeds can be distinguished:

- *Minimum stalling speed* V_S. Equation (11.5-13) shows that the airspeed decreases as the lift coefficient increases. This indicates that in steady flight, where $L = W$, at the maximum lift coefficient $C_{L\max}$. the minimum airspeed is obtained.

- *Minimum power required speed,* $V_{Pr\min}$. This airspeed is higher than the minimum stalling speed V_S and is, according to Equation (11.5-15), obtained at the particular combination of Mach number and lift coefficient for which the ratio C_L^3/C_D^2 is a maximum.

- *Best climb gradient speed,* $V\gamma_{\max}$. This speed is determined by the condition that the excess thrust $(T - D)$ is a maximum (see Equation (11.5-4)).

- *Minimum drag speed,* $V_{D\min}$. The angle between any radius from the origin to a point on the power required curve is a measure for the ratio P_r/V and so for the drag D (see Equation (11.5-9)). The point on the power required curve that corresponds to minimum drag is thus found by drawing a straight line from the origin tangent to the P_r-curve. Hence, the minimum drag speed is always higher than the minimum power required speed. According to Equation (11.5-14), this speed is obtained at the particular combination of Mach number and lift coefficient for which C_D/C_L is a maximum.

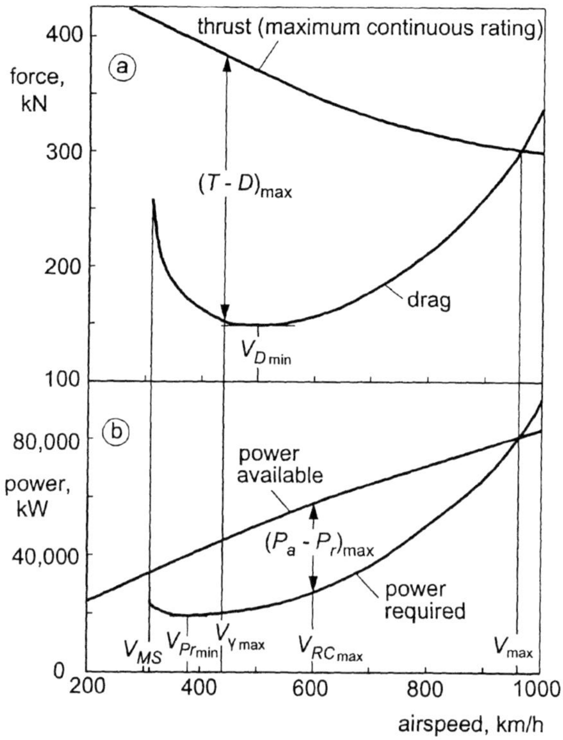

Figure 11.5-3. Performance diagrams for a high-subsonic transport airplane.

- *Best rate of climb speed,* $V_{RC\max}$. According to Equation (11.5-12), the flight speed for $RC_{\max}$ is found as the speed at which the excess power is maximum.

- *Maximum level flight speed,* $V_{\max}$. The maximum forward velocity at which level, steady flight can be maintained is determined by the condition that thrust equals drag ($T = D$) and lift equals weight ($L = W$). These equalities hold for the intersection of the T and D curves or P_a and P_r curves.

At speeds lower than the maximum flight speed $V_{\max}$, thrust exceeds drag for level flight: $(T - D) > 0$. In steady flight the excess thrust is used to increase the potential energy of the airplane.
The RC-values resulting from Figure 11.5-3 are plotted in Figure 11.5-4 as a function of airspeed, showing the optimum airspeed for maximum rate of climb. In the latter diagram, a radius vector from the origin and intersecting the curve has the slope RC/V. Hence, in Figure 11.5-4 the angle between the abscissa and a straight line from the origin to the RC-curve represents the climb angle γ at airspeed V. Hence, the maximum climb angle in steady, symmetric flight at $V\gamma_{\max}$

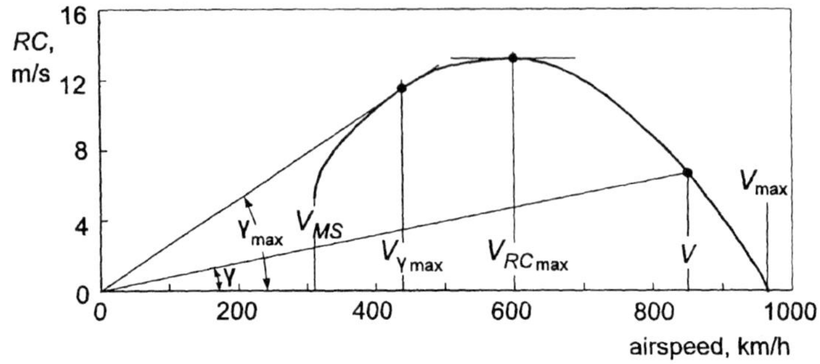

Figure 11.5-4. Climb performance.

is given by the tangent to the curve. Note that the speed for best climb angle is less than the speed for best rate of climb.

Finally, we remark that the effect of the altitude on point performance in steady, symmetric flight arises from the decrease in air density with increasing altitude. The attendant modification of the performance diagram can be found by calculating the two performance curves at various altitudes. Calculation results will show that as altitude increases the maximum rate of climb decreases. The altitude at which for a given airplane configuration, weight, and thrust setting, the maximum rate of climb becomes zero, is called the *theoretical ceiling*. At this altitude the high-speed and the low-speed intersections of the performance curves in Figure 11.5-4 coincide and no excess thrust or power exists.

11.6 PATH PERFORMANCE IN CLIMBING AND DESCENDING FLIGHT

With regard to the continued climb to the cruise altitude, the climb techniques yielding minimum time and minimum fuel are of paramount importance.

In the absence of wind, the time rate of change of altitude is the rate of climb of the airplane, which equals the vertical component of the airspeed:

$$\frac{\mathrm{d}h}{\mathrm{d}t} = RC = V\sin\gamma \ . \tag{11.6-1}$$

Using the approximation that $\mathrm{d}h \approx \mathrm{d}H$, we have:

$$\mathrm{d}t = \frac{\mathrm{d}H}{RC} \ . \tag{11.6-2}$$

The time to climb from a geopotential altitude H_1 to a geopotential altitude H_2 is obtained by integrating Equation (11.6-2) between H_1 and H_2:

$$t = \int_{H_1}^{H_2} \frac{dH}{RC} . \tag{11.6-3}$$

The weight of the fuel consumed is given by:

$$W_f = \int_{t_1}^{t_2} F dt = \int_{H_1}^{H_2} \frac{F}{RC} dH , \tag{11.6-4}$$

and the related horizontal distance in the absence of wind by:

$$s = \int_{t_1}^{t_2} V \cos\gamma dt = \int_{H_1}^{H_2} \frac{V}{RC} dH . \tag{11.6-5}$$

In Equations (11.6-3) and (11.6-4), rate of climb and fuel weight flow rate are functions of airplane weight, engine thrust setting, selected flight speed, and altitude. When we are interested in obtaining minimum time or fuel, of course, RC_{max} and $(F/RC)_{min}$ -values in dependence on the altitude, have to be determined.

Generally, no analytic expressions are obtainable relating the quantity's RC_{max} and $(F/RC)_{min}$ to altitude H, so that integration of Equations (11.6-3) to (11.6-5) must be accomplished numerically.
When we look at the step by step calculation procedure to derive, for instance, the minimum time to climb to a given altitude, we have to solve the summation form of Equation (11.6-3):

$$t_{min} = \sum_{i=1}^{N} \frac{\Delta H}{RC_{max\,i}} . \tag{11.6-6}$$

In this equation, $RC_{max\,i}$ is the average value of the maximum rate of climb within each interval ΔH.
The first step may be the determination of the maximum rate of climb for various airplane weights as a function of the altitude, as sketched in Figure 11.6-1a.
From these data, the time to climb can be found by considering increments in the altitude, ΔH. To take into account the effect of the variation of airplane weight due to the consumption of fuel on RC_{max}, at the end of each interval ΔH, the weight of the fuel consumed is given by:

$$\Delta W_{fi} = F_i \Delta t_i = F_i \frac{\Delta H}{RC_{max\,i}} , \tag{11.6-7}$$

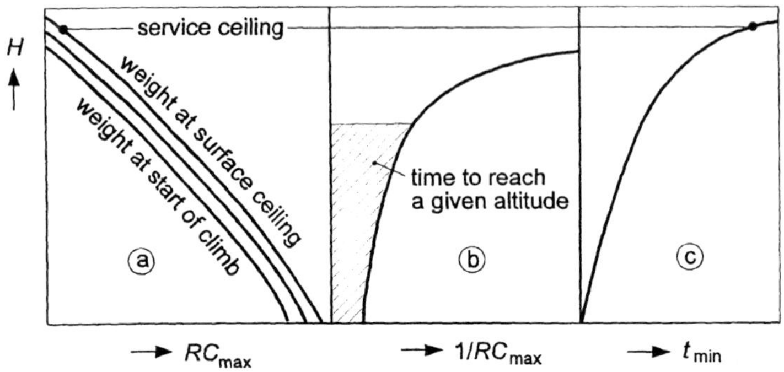

Figure 11.6-1. Determination of time to climb.

where F_i is the average value of the fuel weight flow rate within an interval ΔH, obtained from engine thrust setting and specific fuel consumption data. Then, from Figure 11.6-1a, the maximum rate of climb in the adjacent altitude interval is known at the airplane weight, $W_{i+1} = W_i - \Delta W_{fi}$. The total fuel consumed during this climb is found from the summation:

$$W_f = \sum_{i=1}^{N} \Delta W_{fi} = \sum F_i \frac{\Delta H}{RC_{\min i}} . \tag{11.6-8}$$

By plotting the curve obtained for $1/RC_{\max i}$ against H and determining the shaded area on the left-hand side of the curve, the minimum time to climb was obtained (Figures 11.6-1b and c).

In the same manner, the minimum fuel to climb can be derived graphically by using the summation form of Equation (11.6-4):

$$W_f = \sum_{i=1}^{N} \frac{F}{RC_{\min i}} \Delta H . \tag{11.6-9}$$

A calculation of the climb performance at a number of altitudes will show that the airspeed at which maximum rate of climb occurs increases with height. Accordingly, a portion of the excess thrust must be used to accelerate the airplane along its flight path. This means that to the point performance capabilities in quasi-steady symmetric flight, kinetic energy correction factors must be applied (see Ref. 8).

11.7 RANGE AND ENDURANCE

As depicted in Figure 11.7-1, the term *range* is used for the horizontal straight-line distance that can be flown on a given amount of fuel, applying a given cruise

technique (e.g., constant altitude and airspeed).
The total distance covered in climb, cruise, and descent may be called *total range*, *stage length* or *block distance*. Maximum total range is the distance an airplane can fly between takeoff and landing as limited by its fuel capacity.

According to Equation (11.2-5) the fuel weight flow rate F is given by:

$$F = -\frac{\mathrm{d}W}{\mathrm{d}t} . \qquad (11.7\text{-}1)$$

The range is obtained from the following definite integral:

$$R = \int_{t_1}^{t_2} V \,\mathrm{d}t = \int_{W_1}^{W_2} -\frac{V}{F}\,\mathrm{d}W = \int_{W_2}^{W_1} \frac{V}{F}\,\mathrm{d}W , \qquad (11.7\text{-}2)$$

where V/F is called *specific range* (range per unit weight of fuel). The subscripts "1" and "2" refer to the initial and final conditions at beginning and end of cruise, respectively.

The length of time spent in the cruise, i.e., the *endurance E*, is given by:

$$E = \int_{t_1}^{t_2} \mathrm{d}t = \int_{W_1}^{W_2} -\frac{1}{F}\,\mathrm{d}W = \int_{W_2}^{W_1} \frac{1}{F}\,\mathrm{d}W . \qquad (11.7\text{-}3)$$

It may be noted that for civil transport airplanes, the holding in a stack before landing, perhaps, is the only flight phase in which the endurance is the

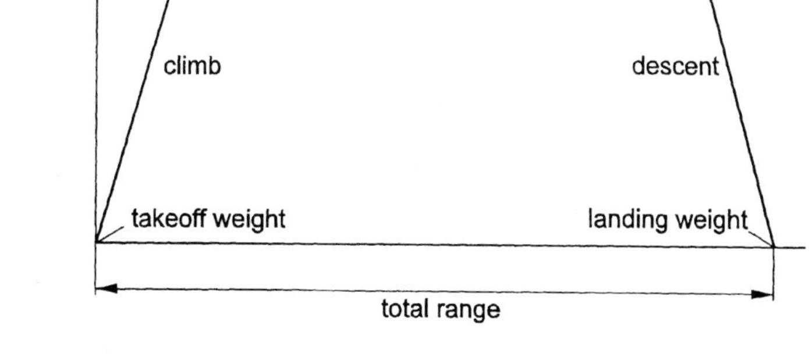

Figure 11.7-1. Mission nomenclature.

performance criterion to be maximized.
Assuming level and steady flight, we may express the fuel weight flow rate as:

$$F = c_T T = c_T \frac{C_D}{C_L} W \, , \qquad (11.7\text{-}4)$$

where c_T is specific fuel consumption (SFC), which varies at constant H and V during the course of the flight (see Section 11.8).
Inserting Equation (11.7-4) into Equation (11.7-2), the cruise range can be written as:

$$R = \int_{W_2}^{W_1} \frac{V}{F} \mathrm{d}W = \int_{W_2}^{W_1} \frac{V}{c_T} \frac{C_L}{C_D} \frac{\mathrm{d}W}{W} \, . \qquad (11.7\text{-}5)$$

The corresponding integral, expressing the endurance is obtained by insertion of Equation (11.7-4) into Equation (11.7-3):

$$E = \int_{W_2}^{W_1} \frac{1}{c_T} \frac{C_L}{C_D} \frac{\mathrm{d}W}{W} \, . \qquad (11.7\text{-}6)$$

To derive an analytic expression for range, we will consider, as a cruise technique of interest for jet-driven airplanes, the flight at constant airspeed and angle of attack. Using the approximation that C_L and C_D have constant values throughout cruise flight, we see from the relationship for level flight $W = C_L \, ½ \, \rho \, V^2 \, S$, that the flight altitude must be steadily increased as fuel is consumed. Therefore, the present cruise technique is commonly referred to as *cruise-climb*. However, the occurring flight-path angle is normally sufficiently small so as to make possible the assumption of quasi-level flight.
Using also the approximation that c_T remains constant, integration of Equation (11.7-5) yields a closed-form solution for the range:

$$R = \frac{V}{c_T} \frac{C_L}{C_D} \ln \frac{W_1}{W_2} \, . \qquad (11.7\text{-}7)$$

The quantity $\frac{V}{c_T} \frac{C_L}{C_D}$ is called *range factor*. Equation (11.7-7) is known as the *Breguet Range Equation* for jet airplanes flying in still air.

Alternatively, it may be convenient to express Equation (11.7-7) in terms of the overall efficiency, η_{tot}, of the propulsion system. According to Equation (7.5-32), we have:

$$\eta_{tot} = \frac{g_0 V}{H c_T} . \tag{11.7-8}$$

Inserting Equation (11.7-8) into Equation (11.7-7) gives:

$$R = \eta_{tot} \frac{H}{g_0} \frac{C_L}{C_D} \ln \frac{W_1}{W_2} . \tag{11.7-9}$$

It should be noted that in the above equations c_T is the *installed specific fuel consumption*. The designation "installed" means that all adverse effects on specific fuel consumption associated with the integration of engine and airplane structures are included. For example, these effects may concern extra pressure losses and interference drags greater than those assumed in the bare engine performance calculations, and power extraction for driving generators and hydraulic pumps. For an important part, turbine installation losses also refer to bleeding compressed air from the engine for air conditioning and cabin pressurization.
As an airplane climbs to cruise altitude, atmospheric temperature and pressure fall to very low values (see Chapter 3). The usual cruise altitudes are far above the highest altitude that will sustain human life. Normally, cabin pressure is maintained at a level equivalent to a geopotential pressure altitude of approximately 6000 feet (1800 m).

11.8 OPTIMUM CRUISE ALTITUDE

The previous Equation (11.7-7) reveals that in order to obtain maximum range for a given altitude, initial weight, and fuel load, a jet airplane should fly at that airspeed and altitude at which the product $V C_L/C_D$ is the maximum, provided that specific fuel consumption remains constant. However, Equation (11.7-7) also unveils that with the assumptions of constant c_T and C_L/C_D, the range has no absolute maximum. Without the effects of compressibility on drag characteristics, a constrained maximum only occurs when airspeed or the initial cruise altitude is specified.
Equation (11.7-7) indicates that in case the airspeed is prescribed, the maximum range is obtained when C_L/C_D is the maximum.

When the initial altitude is dictated, the airspeed can be written as:

$$V = \sqrt{\frac{W_1}{S} \frac{2}{\rho_1} \frac{1}{C_L}} . \tag{11.8-1}$$

Substitution into Equation (11.7-7) gives:

$$R = \frac{1}{c_T}\sqrt{\frac{W_1}{S}\frac{2}{\rho_1}\frac{C_L}{C_D^2}}\ \ln\frac{W_1}{W_2}\ . \qquad (11.8\text{-}2)$$

This Equation shows that now for best range the ratio $\frac{C_L}{C_D^2}$ should be maximum.

However, climbing at constant V and α leads to higher Mach numbers, through which the airplane might penetrate into the drag rise that results from transonic flow (related to speeds close to the speed of sound).

The derivation of the actual flight condition for the finite value of $(V\,C_L/C_D)_{max}$, when the effects of compressibility on airplane drag are taken into account, will be illustrated by means of a numerical example. For that, we consider again the turbofan airplane of which the aerodynamic characteristics are given in the previous Figure 11.5-2. From these data, the curves in Figure 11.8-1a are deduced, which give the maximum lift-drag ratio and the related lift coefficient as a function of Mach number. The graph manifests the characteristic behavior that both quantities drop off sharply at Mach numbers greater than 0.6. In Figure 11.8-1b are plotted the required altitudes (ISA) versus M as computed from the equality $W = C_L\ ½\ \gamma\ p\ M^2\ S$, where C_L is the lift coefficient at which C_L/C_D is the maximum.The related airspeeds follow from $V = M c$.The final result is shown in Figure 11.8-1c, where is plotted the product of V and $(C_L/C_D)_{max}$ against flight Mach number. Clearly, there are an optimum Mach number and an optimum altitude dependent on airplane weight. Figure 11.8-1c also indicates that as airplane weight decreases during the cruise, the airplane should climb in altitude to maintain the optimum flight condition.

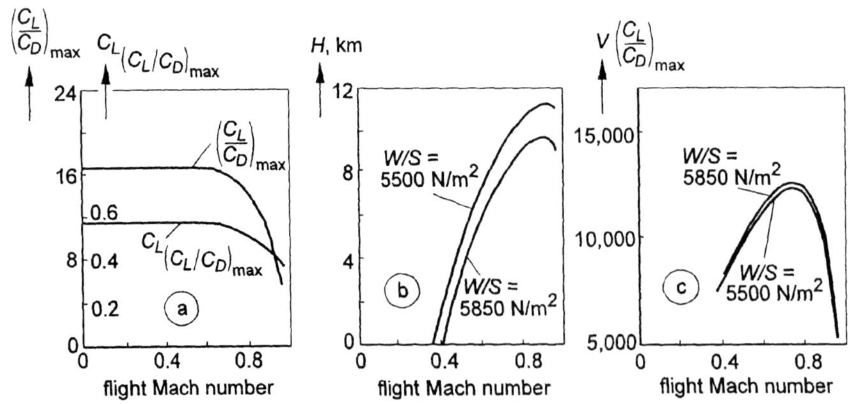

Figure 11.8-1. Derivation of condition for maximum range.

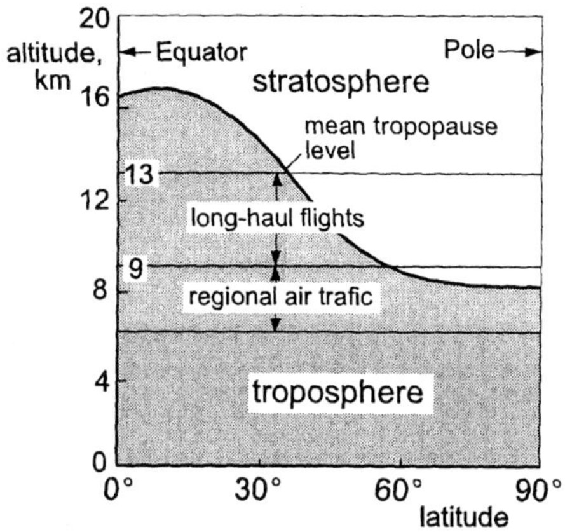

Figure 11.8-2. Cruise flight levels.

It should be remarked in this respect, that each flight from takeoff to landing, normally, is executed under the jurisdiction of *Air Traffic Control* (ATC) regulations. The procedures that a pilot must follow when flying without visual cues from the ground, are called *Instrument Flight Rules* (IFR).

ATC plays a crucial role in ensuring safe and efficient air traffic. There are primarily two kinds of air traffic control: (1) airport ATC, which applies to airplane movements on and in the immediate vicinity of the airport, and (2) airway ATC, which applies to en route flying between the airport of departure and destination. During the cruise, civil airplanes follow the *airways*, i.e., horizontal corridors within which airplanes are protected by internationally laid down rules of separation. Other airplanes must be at least 1000 ft above or below, 10 miles on either side of any airplane flying along one, and 10 minutes flying time in front and behind.

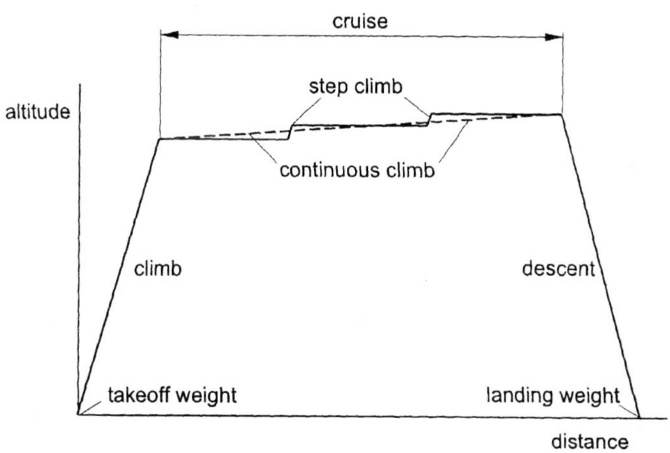

Figure 11.8-3. The execution of step-climb during cruise.

As shown in Figure 11.8-2, the atmosphere is divided into three main altitude layers. High-subsonic civil airplanes use the layer between 29,000 and 45,000 feet (9 - 13 km), and primarily at a mean altitude of about 36,000 ft (11 km). The highest layer, (H_p> 45,000 ft) is available for supersonic airplanes, while the layer below the main jet airways is used by the slower propeller-driven airplanes.
Anyway, when flying along an airway, each change of altitude needs approval. Practically, a *cruise climb* may be approximated by a sequence of level flight segments. This so-called *step climb* is sometimes allowed on long-haul flights (Figure 11.8-3).

11.9 NUMERICAL CALCULATIONS

The cruise range and endurance equations derived in Section 11.7 may be solved by either analytical integration or numerical and graphical procedures. Clearly, accurate predictions of fuel load as functions of range, weight, airspeed, and altitude will demand step by step calculations. In this respect, it is of importance to recognize again that for airplanes propelled by air breathing engines, the fuel flow rate is such small that there is only a slow variation of airplane weight with time. This condition allows us to consider the cruise as a continuous succession of uniform motions under slowly changing conditions. In other words, the instantaneous values of V/F and F in Equations (11.7-2) and (11.7-3) can be determined as though the airplane performs a quasi-steady and quasi-level flight, where the instantaneous flight condition depends on the specification of two control variables.
The graphical procedure for determining F and V/F as a function of airplane weight is shown schematically in Figure 11.9-1, where two different cruise techniques are depicted: 1) a cruise at constant altitude and airspeed, and 2) a cruise at constant

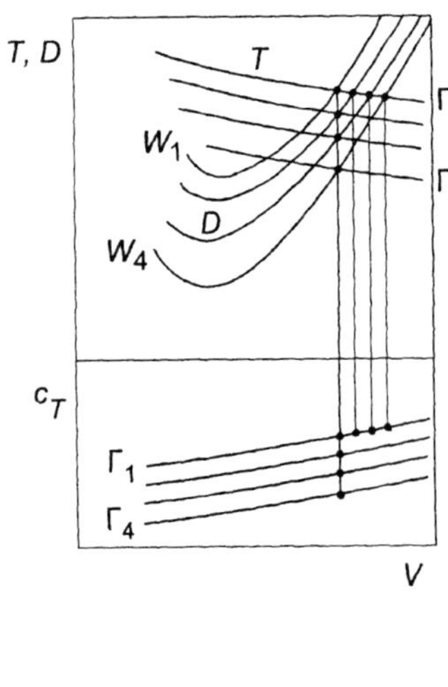

Figure 11.9-1. Determination of range performance.

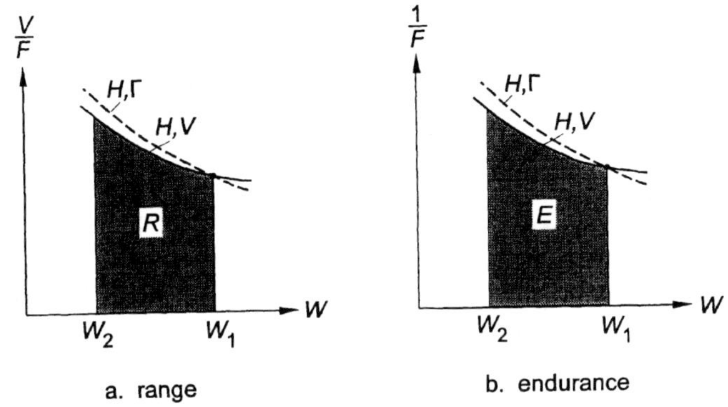

Figure 11.9-2. Calculation of range and endurance.

altitude and thrust setting. Successively, the problem requires the determination of thrust setting and level flight speed as a function of airplane weight. By way of illustration, the airplane-drag curves are sketched for a number of airplane weights W ($W_1 > W_2 > W_3 > W_4$), and the variation of thrust with airspeed for various thrust settings Γ ($\Gamma_1, \Gamma_2, \Gamma_3, \Gamma_4$). At each flight speed and thrust setting, specific fuel consumption, $c_T = c_T(H, \Gamma, V)$, is found from the engine performance data provided by the engine manufacturer.

Next, the associated fuel weight flow rates and specific ranges are obtained from the relationships $F = c_T T$ and $V/F = V/(c_T T)$, respectively.

Knowing these figures, Equations (11.7-2) and (11.7-3) can be evaluated graphically to provide the values of range and endurance for a given cruise technique.

As depicted in Figure 11.9-2a, range follows from plotting V/F against W, and is represented by the shaded area under the curve from the final cruise weight W_2 to the initial cruise weight W_1. Similarly, the endurance follows from the plot of $1/F$

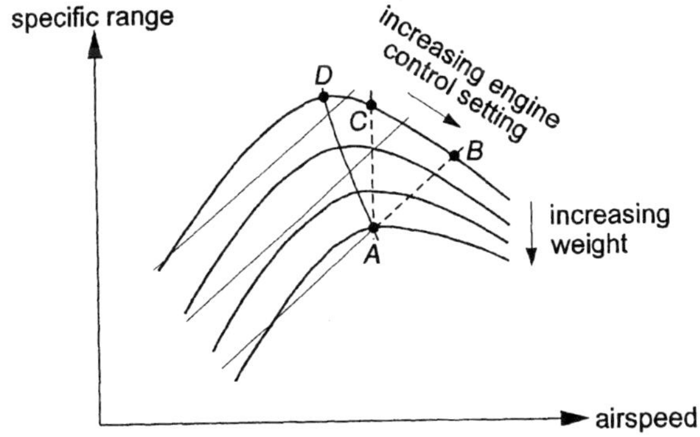

Figure 11.9-3. Typical specific range performance at a given altitude.

against W, as shown in Figure 11.9-2b.

If the preceding calculation procedure is repeated systematically for various cruise techniques, a *cruise chart* can be deduced as exemplified in Figure 11.9-3. This diagram shows the effect of cruise technique by presenting for a given altitude the typical variations of specific range with airspeed, airplane weight, and thrust setting. The constant engine rating program is represented by the line AB, the constant speed program by the line AC, and the maximum range program by the line AD, all for the same initial conditions.

To meet the ATC regulations, an airplane normally cruises at constant speed and pressure altitude (*Flight Level*). In this respect, it may be remarked that, practically, the range achieved in a cruise in which airspeed and altitude are held constant is almost the same as gained by the maximum range program, in which the specific range is maximum for each instantaneous flight condition.

The instantaneous airspeeds which give the greatest distance on a given amount of fuel (line AD), may be called *economic speeds*, V_{ec}.

For a particular cruise program, the relationship between specific range and airplane weight follows directly from the cruise chart and so the required fuel load for a given range.

Cruise charts as sketched in Figure 11.9-3, will be found in the airplane's performance manuals for several typical cruise altitudes.

11.10 TANKERING

Tankering is taking in more fuel than required on the basis of the flight plan. Apart from achieving a rapid turnround, in particular, a much higher fuel price at the destination airport than at the airport of departure will be responsible for the fact that operators may decide to 'tanker' fuel. Obviously, the amount of tankering is restricted to the maximum takeoff and landing weight of the aircraft (see Section 11.13) and, practically, will be limited to the amount of fuel that is required for the return trip. Carrying extra fuel imposes a penalty on the performance of the airplane. For the related fuel consumption and so for the emissions, especially, the

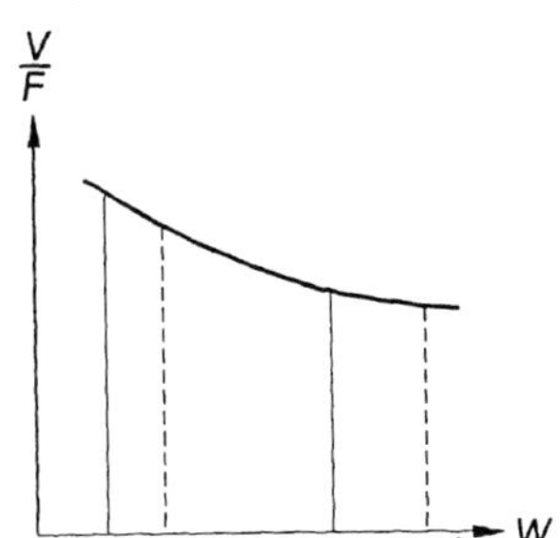

Figure 11.10-1. Tankering and range.

effect of airplane weight on specific range will be of importance.
To explain the latter effect, we turn to Figure 11.10-1, showing the relationship between specific range and airplane weight for a given cruise technique (see also Figure 11.9-2a). To travel a cruise range R, the required fuel weight is $W_1 - W_2$. According to the previous Equation (11.7-2), we have:

$$R = \int_{W_2}^{W_1} \frac{V}{F} \mathrm{d}W \,. \tag{11.10-1}$$

From Section 11.9, we also know that the range R is represented in Figure 11.10-1 by the area under the curve from the final cruise weight W_2 to the initial cruise weight W_1.
If, due to tankering, the initial weight is increased to W_1*, also the final weight after covering the distance R will have a higher value $W_2^* > W_2$, so that now the range is given by:

$$R = \int_{W_2^*}^{W_1^*} \frac{V}{F} \mathrm{d}W = \int_{W_2^*}^{W_1} \frac{V}{F} \mathrm{d}W + \int_{W_1}^{W_1^*} \frac{V}{F} \mathrm{d}W \,. \tag{11.10-2}$$

From Figure 11.10-1 we find that the fuel weight available for tankering is equal to the weight difference: $W_2^* - W_2$.
Evidently, the additional fuel burn due to the higher airplane weight is: $(W_1^* - W_1) - (W_2^* - W_2)$.

Note that the quantity just mentioned has a positive sign because the slope of the curve in Figure 11.10-1 at the lower weights is steeper than at the higher weights. In other words, the weight difference necessary to encompass the same area under the curve is larger at the higher airplane weights than at the lower weights.

Before leaving the subject of tankering, it is worth mentioning the current health problem of human overweight and obesity. With reference to aviation, recent studies reveal that airlines increasingly have to worry about additional passenger weight. Apart from higher operating costs, the resulting enlarged fuel consumption has also an environmental impact as extra combustion products are released into the atmosphere.

11.11 EFFECT OF WIND

Often, the effect of wind, i.e., the horizontal displacement of air relative to the Earth, on range is pronounced. An expression for range with head or tailwind, is the following integral:

$$R_w = \int_{t_1}^{t_2} V_g \, dt = \int_{W_2}^{W_1} \frac{V_g}{F} dW \,, \tag{11.11-1}$$

where R_w is the range with wind, and V_g the speed of the airplane relative to the ground, the so-called *ground speed.*
In steady, level flight the relationship between ground speed V_g, airspeed V, and wind velocity V_w is given by:

$$V_g = V - V_w \,, \tag{11.11-2}$$

where a headwind is taken positive and a tailwind negative.

Substitution of Equation (11.11-2) into Equation (11.11-1) yields with V_w = constant:

$$R_w = \int_{W_2}^{W_1} \frac{V - V_w}{F} dW = \int_{W_2}^{W_1} \frac{V}{F} dW - V_w \int_{W_2}^{W_1} \frac{dW}{F} \,, \tag{11.11-3}$$

or with Equations (11.7-2) and (11.7-3):

$$R_w = R - V_w E \,. \tag{11.11-4}$$

This equation shows that for a given fuel load, the range is decreased by a headwind and increased by a tailwind. Clearly, the actual fuel consumption not only is determined by block distance, but also by the prevailing wind velocities along the flight path.
Inspection of the specific range in Equation (11.11-1), learns us that for maximum range the airplane should fly at that altitude and airspeed at which the ratio V_g / D is the maximum, provided that specific fuel consumption is constant:

$$\frac{V_g}{F} = \frac{V_g}{c_T T} = \frac{V_g}{c_T D} = \frac{1}{c_T} \frac{(V - V_w)}{D} \,. \tag{11.11-5}$$

From Equation (11.11-5) and the performance diagram sketched in Figure 11.11-1, it can be seen that in the presence of wind the economic speed relative to the

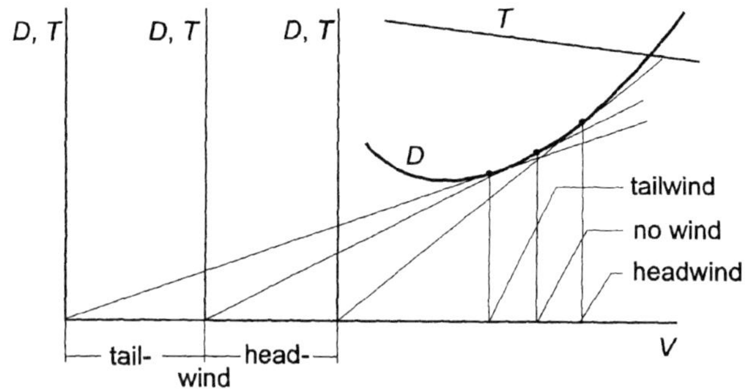

Figure 11.11-1. Effect of steady wind velocity on airspeed for maximum specific range.

ground is found by determining a new origin on the airspeed axis and drawing the tangent to the drag curve. Apparently, the airspeed for maximum specific range with a headwind is greater than in still air conditions. The opposite is true when flying with a tailwind.

11.12 EFFECT OF CLIMB AND DESCENT ON RANGE PERFORMANCE

The block distance includes cruise range plus distances traveled in climb and descent. Accordingly, the block fuel includes the fuel consumed in the cruise plus the fuel required for climb and descent. In addition, fuel is used in taxi, takeoff, maneuvering in the vicinity of the airport, holding and landing. Range studies, therefore, always require the calculation of distance, fuel, and time to climb to and descend from the cruise altitude.

Returning to the typical division of the block fuel in Figure 11.1-2, we recognize again the importance for long-range transport airplanes to minimize the cruise fuel. Nevertheless, the effects of climb and descent on fuel load remain of significance. Lifting the airplane in the climb requires extra fuel, so that the mean value of the specific range in climb, $(V/F)_c$, will be less than the mean value in the cruise, $(V/F)_{cr}$. On the other hand, in descent $(V/F)_d$ may be greater than $(V/F)_{cr}$. However, this only partially can compensate for the effect of the climb, so that the overall effect of climb and descent is a loss of range at a given fuel load, in comparison to the situation that the whole range could be flown under cruise conditions. Obviously, detailed calculations are required to determine distance, fuel, and time in the various flight phases.

Apart from the step by step calculation procedures indicated briefly in Section 11.6, estimations of distance and fuel for climb and descent may be made by the simple approximate method given in Ref. 66, where the concepts of *lost range* and *lost fuel* are introduced.
The lost range is defined as the difference between the range obtained by assuming that all the available fuel is consumed in the cruise and the total range traveled in the real flight. On the other hand, to obtain the true necessary total fuel load for a given total range, the lost fuel must be added to the mass of fuel calculated by assuming cruise conditions for the whole distance. As far as possible, the quantities lost range and lost fuel must include the contributions from all flight phases other than continued climb and descent (see the previous Figure 11.1-1). In this respect, we have to mention the use of a *detour factor*. Detouring involves the deviation from the great circle distance between the airport of departure and of arrival due to airspace routing and ATC requirements. Then, the true distance to fly equals the great circle distance corrected by a detour factor.

The availability of the above-mentioned correction quantities provides estimates of still-air total range and fuel load, based on cruise performance alone to be adjusted for the net loss of energy during the non steady and steady climb and descent. Since, especially, we are interested in the cases where the distances flown in climb and descent are not a substantial part of the total range, the method may be of advantage to studies on fuel conservation for long-range flights.
For further details of the method, the reader may turn to the original text in Ref. 66, where, also expressions for the *lost time* due to climb and descent of airplanes with turbofan engines are derived.

11.13 PAYLOAD-RANGE RELATIONSHIP

Referring to the aircraft's weight build-up discussed in Section 11.3, the takeoff weight is given by:

$$W_{to} = W_b + W_p + W_f + W_{fr} \,, \qquad (11.13\text{-}1)$$

where W_{to} is the takeoff weight (*TOW*) of the aircraft, W_b is the *operational empty weight* (*OEW*), W_p the payload, and W_{fr} the weight of the fuel reserves.
We readily derive an analytic expression for the fuel weight fraction, W_f/W_{to}, when we use Equation (11.7-7), which describes a cruise at constant airspeed and angle of attack. Assuming that the entire distance is performed in cruise, we find with $W_1 = W_{to}$ and $W_2 = W_{to} - W_f$:

$$\frac{W_f}{W_{to}} = 1 - e^{-\frac{R}{\frac{V}{c_T}\frac{C_L}{C_D}}} = 1 - e^{-\frac{R}{\eta_{tot}\frac{H}{g_0}\frac{C_L}{C_D}}}, \tag{11.13-2}$$

where R is the total range.

Equation (11.13-2) indicates that the aim for a minimum fuel weight fraction, obviously, is to optimize the range factor: $\frac{V}{c_T}\frac{C_L}{C_D} = \eta_{tot}\frac{H}{g_0}\frac{C_L}{C_D}$.

The weight fraction W_b/W_{to} in Equation (11.13-1) can be regarded as the structural efficiency since the lighter the airplane is designed and built, the greater is the *useful load fraction*, $W_u/W_{to} = W_p/W_{to} + W_f/W_{to}$.

At a given useful load fraction, obviously, a greater payload fraction is obtained as less fuel is consumed for a given total range.

We also conclude from Equation (11.13-2) that at a given range factor, a larger fuel weight fraction is required as the total range increases. Consequently, for a given airplane, the payload fraction will be dependent on total range.

In Figure 11.13-1a is sketched, the weight breakdown with respect to total range in still air for a typical transport airplane.

The line *AB* in Figure 11.13-1a gives the takeoff weight (TOW) at the maximum payload that can be transported. The line shows that as range increases, the fuel load must be larger. This can continue until the takeoff weight reaches its maximum value at point *B*. The range at maximum takeoff weight (MTOW) is called *maximum payload range.* To fly farther than this range requires that payload is replaced by fuel. This is represented by the line *CD*. Point *D* represents

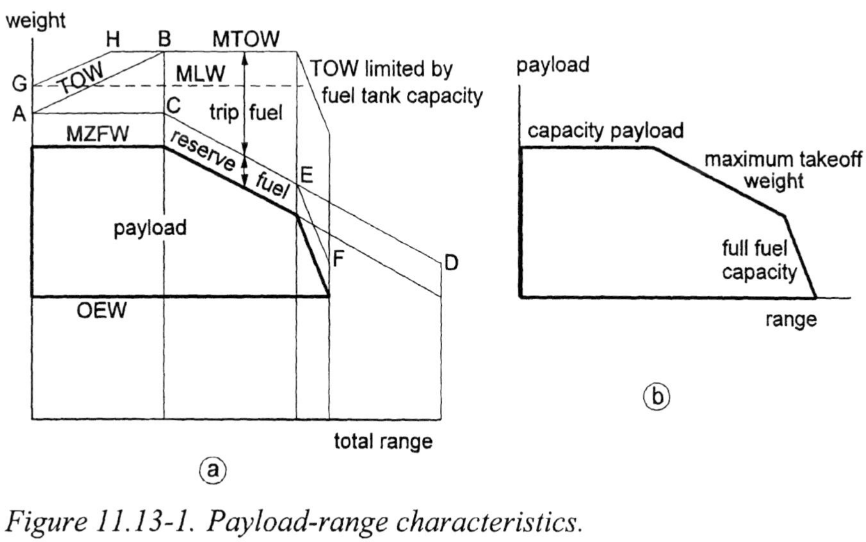

Figure 11.13-1. Payload-range characteristics.

the ultimate range, at which no payload can be carried (unconsumed reserve fuel). Normally, the fuel tank capacity is such that the range cannot be increased beyond point *E*. The line *EF*, finally, indicates that some extra increase of range is possible by reducing the takeoff weight through decreasing further the payload. However, transport airplanes seldom operate in this uneconomical mode.

Since the payload-range relationship in Figure 11.13-1a may be regarded as the basis of the economic value of commercial transports, their payload-range characteristics, usually, are shown separately (Figure 11.13-1b).

For the sake of completeness, in Figure 11.13-1a also the line *GH* is drawn, which represents a limit to the takeoff weight that on short ranges may be dictated by the maximum allowable weight of the airplane at the landing. The *maximum landing weight* (MLW) is imposed by structural design requirements.

Another weight limit of importance is the *maximum zero fuel weight* (MZFW), the maximum allowable weight of the airplane without fuel. Normally, the maximum landing weight is greater than the maximum zero fuel weight plus the weight of the reserve fuel.

11.14 ECONOMIC PERFORMANCE PARAMETERS

In this section we summarize some valuable parameters describing the commercial merit of transport airplanes. The economic performance items considered here are:

- block distance, R,
- block time, E_B,
- block speed, V_B,
- transport product, P_r,
- efficiency of transportation, P_t,
- transport productivity, P_h,
- revenue-earning capacity, P_y.

The *block distance*, total range or stage length is, as we already know, the horizontal distance traveled in climb, cruise, and descent.

The *block time* is the time elapsing from starting engines at the departure airport to engines off at the destination airport. Thus, the block time includes taxi time from the loading point to the takeoff runway, checks, takeoff, initial and continued climb, cruise, descent, holding, approach flight, landing, and taxi time to the unloading point.

The *block speed* is the block distance divided by the block time:

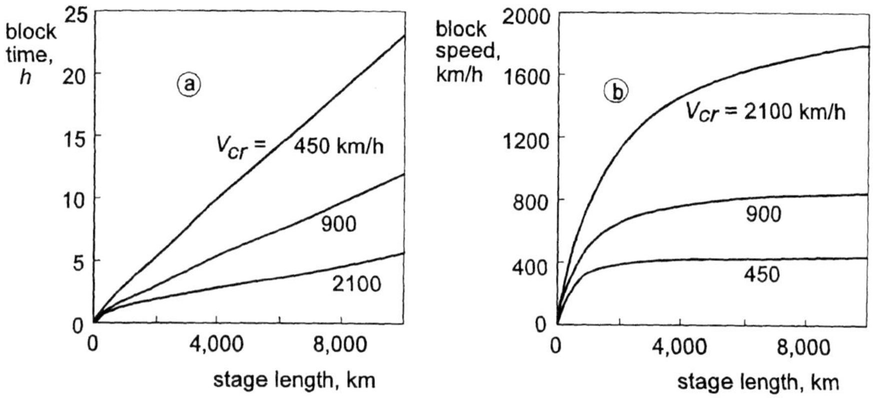

Figure 11.14-1. Block time and block speed versus block distance.

$$V_B = \frac{R}{E_B} \; . \tag{11.14-1}$$

Of course, the block speed is lower than the cruise speed, V_{cr}. The relationship between block time, block distance and cruise speed can be written as (Ref. 8):

$$E_B = \frac{R}{V_{cr}} + \Delta t \; , \tag{11.14-2}$$

where Δt is the time that accounts for the field operations and the lower airspeeds in flight phases other than the cruise.
Combining Equations (11.14-1) and (11.14-2) delivers the following expression for the block speed:

$$V_B = \frac{R}{\dfrac{R}{V_{cr}} + \Delta t} \; . \tag{11.14-3}$$

For a number of cruise speeds, Figures 11.14-1a and b illustrate the dependence of block time and block speed on block distance, using $\Delta t = 50$ minutes.
The graphs show that at a given cruise speed, block time and block speed increase with increasing block distance, and that raising the cruise speed is especially beneficial as the block distance is greater.
The revenues gained by transport airplanes are given by the transport product, that is, the product of payload and distance:
where P_r may be expressed in the units *ton.km* or *passenger.km*.

$$P_r = W_p R \; , \tag{11.14-4}$$

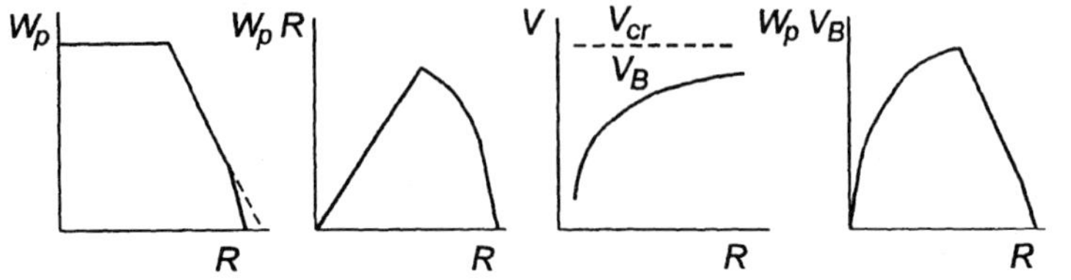

Figure 11.14-2. Economic parameters for commercial aviation.

Important to the transport product is the *load factor*, which is defined as the ratio between the transported payload and the maximum payload.
Related to the transport product is the *efficiency of transportation*, P_t, which is the fuel required to transport one passenger over a distance of 1 km, or the energy per passenger.km.
The *transport productivity* is defined as the transport product delivered per unit block time:

$$P_h = \frac{W_p R}{E_B} = W_p V_B . \tag{11.14-5}$$

Based on the typical variations of payload and block speed with range, the quantity's P_r and P_h are portrayed in Figure 11.14-2. Note that both the transport product and the transport productivity peak at the design range (see Section 11.13).
Finally, the *revenue-earning capacity* is the transport product per year. If U is the annual flight utilization of the airplane in hours, we obtain:

$$P_y = P_h U = W_p V_b U . \tag{11.14-6}$$

For intercontinental operations, U may reach a value of 4500 hours. This figure gradually decreases as the block distance becomes shorter.

11.15 DIRECT OPERATING COST

Naturally, the costs of operating a commercial airplane are of vital importance to its economic value. The operating costs are conveniently broken down into indirect and direct costs.
The *indirect operating cost* (IOC) are independent of the characteristics of the airplane since they are associated with the costs of operating an airline. They may include training, management, administration, sales, housing, and depreciation of ground properties and equipment.

The *direct operating cost* (DOC) are those cost which are associated with flight operations and direct maintenance. These may comprise the costs of fuel and oil, crew, landing fees, depreciation of capital invested, insurance, interest, and maintenance. On long-range flights the fuel price may account for some 30% of the direct operating cost (Ref. 63).
However, when considering the economic value of an airplane, the direct operating costs are of primary interest since these are directly related to the operational characteristics of the airplane.
Besides fuel and oil, the above-mentioned expenses can be considered to be time costs as they are directly dependent on the time that the airplane is airborne.

Each airline uses its own method for estimating operating costs related to their particular circumstances and requirements. However, for aircraft manufacturers and also for research activities it is necessary to employ a standardized method for assessing operating costs. To this end, the method published by the Association of European Airlines (AEA) in 1989 may serve as a guideline (Ref. 64).

For our cost analysis, we divide the direct operating cost into costs of time and fuel. Then we can write:

$$\mathrm{DOC} = CT \times E + CF \times W_f \,, \tag{11.15-1}$$

where CT designates the cost of time in \$/h and CF the cost of fuel in \$/N, E is the endurance, and W_f the weight of fuel consumed for a given range.
A criterion employed to express the relative costs of time and fuel is the *cost index*, *CI,* defined as the ratio of time cost to fuel cost (Ref. 58):

$$CI = \frac{CT}{CF} \,, \tag{11.15-2}$$

where CI is expressed in (\$/h)/(\$/N) = N/h, which are units of fuel weight flow rate. Insertion of the cost index (11.15-2) into Equation (11.15-1), enables us to reduce time cost and fuel cost to an equivalent fuel weight, $W_{f,eq}$, which can be used as a measure for the total cost:

$$W_{f,eq} = CI \times E + W_f \,. \tag{11.15-3}$$

Since DOC = $W_{f,eq} \times CF$, the weight $W_{f,eq}$, in fact, is the fuel weight that could be purchased for the direct operating cost, DOC.
In order to investigate the minimum direct operating cost for a given range, the economic specific range, SER, is introduced, which parameter is defined as the ratio of the rate of change of distance to the rate of change of equivalent fuel weight (Ref. 58).
In still air, the rate of change of distance is the airspeed V, and the rate of change

of equivalent fuel weight follow from Equation (11.15-3):

$$\frac{\mathrm{d}W_{f,eq}}{\mathrm{d}t} = CI + F \ , \tag{11.15-4}$$

where $F = \dfrac{\mathrm{d}W_f}{\mathrm{d}t}$ is fuel weight flow rate in N/h.

This results in the following expression for the economic specific range:

$$\mathrm{SER} = \frac{V}{CI + F} \ , \tag{11.15-5}$$

where the economic specific range has the dimensions km/N.

Similar to the condition for minimum fuel, which is to say, maximum specific range, $(V/F)_{max}$, the condition for minimum DOC in cruise flight is that SER should be maximum for each instantaneous flight condition. Then the equivalent fuel weight is minimized, and therefore the total cost for the distance flown.

11.16 POLLUTION NUMBER

From the previous Equations (11.7-1) and (11.7-2), it follows that the fuel required to travel a particular range *R,* is given by:

$$W_{f_{cr}} = W_1 - W_2 = \int_0^R \frac{1}{V/F}\,\mathrm{d}R \ , \tag{11.16-1}$$

where the suffix "*cr*" designates cruise.

The emitted amount of pollutant during the flight is determined by:

$$Q = \int_0^{W_{f_{cr}}} EI\,\mathrm{d}W_f = \int_{W_2}^{W_1} EI\,\mathrm{d}W \ . \tag{11.16-2}$$

Combination of Equations (11.16-1) and (11.16-2) furnishes:

$$Q = \int_0^R \frac{EI}{V/F}\,\mathrm{d}R \ . \tag{11.16-3}$$

This equation indicates that for a given distance *R*, the amount of emitted substance will be minimum if the quantity *EI/(V/F)* is minimum at each instant during the flight. This requirement certainly forces engine manufacturers to develop fuel-efficient combustors, that produce only low amounts of pollutant substances.

However, as remarked above, a high fuel efficiency resulting in a high specific range, may nevertheless lead to a high amount of pollutant emission, due to a high emission index. In other words, the solution of the problem is not as simple as it might appear, as the emission index for nitrogen oxides is, among others, a function of flame temperature and rapidity of combustion. For example, the lower the temperature and the greater the rapidity, the less NO_X is produced, whereas slowing combustion time and increasing flame temperature reduces fuel consumption and so carbon dioxide and water vapor emissions, but increases NO_X production (see also Chapter 10). Clearly, improvements in one area may have a negative influence in another.

For any pollutant substance, the parameter $[(EI)/(V/F)]$ in Equation (11.16-3) may be called *pollution number*:

$$P = \frac{EI}{V/F} , \qquad (11.16\text{-}4)$$

where P is expressed in the units g / km (Ref. 68).

E.g., for the emission of NO_X we can write:

$$P_{NO_x} = \frac{EI_{NO_x}}{V/F} , \qquad (11.16\text{-}5)$$

and

$$Q_{NO_x} = \int_0^R \frac{EI_{NO_x}}{V/F} \mathrm{d}R = \int_{W_2}^{W_1} \frac{EI_{NO_x}}{V/F} \frac{V}{F} \mathrm{d}W = \int_{W_2}^{W_1} EI_{NO_x} \mathrm{d}W . \qquad (11.16\text{-}6)$$

Apparently, *NO_x-emission efficiency* may be described by the amount of pollutant produced per passenger.km. It incorporates emission index, fuel consumption, and productivity.

12 OPERATION AND DESIGN FOR LOW EMISSIONS

12.1 HIGH ALTITUDE EMISSIONS FROM TRANSPORT AIRCRAFT

Generally speaking, measured data of pollutant emissions from flight operations are only available for the ICAO-LTO cycle. This simulated flight procedure controls the level of atmospheric pollution occurring at and in the vicinity of airports (see Chapter 10, Section 10.5). LTO-cycle data are based on static engine tests at ground level yielding for particular thrust settings, which correspond to those used in the various segments of the flight profile, the emission indices of the species carbon monoxide CO, unburned hydrocarbons UHC, nitrogen oxides NO_X, and smoke (soot). The last substance is expressed in terms of smoke number. The quantification of the level of pollutants is obtained by adopting an average duration in each of the segments in the LTO cycle (see the previous Table 10.5-2).

By way of illustration, for the airplane type Boeing 747-400, the fuel mass flow rate and the emission index values in each LTO-cycle segment are given in Table 12.1-1(cf. Table 10.2-1). A three-view diagram of this typical wide-body airplane powered by turbofan engines is shown in Figure 12.1-1, while some characteristic data of the airplane are given in Table 12.1-2.

In Table 12.1-3 are stated the obtained fuel burn and emission quantities in the various flight segments. The results demonstrate the large difference between the gas emissions in flight and those during ground maneuvering.

During taxiing the engines operate at a very low thrust setting with imperfect combustion, through which large amounts of CO and UHC are released from the engine exhaust.

flight phase	duration, minute	fuel mass flow rate, kg/s	emission index, g/kg of fuel		
			UHC	CO	NO_X
takeoff	0.7	2.342	0.06	0.4	28.0
climb out	2.2	1.930	0.01	0.6	22.9
approach	4.0	0.658	0.13	2.0	11.6
taxi	26.0	0.208	1.92	21.9	4.8

Table 12.1-1. LTO-cycle data for Boeing 747-400 airplane (Ref. 63).

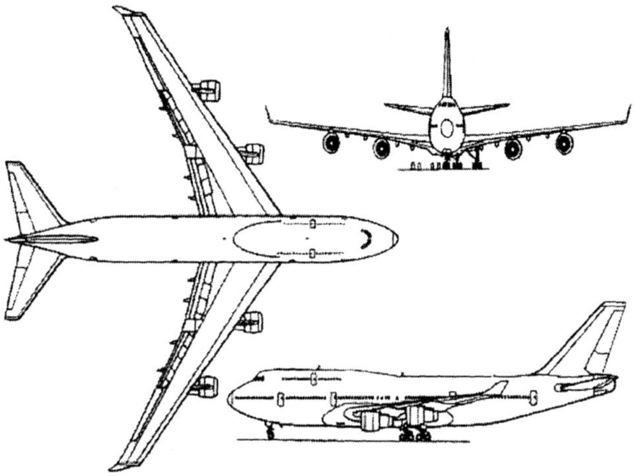

Figure 12.1-1. Three-view diagram of Boeing 747-400 airplane (Jane's All the World's Aircraft, 1989).

On the other hand, when the airplane is airborne, high thrust settings are selected. This is coupled with high turbine entry temperatures, leading to the generation of relative large quantities of NO_x.
With regard to the LTO-cycle emissions, we find that since the introduction of these requirements in 1981 (Ref. 6), the amounts of smoke, unburned hydrocarbons, and carbon monoxide emitted near ground level have been greatly reduced. As a departure from this development, the emission of NO_x appears to be increased (Ref.18). The reasons for this exception are the high

year of introduction	1989
wing loading	7151 N/m^2
wingspan	64.92 m
wing area	541.16 m^2
maximum thrust	4×258 kN
cruise Mach number	0.85
maximum takeoff weight	3870 kN
operating empty weight	1792 kN
fuel capacity	1603 kN
range	13157 km
initial cruise altitude	10030 m
takeoff field length	3352 m
landing field length	2072 m
maximum lift-drag ratio	18

Table 12.1-2. Characteristic data of Boeing 747-400 airplane.

flight phase	fuel burn, kg	emission constituent CO_2, kg	H_2O, kg	UHC, g	CO, g	NO_X, g
takeoff	98.36	309.83	122.95	5.90	39.34	2754.08
climb out	254.76	802.49	318.45	2.55	152.86	5834.00
approach	157.92	497.45	197.40	20.53	315.84	1831.89
taxi	324.48	1022.11	405.60	623.00	7106.11	1557.50
total	835.52	2631.88	1044.40	651.98	7614.15	11977.45

Table 12.1-3: Boeing 747-400 fuel consumption and gas emission during LTO cycle.

overall pressure ratios and the high combustion temperatures of the modern high-bypass ratio turbofan engines.

Given the anticipated growth of air traffic, the high-altitude gas emissions from aviation will play an important role in future tropospheric ozone production and stratospheric ozone depletion. Table 12.1-4, taken from Ref. 29, presents in this respect the percentage distribution of the pollutants over the various flight phases of a medium-haul airliner performing a stage length of 500 nautical miles. Indicated are the four most significant pollutant substances, CO, UHC, NO_X, and soot. The unit of distance nautical mile (nm) is equal to 1.852 km. The predictions agree with the results given in Table 12.1-3 that NO_X, by far, contributes most to the total emission of undesired pollutants. For the landing and takeoff cycle, 77.4% of these emissions consist of NO_X, while the contribution from the climb, cruise and descent is even 89.0%. For the whole flight, the NO_X emissions comprise 85.9% of the total production of pollutants.

constituent	takeoff and landing	climb, cruise, and descent	total flight
CO	5.4	7.0	12.4
NO_X	20.6	65.3	85.9
UHC	0.6	1.0	1.6
soot	-	0.1	0.1
total	26.6% (77.4% NO_X)	73.4% (89.0% NO_X)	100.0% (85.9% NO_X)

Table 12.1-4: Percentage distribution of emissions per flight phase for a flight distance of 500 nautical mile.

Obviously, if a reduction of the emissions would be desirable, NO_x is the most appropriate constituent to suppress.

A similar study on the emission of NO_x for a flight between London and Tokyo with a Boeing 747-400 airplane is reported in Ref. 34. Considered is a *cruise climb* in International Standard Atmosphere (ISA), with an initial cruise altitude of 9845 m and at a flight Mach number of $M_0 = 0.85$.
Results of the investigation are specified in Table 12.1-5, indicating that on this long-haul flight the airplane produces 1321 kg of NO_x, of which around 85% is emitted during the cruise. The NO_x emission at higher altitudes even constitutes about 97% of the total production. These numbers demonstrate again that for a reduction in NO_x emission the cruise will be the most relevant flight phase to achieve this.
In Table 12.1-5, the amounts of NO_x emitted during climb, cruise, and descent, are determined according to the previous Equation (10.6-4). The other NO_x emissions are taken from LTO-cycle data.
Correctly, it is argued in Ref. 34 that for the mode of operation in question the flight altitude increases when the airplane becomes lighter through the consumption of fuel, through which airspeed and thrust required decrease (see the equations in Chapter 11, Section 11.7). This leads to a reduced fuel flow rate into the combustors, which is responsible for a proportionate reduction in NO_x generation. At the same time, lower combustion temperatures and thus lower NO_x reaction rates occur, resulting in lower NO_x-emission indices. This beneficial dual effect originates from the possibility for the airplane to increase its altitude continuously as fuel is burned off during the cruise. Unfortunately, the execution of a cruise climb may at variance with spacing requirements from current air traffic control (ATC). With the view of the reduced NO_x emissions,

flight phase	NO_x emissions	
	kg	% total
takeoff	15	1.1
climb out to 457 m (1500 ft)	15	1.1
continued climb (457 m to 9500 m)	158	12.0
cruise climb (9.5 km to 11.8 km)	1113	84.3
descent	5	0.4
approach	7	0.5
taxi	8	0.6
total	1321	100.0

Table 12.1-5: NO_x emissions per flight phase for B 747-400 airplane on the London - Tokyo route (Ref. 34).

it seems important that ATC does not prevent air traffic to increase the altitude during the cruise.
Regrettably, such a flight profile is rarely allowed by the controlling authorities. As discussed already in Chapter 11, aircraft sometimes receive permission to change the cruise altitude by a fixed increment. This mode of operation is the *step climb*, as already depicted in the previous Figure 11.8-3.
Ref. 34 also reports that optimizing the initial cruise altitude and cruise Mach number, can lead to a reduction of NO_X emissions into the atmosphere of up to 10% (see also Section 12.2). However, although more fuel efficient, optimized cruise techniques with lower cruise Mach numbers can penalize flight time and associated direct operating cost, and also may influence passenger appeal.

Certainly, the methods used for the prediction of cruise-level emissions need validation by measurement. Such validation can be carried out by testing engines in chambers in which the same rarefied high-altitude conditions prevail as occurring in the cruise. Another possibility is the execution of in-flight measurements in the exhaust plume of the airplane (Ref.1). Also information may be derived from LTO-cycle emissions, provided that available ground test data can be used for predicting altitude-emissions.

As remarked already in Chapter 10, Section 10.5, from the above statements there is reason to expect that also for flight operations near the tropopause, the introduction of supplemental legislation may be anticipated. This may have considerable influences on airplane and engine design, and airplane operation. Especially, the increasing awareness of the harmful effects of nitrogen oxides, and the fact that the emission index of this substance can be affected, has led to the determination to achieve reduced NO_X-emission levels.
It was pointed out in Chapter 10 that currently much effort is put into new combustor technology for low-NO_X emissions. Obviously, another approach to the problem is investigating the potentials for reducing engine emissions through changes in the way civil airplanes are operated. In this connection, it is important to notice, as may be seen from the previous Figure 11.8-2, that due to the variation of the tropopause level with latitude, it will be less simple to keep away from the stratosphere when flying over the Poles than over the Equator. Furthermore, it is of importance to recognize that the potential effects of NO_X and H_2O emissions not only vary with the altitude, but also are dependent on latitude and time of year (see also Section 12.5).
Several theoretical studies have been undertaken on the effect of the cruise technique on pollutant emissions, by performing so-called *low-emission flight paths*. See, for example, Ref. 68, of which some results will be reviewed in the next section.

12.2 EFFECT OF CRUISE CONDITIONS ON NO_x EMISSIONS

In Ref. 68 a *Cruise NO_x Simulation Model* has been developed. It is a computer simulation by which the effects of varying cruise conditions on the emission of nitrogen oxides and the accompanying effects on fuel consumption and direct operating cost (DOC) can be predicted.

The results presented in this section are based on the airplane type Boeing 747-400, powered by a fictitious turbofan engine model, using state-of-the-art engine technology. Varying cruise conditions includes variation of flight altitude, cruise Mach number or airspeed, and ambient conditions.

The calculation procedure accomplishes the determination of the relationship between atmospheric conditions, airplane-aerodynamic characteristics, engine performance, and NO_x formation.

Firstly, two point performance capabilities in ISA are presented, namely, specific air range V/F and specific economic range $SER=V/(CI+F)$, where F is the fuel flow rate and CI is the cost index (see Chapter 11, Section 11.15). The matching instantaneous flight conditions are defined by geopotential altitude H, flight Mach number M, and airplane weight W. The emission indices EI_{NOx} are calculated according to the previous Equation (10.6-2).

Figure 12.2-1shows the effects of the flight Mach number and cruise altitude on specific range. Obviously, at a given weight (W = 2800 kN) an optimum Mach number as well as an optimum cruise altitude occurs for maximum specific range, $(V/F)_{max}$.

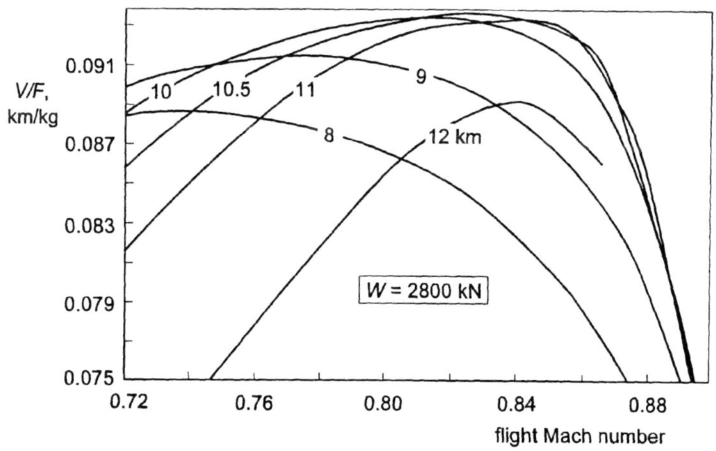

Figure 12.2-1. Specific range as functions of Mach number and altitude.

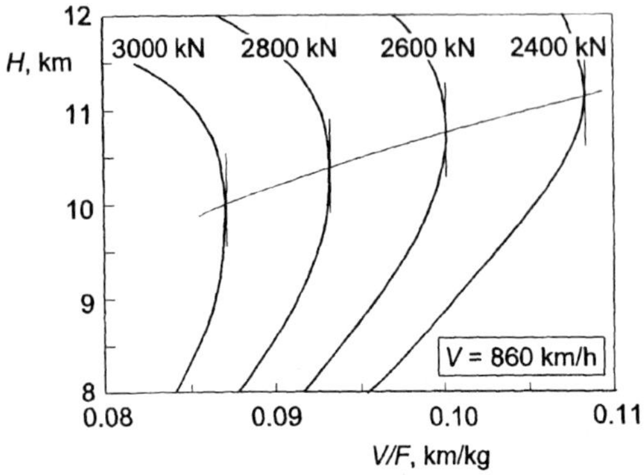

Figure 12.2-2. Specific range as functions of altitude and weight.

Figure 12.2-2 indicates that at a given cruise speed (V = 860 km/h), the optimum altitude for maximum specific range increases with decreasing weight. Hence, when the cruise with minimum fuel consumption is aimed for, the airplane should ascend in altitude during the course of the flight.

Figure 12.2-3 displays the effects of Mach number and cruise altitude on specific economic range (W = 2800 kN). The quantity $(SER)_{max} = [(V / (CI + F)]_{max}$, defines the condition for maximum specific economic range. Used is a cost index of $CI = CT/CF$ = 3220 kg/h, where the time cost CT = 1288 $/h and the fuel cost CF = 40 $ ct/kg fuel.

Notice that, different from the definitions given in Chapter 11, Section 11.15, the various quantities are expressed in mass instead of weight.

If a cruise for minimum cost is aimed for, then similar to minimum fuel consumption, the airplane should ascend in altitude. The graph also emphasizes

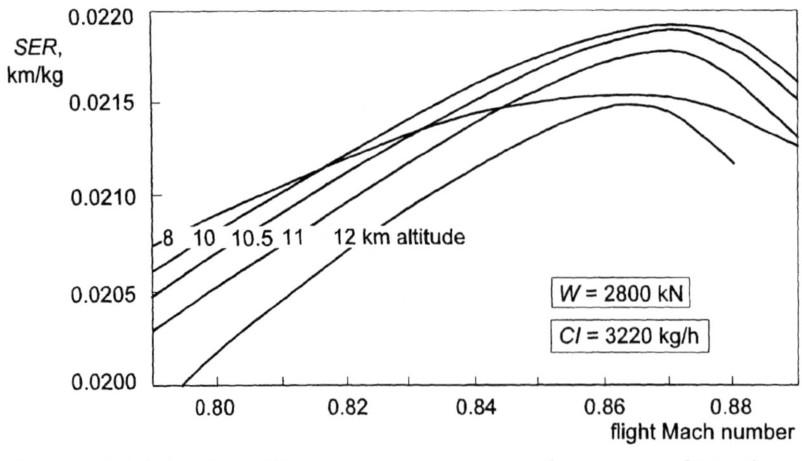

Figure 12.2-3. Specific economic range as functions of Mach number and altitude.

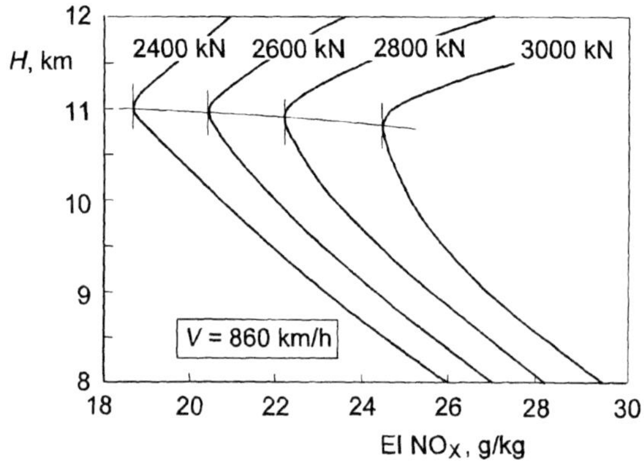

Figure 12.2-4. NO_x-emission index as functions of altitude and weight.

the importance of both a high flight Mach number and a high cruise altitude for an economic operation of the airplane.

In Figure 12.2-4 are presented, the effects of cruise altitude and airplane weight on NO_x-emission index for given airspeed (V = 860 km/h). The plot reveals that at a given weight and airspeed, an optimum cruise level occurs for minimum emission index, $(EI_{NOx})_{min}$. The curves also manifest that when at a given airspeed and altitude, weight is reduced, a lower NO_x production is obtained due to the lower thrust setting required. Clearly, empty weight, mission fuel, and excess fuel, are challenging parameters for aircraft designers and airline operators, since restricting them will have a direct effect on fuel consumption as well as on NO_x-emission.

Another observation from Figure 12.2-4 is that the optimum altitude at which $(EI_{NOx})_{min}$ occurs, is almost independent of airplane weight.

Figure 12.2-5 shows that when at a given airplane weight the higher is the

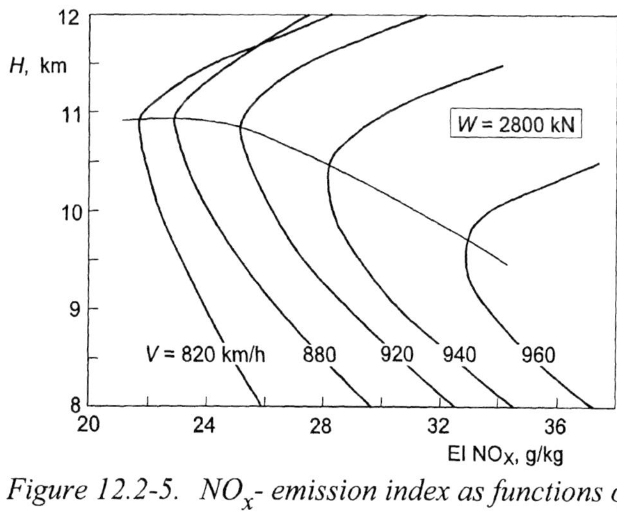

Figure 12.2-5. NO_x- emission index as functions of altitude and airspeed.

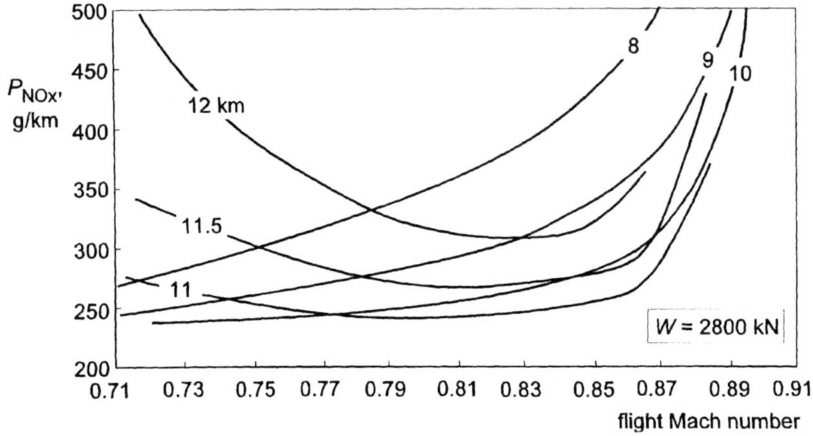

Figure 12.2-6. NO_x- pollution number as functions of Mach number and altitude.

airspeed, the higher is the NO_x production and the lower the optimum cruise altitude.

In accordance with the definition of the pollution number, the effect of varying cruise conditions on P_{NOx} is the combined effect of NO_x-emission index and specific range: $P_{NOx} = EI_{NOx} /(V/F)$ (see Chapter 11, Section 11.16).

From the two influences the curves shown in Figure 12.2-6 result. It can be seen that at an altitude of approximately 11 km, a minimum for P_{NOx} occurs. Apparently, hardly any variation of $(P_{NOx})_{min}$ with Mach number is found. It appears that the altitude yielding the lowest pollution number is virtually independent of cruise Mach number. Beyond a Mach number of 0.86, the pollution number increases sharply.

When a cruise with minimum emission of NO_x is aimed for, P_{NOx} should be minimum for each momentary airplane weight. This can be achieved by choosing the optimum altitudes for minimum P_{NOx}.

The resulting effects of the flight Mach number on range performance for the low-NO_x flight, are shown in Figure 12.2-7. The required fuel, DOC, and emitted NO_x, are presented for an example cruise range of 5800 km, which is approximately equal to the distance Amsterdam-New York. The cruise is taken as a continuous succession of quasi-steady state motions. The NO_x reduction is determined by considering the cruise part of the flight only and assuming that the airplane is allowed by ATC to perform a cruise climb. Also, the additional fuel consumption, NO_x emission, and cost change due to an extended climb to a higher initial cruise altitude, are neglected.

In Figure 12.2-7 the NO_x reductions, and the associated changes in cost and fuel are given in terms of percentage. The diagram shows that NO_x reductions

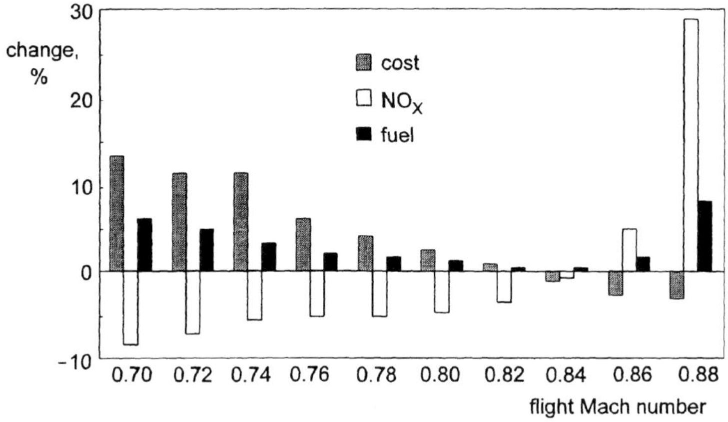

Figure 12.2-7. Effect of flight Mach number on changes in NO_x, fuel, and cost for a low-NO_x flight of 5800 km.

up to almost 10% are achievable as the cruise Mach number is lower. However, this measure is accompanied by a considerable cost increase and also penalizes the flight time. Obviously, a reduction in NO_X is also accompanied by an increase in fuel consumption, which leads to an increase in CO_2 and H_2O emissions. The harmful effect of these pollutant species should be carefully weighted against the effects of NO_X upon the atmosphere.

In conclusion, it can be said that for the reduction of NO_X emissions into the upper troposphere, the cruise performance of airplanes deserves constant attention. However, it seems that the potential benefits in combination with an improved exploration of the existing potential of air traffic control (ATC) will never fully offset the effects of increased emissions resulting from the expected growth in air traffic.

12.3 FUEL CONSERVATION

As mentioned already in Chapter 1, Section 1.6, the global annual consumption of aviation fuel (kerosene) is only a small proportion of the world's energy consumption. Accordingly, also the emissions from aircraft will represent a small fraction of total global emissions produced by the combustion of fossil fuels.
However, due to the growth in air transport, the annual aviation fuel consumption is expected to increase about 3% per year, in the period to 2015, provided that improved technology is applied. Owing to this development,

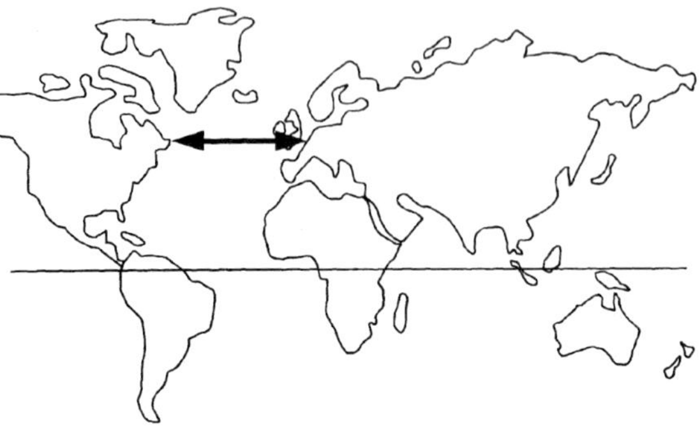

Figure 12.3-1. North Atlantic flight corridor.

inevitably, airplane engine emissions will significantly increase, which could possibly change the relative contribution of aviation to the total air pollution.
Of importance to the problem is the fact that air traffic mainly is executed on the northern hemisphere. The arrows in Figure 12.3-1 represent the North Atlantic flight corridor, along most long-haul air transport takes place. Moreover, it is forecasted that the number of flights on this route will increase further in the future. Another thing is that 30 to 50 percent of these flights take place above the tropopause (Ref. 32). Therefore, as stated before in Chapter 5, Section 5.5, the radiative imbalance from aviation is certainly not globally distributed evenly, strengthening the requirement for reducing the level of pollutants from aero-engine emissions.

As pointed out in Chapter 11, a first approach to the problem deals with the optimization of the efficiency of air transportation, defined as the distance that a given payload can be transported while burning a given mass of fuel. Therefore, any change in the design and way of operation of airplanes, which leads to a reduction of the fuel needed for a specified flight distance is always desirable. The reduction of fuel consumption in itself is obviously beneficial in terms of the conservation of resources but is also highly significant in attempting to reduce the emissions of carbon dioxide and water vapor, which are the natural and inevitable results of burning fossil fuels. Apparently, fuel conservation by engine design has the potential for affecting more than one aspect of the problem.
It will be clear that, especially, specific fuel consumption (SFC) has a considerable influence on block fuel and so on the takeoff weight of the airplane, and is therefore of great importance to the reduction of fuel and gas emissions. Moreover, block fuel is a significant cost factor in the operation of airplanes and may affect the payload capability (see Chapter 11, Section 11.12).

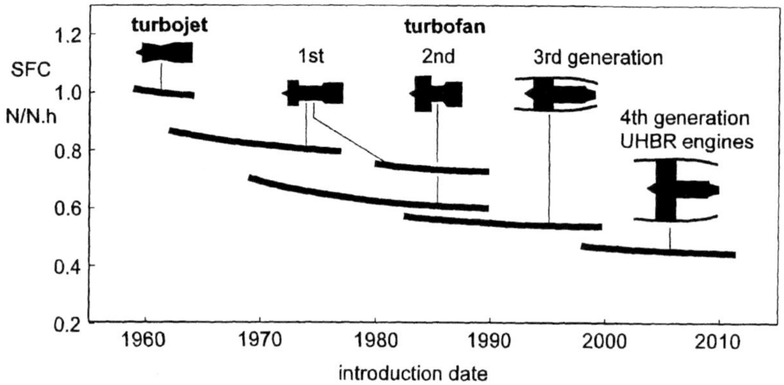

Figure 12.3-2. Development history of specific fuel consumption of engines for commercial aircraft in cruise.

In the light of this statement, it is not surprising that fuel conservation has always been important in aviation. This can be seen from Figure 12.3-2, where the effect of aero-engine technology on the development of specific fuel consumption over the past forty-five years is portrayed. The most fuel-efficient propulsion systems for today's transport airplanes are turbofan engines with bypass ratios up to about nine. These so-called ultra high bypass ratio (UHBR) engines employ high pressure ratios in combination with high turbine entry temperatures (TET) through the use of improved hot section materials and blade cooling. Plans for the introduction of engines with a bypass ratio up to eleven or more have been announced. Unfortunately, there are also concealed difficulties attached to these engine types in the form of extra weight and nacelle drag, which are becoming more significant factors as the bypass ratio goes up. Furthermore, ultra high bypass ratios not only require enlarged fan diameters, but also a gear box between power turbine and fan. Engines with bypass ratios above fifteen are likely to require the removal of the duct to limit weight and drag. Unducted fan (UDF) engines, using small-diameter propellers with blades

Figure 12.3-3. Experimental propeller.

shaped like scimitars (Figure 12.3-3), provide bypass ratios of more than thirty. This engine type has been flight tested and offers significant improvements in propulsive efficiency (see Ref. 69). However, interior and exterior noise problems appertaining to these so-called *propfan engines* seem to remain largely unsolved. Thus, as frequently occurring, improvements in one area may have a negative influence in another.

From previous discussions we also know that consistently with their improved fuel efficiency, the modern turbofan engine types show high combustion temperatures and pressures, through which their NO_x-formation rates at takeoff and at cruise conditions even have been increased. Figures 12.3-4a and b give an impression of the increase with time of the overall pressure ratio and the temperatures at the turbine entry occurring at takeoff thrust setting.
At this point, it may be noted that the thermal efficiency of *internal combustion engines* increases with increasing compression ratio ε (see the previous Equation (9.3-5)). If ε is too high, "knocking" occurs due to irregular burning and detonation. It is avoided by burning fuel of high octane number, and using antiknock additives. In order to obtain better specific fuel consumption and lower emissions of toxic or acidic species, future engines may run at lean fuel mixtures of about $\Phi \approx 0.65$, instead of burning a stoichiometric mixture ($\Phi = 1$).

An approach to the reduction of the exhaust of NO_x concerns the chemistry of the combustion process. This possibility, which involves the improvement of

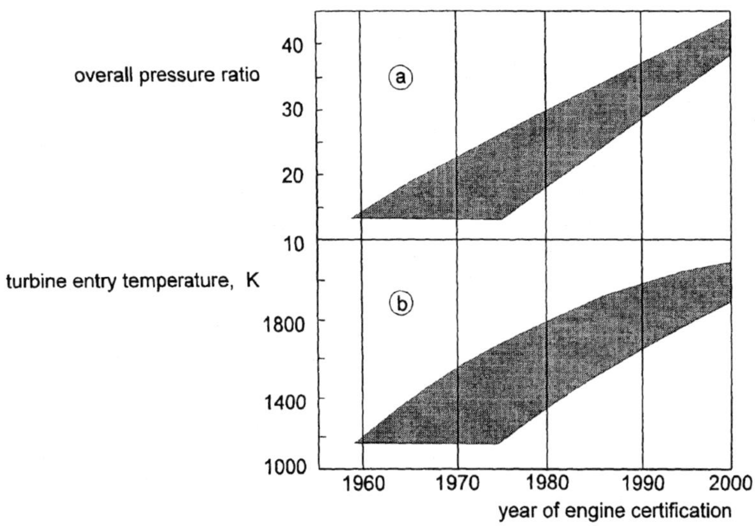

Figure 12.3-4. Overall pressure ratio and turbine entry temperature trends versus time (Ref. 1).

the technology of current combustion chamber design, has been discussed briefly in Chapter 10, Section 10.7.

Also progress in aerodynamics may play an important role in influencing fuel consumption by measures which are related to the drag of the airplane.
The various design parameters of airplanes and aero-engines affecting block fuel, may become apparent by considering the previous Equation (11.13-2). This expression, repeated below for convenience, governs the fuel weight fraction for a flight at constant airspeed and angle of attack, assuming that the entire distance is performed in a cruise:

$$\frac{W_f}{W_{to}} = 1 - e^{-\frac{R}{\frac{V}{c_T}(C_L/C_D)}} = 1 - e^{-\frac{R}{\eta_{tot}\frac{H}{g}(C_L/C_D)}} , \qquad (12.3\text{-}1)$$

where R is the total range and $\frac{V}{c_T}(C_L/C_D)$ is the range factor.

From Equation (12.3-1) we can conclude that at the selected cruise technique (V and α being constants), the fuel load required for traveling a specified distance depends on the following main factors: (1) selection of airspeed, (2) engine control setting, (3) airplane lift-drag polar, and (4) airplane initial weight and altitude.
The effects of airplane drag and weight will be discussed in the Sections 12.4 and 12.6, respectively.

As a final note in this section, we mention the role of future air traffic management (ATM) systems in relation to the optimization of flight profiles.
Improved ATM as anticipated, will integrate traditional elements of air traffic services (ATS), air traffic control (ATC), air traffic flow management (ATFM), and airspace management (ASM) into a total system (Ref. 1). The main purpose of ATC is to provide safe air transport by keeping airplanes separated. An integrated global ATM system not only will ensure high levels of flight safety, but also will coordinate airplane movements such that airline operators are able to meet planned times of departure and arrival. In some airspace systems, air transportation growth and capacity constraints have resulted in increased congestion in the airspace around the airports, leading to increased flight times spent in holding, in recent years. Therefore, although the majority of aviation pollution still occurs during the cruise, the emissions in terminal areas, perhaps, may grow faster than cruise emissions. For example, development of revised standardized departure and arrival routes, which shorten the curved flight paths executed in the vicinity of airports, may reduce fuel consumption. Also,

reduction of final approach spacing and/or use of parallel runways may minimize fuel-consuming airborne delays.

A noticeable measure is the application of the reduced vertical separation minima (RVSM) in the altitude region above 30,000 ft, where the altitude differences of flight levels is reduced from ΔH_p = 2000 ft to 1000 ft (see Chapter 3, Section 3.2). Additional measures to enhance fuel conservation are the implementation of reduced longitudinal and lateral separation standards to allow a larger still number of airplanes to utilize and retain their minimum fuel cruise altitudes and perhaps offering airplanes the possibility to execute a quasi-continuous cruise climb.

Unfortunately, much airspace is restricted to military use, and civil airways are often channeled into the narrow gaps between them. Airplanes mostly follow designated airways, standard instrument departures (SID's) and standard terminal arrival routes (STAR's). A global ATM system must provide the ability to follow great circle routes during the cruise as near as possible with no compromise on flight safety. In the absence of wind, these preferred flight routes yield shorter cruise ranges and less fuel for a given great circle distance between the points of departure and destination.

12.4. LIFT AND DRAG

The aerodynamic force, i.e., the vector sum of lift and drag, which acts on a moving airplane is produced by pressure and viscous forces. The pressure forces arise due to the asymmetric pressure distribution about the wing and other airplane component parts, whereas the viscous forces occur because of shearing stresses in the boundary layer flow near the outer surface of the airplane.

Figure 12.4-1 shows the pressure distribution over a wing section slanted at a low incidence angle (angle of attack) to the free-stream direction in a steady flow. According to Bernoulli's equation for compressible isentropic flow in Chapter 7, Equation (7.4-16), we have:

$$p_t = p\left[1 + \frac{\gamma - 1}{2\gamma}\frac{p}{\rho}V^2\right]^{\frac{\gamma}{\gamma - 1}} = \text{constant}\,. \qquad (12.4\text{-}1)$$

Along the streamline which follows the surface of the wing section, the velocity lowers from the free stream value ahead of the wing till zero at the stagnation point on the nose of the wing, where static pressure p becomes equal to free stream total pressure p_t.

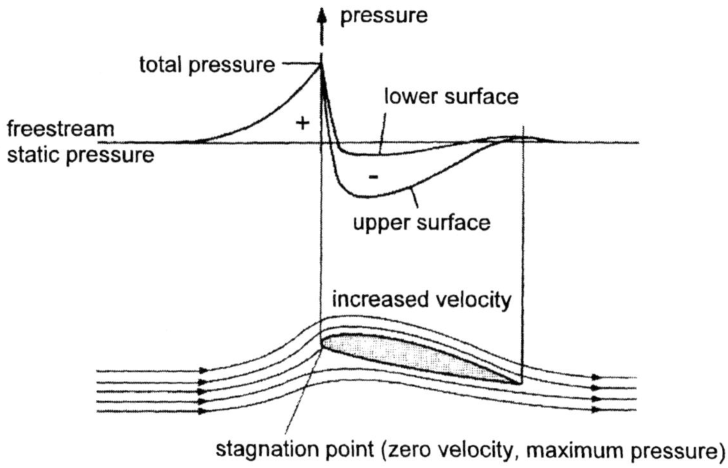

Figure 12.4-1. Pressure distribution over a wing section.

Following the streamline from the nose along the upper surface, the velocity of the air increases, and the local static air pressure decreases. At some point the velocity reaches its highest value and the static pressure its lowest value. Beyond this point the velocity decreases again and the local static pressure comes back to the free-stream static pressure. Similar variations of local velocity and pressure occur along the lower surface. Owing to the difference between the amount of upper and lower surface camber, and also because the wing section is at an angle of incidence, the velocity of the air flowing over the upper surface is greater than the velocity along the lower surface. Consequently, the pressure on the upper surface will be lower than that acting on the lower surface. As a result, an aerodynamic force is produced due to the shape and attitude of the wing section.

In addition, the existence of viscosity introduces a shearing stress at the surface τ_0, which is the force per unit area acting tangentially on the surface of a body due to the frictional effects between body and surrounding airflow. Figure 12.4-2 illustrates the cause of shearing stress, namely, the fact that the air particles

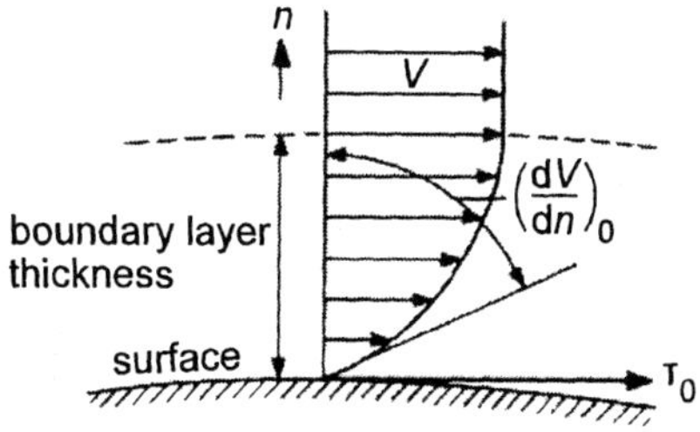

Figure 12.4-2. Velocity profile across a laminar boundary layer.

adjacent to a solid boundary are brought to rest and those close to it are slowed down markedly due to the viscosity of the air. It allows motion only by a sliding action between adjacent layers, which causes shearing forces. This effect decays quickly across the flow so that always there is a velocity gradient in the flow near the body. The very thin layer in which the velocity of the air particles increases from zero at the surface to the local velocity about the body is called the *boundary layer*.

The effect of the boundary layer is referred to as the skin-friction drag. Apparently, the concept of boundary layer flow allows the flow field around a body to be divided into a thin layer (thickness < 1 cm) in which friction is important, and a region in which the air shows zero viscosity.

Experimentally, it has been observed that the shearing stress at the surface of the body, τ_0, is given by the Newtonian relation:

$$\tau_0 = \mu \left[\frac{\mathrm{d}V}{\mathrm{d}n} \right]_0 , \qquad (12.4\text{-}2)$$

where $[\mathrm{d}V/\mathrm{d}n]_0$ is the slope of the velocity profile at the surface and the proportionality constant μ is the coefficient of dynamic viscosity of the air (see Chapter 3).

Besides causing skin friction drag, the boundary layer also affects the pressure drag. The presence of the boundary layer requires that some of the outer air moves away from the wing surface to accommodate the boundary layer. In this respect, one considers the thickness of a fictitious boundary layer with zero velocity that would displace the flow outside the boundary layer by the same amount as the actual boundary layer. The thickness of the fictitious boundary layer is termed the *displacement thickness,* which, naturally, is less than the thickness of the actual boundary layer. Clearly, the occurring velocities and pressure distributions are equal to those of a wing section with an increased thickness (Ref. 61).

Figure 12.4-3 depicts the most important parameters defining the wing geometry used for long-haul jet airplanes. These airplane types have tapered and swept back wings in order that they can reach high-subsonic flight Mach numbers before drag becomes excessive.

The geometrical parameter defining the sweep is the *sweep angle* Λ, which is the angle between the Y-axis of the body-axis system, and the projection of the wing leading edge onto the XY-plane.

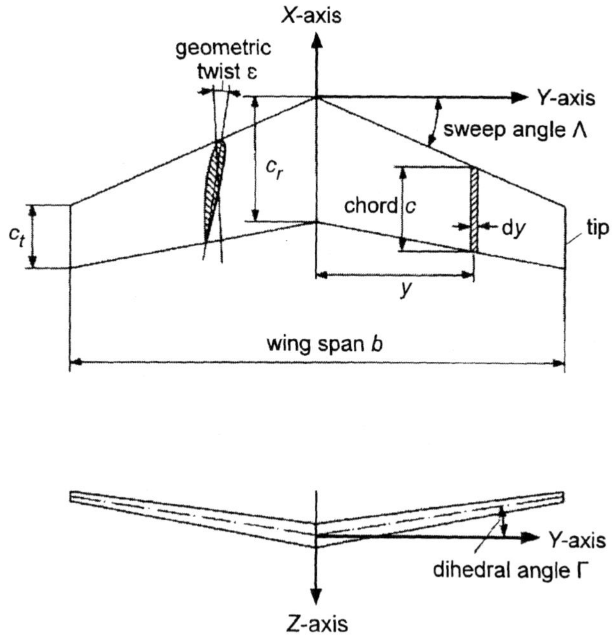

Figure 12.4-3. Wing geometry.

The wingspan is the distance in Y-direction between the wing tips and the wing area, S, is the area of the wing projected onto the XY-plane of the airplane's body axis system.
Since the chord c varies with the Y-coordinate, the wing area is given by:

$$S = \int_{-\frac{b}{2}}^{\frac{b}{2}} c(y)\mathrm{d}y \; . \tag{12.4-3}$$

From Figure 12.4-3 it is clear that the wing area, arbitrarily, also comprises the central part of the wing covered by the fuselage. Knowledge of the wing area is a necessity because it is used to describe the airplane's *wing loading* W_{to} / S.
As a measure of the slenderness of the wing use is made of the *aspect ratio A, which* is the ratio of the wing span to the mean geometric chord $c_{mg} = S / b$:

$$A = \frac{b}{c_{mg}} = \frac{b^2}{S} \; . \tag{12.4-4}$$

Another wing parameter of importance is the mean aerodynamic chord $\bar{c}$, which, for instance, is used for the quantification of the Reynolds number of the flow around the airplane:

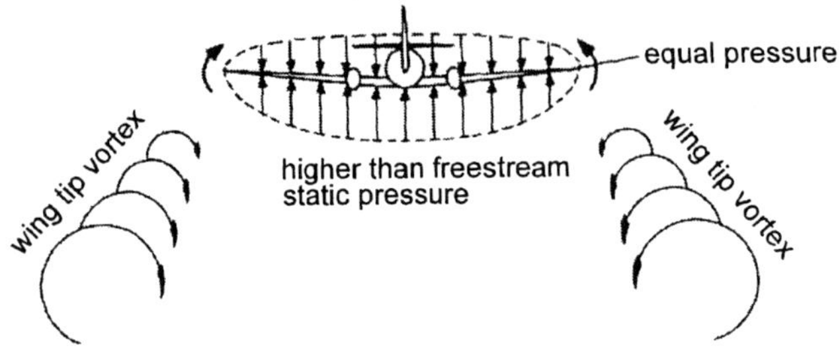

Figure 12.4-4. Three-dimensional wing flow.

$$\bar{c} = \frac{1}{S}\int_{-\frac{b}{2}}^{\frac{b}{2}} c^2(y)\mathrm{d}y \ . \tag{12.4-5}$$

Since the pressure on the bottom surface of a wing is greater than that on the top surface, a decrement of the lift coefficient at a given angle of attack will occur compared to the two-dimensional situation. In exchange for this loss of lift due to the finite width of the wing, we get from the span wise inflow above the wing and the outflow below the wing, two discrete wing tip vortices in the flow downstream of the wing (Figure 12.4-4).

In Figure 12.4-5 is shown the condition for the angle of attack in the case of an equal lift coefficient C_L obtained from a wing placed in a three-dimensional flow with using the two-dimensional aerodynamic properties of the composing wing sections. We see that this equality is achieved by increasing the angle of attack of the wing in the actual three-dimensional flow by a small value $\Delta\alpha$ over that for the wing in two-dimensional flow. The latter flow condition is the same as would exist for a wing of an infinite span.

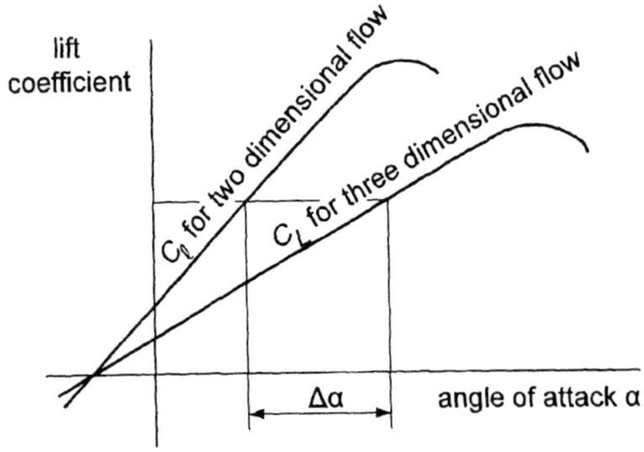

Figure 12.4-5. Lift coefficient versus angle of attack.

The effect of the enlarged angle of attack is expressed by the drag for the wing in three-dimensional flow, D_w, which is equal to the two-dimensional drag D_p plus the induced drag D_i:

$$D_w = D_p + D_i , \tag{12.4-6}$$

where D_p is called the *profile drag*, which force is due to pressure drag and skin-friction drag of the two-dimensional wing operating at the lift coefficient C_L.

When successive layers of air in the boundary layer slide over one another in the form of parallel layers, we speak of a *laminar flow*. On the other hand, a turbulent flow may prevail, where oscillations of the air particles across the boundary layer occur, and through which there is an exchange of kinetic energy between the laminae and a transfer of energy from the free stream to the boundary layer.

Figure 12.4-6 shows the velocity profiles across the boundary layer for both laminar and turbulent flow. Assuming the same flow conditions, the turbulent boundary layer is thicker than the laminar boundary layer. Another important difference is that near the surface the velocity gradient of the turbulent layer is much greater than that of the laminar boundary layer. Then, the previous Equation (12.4-2) indicates that the occurrence of turbulence in the flow will result in increased shearing forces and so in skin-friction drag.

The transition from laminar to turbulent flow, is dependent on: (1) Reynolds number, (2) surface roughness, and (3) pressure gradients in the flow field.

As stated before in Chapter 11, Section 11.2, the Reynolds number has an important effect on the flow in the boundary layer. The higher this parameter, the further upstream along the wing chord the boundary layer becomes turbulent. For airplanes in the cruise, the Reynolds number range is such that laminar as well as turbulent flows are present.

The effect of *surface roughness* of a wing is that it also may force the flow near the surface to change from laminar into turbulent. As the surface roughness increases, the point of first occurrence of turbulent flow will move upstream

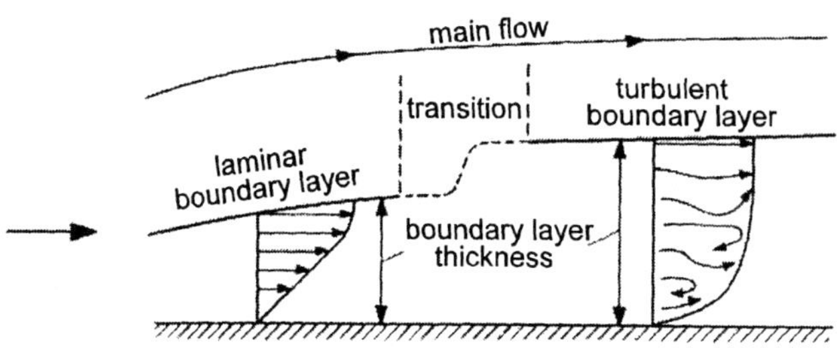

Figure 12.4-6. Boundary layer velocity profiles.

along the wing section. Therefore, surface roughness and Reynolds number both determine the location at which the flow turns from laminar into turbulent. The third factor affecting the transition from laminar to turbulent flow is the pressure gradient in the flow along the wing section. If the local pressure increases with increasing distance to the nose, disturbances in the laminar flow will be intensified and turbulent flow will result. If the pressure decreases with distance to the nose, disturbances in the laminar flow deaden and the flow will tend to remain laminar. Note that over an airfoil, first the pressure decreases up to the point of maximum thickness, so that laminar flow will be found in this region. After the point of maximum thickness the local pressure decreases and the flow will become turbulent at some point before the trailing edge.

The total drag D of an airplane may be divided into the drag of the wing D_w and the sum of the component drags D_n:

$$D = D_w + D_n \ . \qquad (12.4\text{-}7)$$

Referring to Equation (12.4-6), we can write the wing drag as the sum of the induced drag (drag due to lift) D_i and the profile drag D_p. Using this in Equation (12.4-7) gives:

$$D = D_i + D_p + D_n \ . \qquad (12.4\text{-}8)$$

The profile drag is composed of pressure drag and skin friction drag. Evidently, pressure drag and skin friction drag together, form also the drag of the airplane components.

Since the drag coefficient of each part, C_{Dn}, is based on some reference area S_n, the total airplane drag can be expressed as:

$$C_D \frac{1}{2} \rho V^2 S = C_{Di} \frac{1}{2} \rho V^2 S + C_{Dp} \frac{1}{2} \rho V^2 S + (\Sigma C_{Dn} S_n) \frac{1}{2} \rho V^2 \ . \qquad (12.4\text{-}9)$$

Accordingly, the drag coefficient of an airplane is (Figure 12.4-7):

$$C_D = C_{Di} + C_{Dp} + \frac{\Sigma C_{Dn} S_n}{S} \ , \qquad (12.4\text{-}10)$$

where S is the wing area and the quantity $(\Sigma C_{Dn} S_n)/S$ is called the *parasite drag coefficient*.

Theoretical aerodynamics predicts that the *induced drag coefficient* C_{Di} is

$$C_{Di} = \frac{C_L^2}{\pi A \varphi} \ . \qquad (12.4\text{-}11)$$

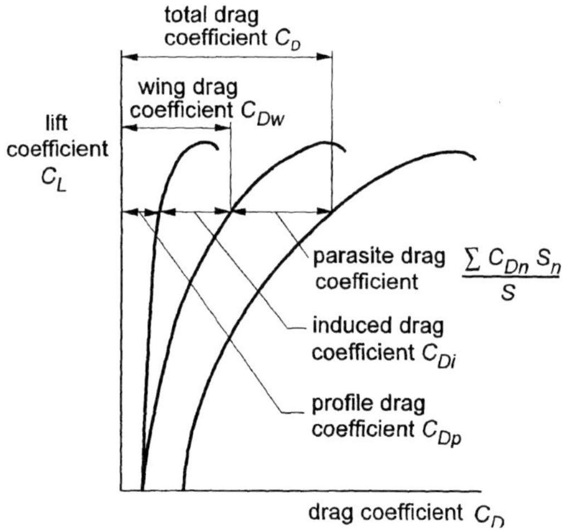

Figure 12.4-7. Components of airplane drag.

directly proportional to the square of the lift coefficient C_L, and inversely proportional to the aspect ratio A, and a wing efficiency factor φ:
The factor φ depends primarily on the planform and geometric twist of the wing since it indicates how close the elliptic spanwise lift distribution is obtained. For an elliptic lift distribution $\varphi = 1$ (minimum induced drag coefficient). In all other cases φ will be less than one.
Thus, the drag coefficient C_D of an airplane is given by:

$$C_D = \frac{C_L^2}{\pi A \varphi} + C_{Dp} + \frac{\Sigma C_{Dn} S_n}{S} . \tag{12.4-12}$$

Since also the profile and parasite drag coefficients are dependent on the angle of attack, Equation (12.4-12) can be written as:

$$C_D = \frac{C_L^2}{\pi A \varphi} + X C_L^2 + \left[C_{Dp} + \frac{\Sigma C_{Dn} S_n}{S} \right]_{C_L = 0} . \tag{12.4-13}$$

The term $XC_L{}^2$ represents the assumed parabolic change of the profile and parasite drag coefficients with lift coefficient. The quantity in parentheses is termed *zero-lift drag coefficient* and given the symbol C_{D0}. Then, the so-called *parabolic lift-drag polar* (12.4-13) can be expressed in the form:

$$C_D = C_{D0} + \frac{C_L^2}{\pi A e} . \tag{12.4-14}$$

The factor e is the *Oswald* or *wing efficiency* factor, which follows from the relationship:

$$\frac{1}{e} = X\pi A + \frac{1}{\varphi}\ . \tag{12.4-15}$$

Apparently, the factor e accounts for the variation of the profile and parasite drag coefficients with lift coefficient, and the effect of the actual span wise lift distribution on induced drag coefficient. For most airplane types the value of e varies, dependent on configuration, between 0.6 and 0.9.
Figure 12.4-8 gives a schematic representation of the parabolic lift-drag polar, where also is shown the variation of the lift-drag ratio with C_L. The maximum value of this ratio may be seen as a measure of the aerodynamic efficiency of the airplane.

Using the lift-drag polar, we recall that the drag coefficient not only is dependent on the airplane's configuration, angle of attack, and Reynolds number, but also on the flight Mach number. The effect of Mach number has to do with the formation of shockwaves, which convert a substantial part of the free stream kinetic energy into heat. Here, it suffices to say that a shockwave is an almost sudden increase in pressure, temperature, and density, and a decrease in velocity and total pressure. Consequently, there is a change in pressure distribution over the surface from that experienced at low-subsonic airspeeds with the result that a certain drag rise occurs at the usual cruise Mach numbers for conventional subsonic airplanes ($0.82 \leq M \leq 0.85$).
Figure 12.4-9 shows the variation of C_D with flight Mach number for fixed values of C_L. Up to a Mach number of 0.7 the drag coefficient C_{D0} remains virtually constant. The flight Mach number at which the zero-lift drag coefficient

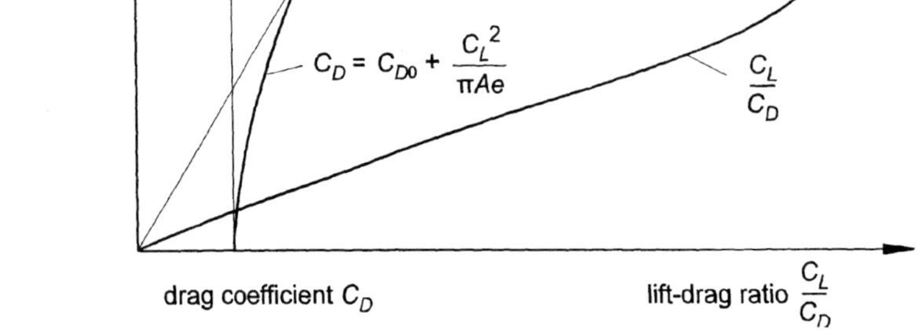

Figure 12.4-8. Parabolic lift-drag polar.

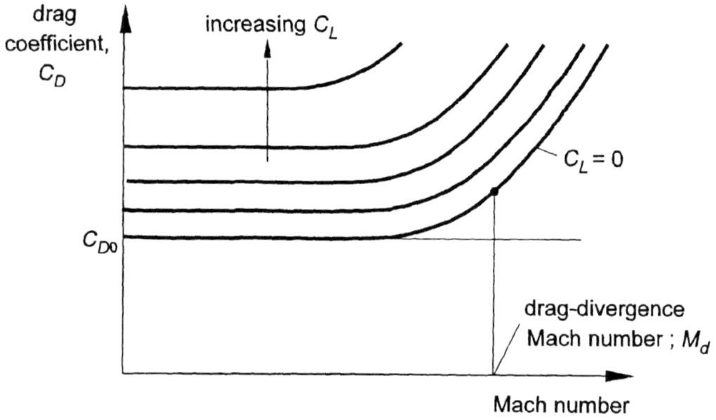

Figure 12.4-9. Variation of drag coefficient with flight Mach number.

increases markedly is termed the *drag-divergence Mach number*, M_d.

It will be appreciated that the parabolic lift-drag polar can be used in the whole region of subsonic airspeeds, if in Equation (12.4-14) the values of C_{D0} and πAe are adjusted in terms of Mach number.

So far we have considered the common division of airplane drag into wing and parasite drag. In this connection, it is important to be aware of the fact that the drag components D_p and D_n both are composed of skin friction and pressure drag. Hence, for our purpose, it is more convenient to express the total airplane drag in terms of friction, pressure, and induced drag. In terms of drag coefficients, this drag breakdown is illustrated by the curves drawn in Figure 12.4-10.
According to this sketch, the most effective way to reduce airplane drag is to

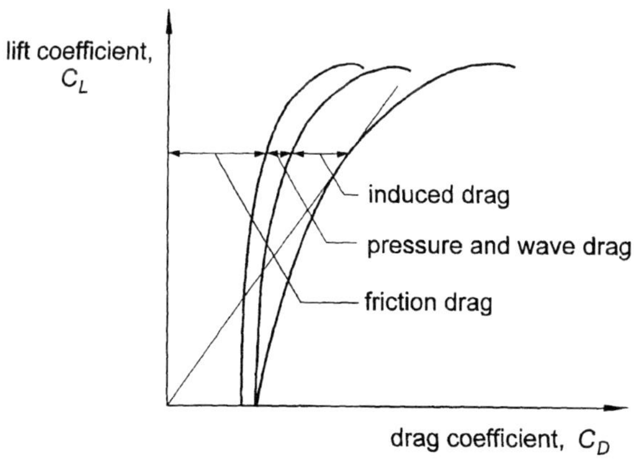

Figure 12.4-10. Airplane drag breakdown.

decrease both friction and induced drag, which together represent the majority part of the total drag at the angle of attack for which the lift-drag ratio attains its maximum value.

However, it seems unlikely that in the foreseeable future an appreciable effect on the induced drag is obtained, in view of the current high-subsonic turbofan-airplanes with their *high-aspect ratio wings*, *supercritical airfoils* and *winglets.* An airplane with supercritical wings can fly closer to Mach 1 before drag divergence due to shockwave formation is encountered (Ref. 8). Winglets are plates mounted on the wing tips to enhance a two-dimensional flow near the tips by avoiding the formation of tip vortices (see Figure 12.1-1).
The application of wings with even higher aspect ratios to decrease the induced drag, may require much added structural mass to oppose the higher wing-root bending moments.
Therefore, it may be expected that most improvement of the aerodynamic efficiency of the airplane must be realized by decreasing friction drag from the airframe skin, in combination with a reduction of nacelle interference drags.
The magnitude of the skin friction drag of the airplane depends strongly on the nature of the boundary layer, i.e., whether this layer is laminar or turbulent. Turbulent boundary layers produce a much larger friction drag, so that a delay of transition from the laminar to the turbulent state can reduce skin friction substantially.
In the first place, this may be obtained by improving wing and fuselage surface smoothness, through the development of advanced manufacturing techniques. Further, a possible method for application of laminar flow control (LFC) in the design of future subsonic long-range transports is the technique of suction through the airplane skin (Figure 12.4-11). Likewise, turbulent skin friction can be reduced by air injection into the boundary layer through slots or porous surfaces.
Successful application of these techniques would also reduce the *aerodynamic*

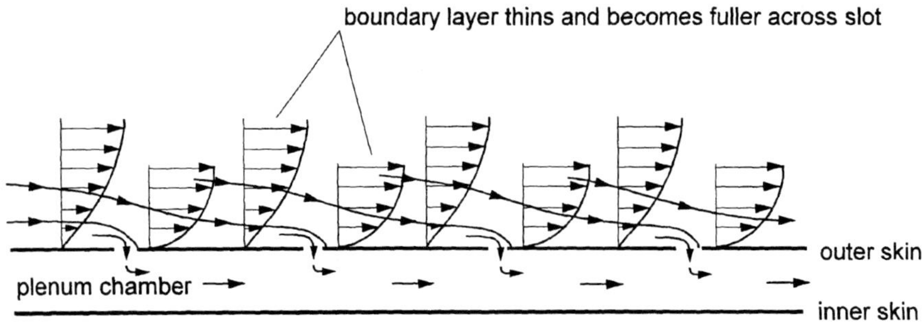

Figure 12.4-11. Laminar flow control by suction (ref. 61).

noise generated by boundary-layer turbulence.

The savings in drag, of course, must be balanced against the weight penalty due to the suction system and the double skin necessary to remove the boundary layer air.

Certainly, any application of this technique is likely to require a long development time period.

A not yet well understood measure in the reduction of turbulent flow areas is the application of a film with micro grooves, termed *riblets*, to the fuselage skin, wing, and horizontal tail, in the direction of the airflow (Ref. 65).

12.5 WING DESIGN FOR LOW POLLUTION

Airbus has designed its new A 380-800 airplane within the ICAO 80 m span limit, in order to assure compatibility with the available space at the major airports around the world (Ref. 71). This limitation has resulted in a relatively low aspect ratio of $A = b^2/S = 7.53$ (see Figure 12.5-1).

A lower aspect ratio often leads to higher fuel consumption. In view of this, the objectives of the studies described in References 70 and 71 are to determine for the above airplane type the economic and environmental consequences of having a wingspan restriction, assuming that the present span may be exceeded.

Here we limit our review to Ref. 71, where is studied the combined change of the wingspan and aspect ratio, employing an updated version of the parametric design model developed in Ref. 70. For this purpose, a A380 aircraft has been analyzed, defining the baseline for the research.

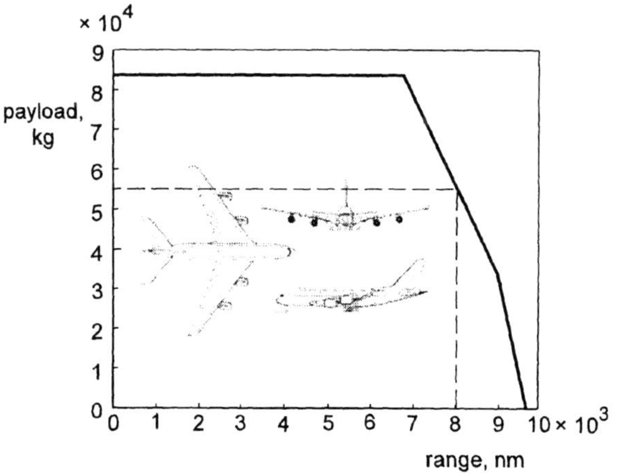

maximum takeoff mass	560,000 kg
max. payload mass	84000 kg
design range	8000 nm
matching seating capacity	555
cruise mach number	0.85
initial cruise altitude	FL 350 (10668 m)
wing area	845 m^2
wingspan	79.75 m
total takeoff thrust	4 × 356 kN
bypass ratio	7.8
overall pressure ratio	41

Figure 12.5-1. Airbus A380-800 specifications (Ref. 71).

Used are the simplifications that the size of the fuselage is fixed and that flight takes place in the International Standard Atmosphere. Also, the wing area is to remain unchanged in order to maintain field performance and fuel tank volume. The higher aspect ratio is thus to be established through wingspan increase and chord reductions.

Only harm is considered caused by the greenhouse gases CO_2, H_2O, and NO_X. The contribution of water to the greenhouse effect through contrail formation, is not examined.

Two types of optimization in terms of direct operating cost and environmental harm have been performed:

a. Variation and optimization of the aspect ratio only.
b. Simultaneous variation of and optimization for the aspect ratio and cruise altitude.

The calculation of the performance includes the determination of empty weight, engine performance, and aerodynamic characteristics of the airplane. The takeoff weight is made up of operational empty weight, fuel weight (trip fuel and reserve-fuel), and payload. Of these weights, the payload weight is fixed. The airplane drag coefficient is represented by a parabolic lift-drag polar (see Equation (12.4-14)). The actual lift-drag ratio in the cruise is determined from flight condition and lift-drag polar.
Engine performance is calculated by employing the method described in Ref. 72. The required fuel load W_f for a flight distance R, follows from the Breguet range formula expressed by Equation (12.3-1). Since the fuel weight determines the takeoff weight, and the takeoff weight itself is an input parameter for the empty weight prediction, mission-fuel weight and takeoff weight are computed by iteration.

The calculation of the direct operating cost (*DOC*) is performed by using the *DOC*- calculation method published in Ref. 64, where the *DOC* is divided into costs for finance, crew, fuel, maintenance, and fees and charges for landing, ground handling and navigation (see also Chapter 11, Section 11.15).

The evaluation of the enhanced greenhouse effect, considering the gases CO_2, H_2O, and NO_X, is based on the masses of the species, just mentioned, which result from the product of their emission index and the cruise fuel mass.
The emission indices of CO_2 and H_2O have the constant values of 3150 and 1250 g/kg, respectively (see the previous Table 10.2-1).

The emission index of NO_x is based on a direct relationship between emission index and fuel flow, which is obtained by application of the so-called "fuel flow method" reported in Ref. 73, and using the *EI* data given in Ref. 53.
In the absence of Airbus A380-800 data, the above relationship has been defined by using the data of a number of comparable engines.
The environmental harm caused by an exhaust gas is represented by the product of the amount of pollutant and its Global Warming Potential (*GWP*) value as a weighing factor (see Chapter 5, Section 5.6).
The *GWP* value of CO_2, by definition, is equal to one. Emissions of NO_x affect global warming indirectly through tropospheric ozone formation, which process is highly dependent on altitude, latitude, and season. Different chemical regimes will produce different amounts of ozone for the same injection of NO_x, so that the *GWP* value of the ozone perturbation will vary by location and point in time. The *GWP* values of H_2O exhibit a similar dependence on the altitude and season.
In the present study, the curves shown in Figures 12.5-2 and 12.5-3 are used. These *GWP* values are established originally in Ref. 74, where is stated that they must be seen as approximations only, which are valid at mid-latitudes and summer atmospheric conditions.

Then, as shown in Figure 12.5-2, the *GWP*-NO_x value increases with the altitude and peaks at an altitude of about 12 km, while at altitudes beyond 15 km only depletion of ozone may take place. There are two reasons for the maximum at about 12 km. The first reason is that NO_x is more effective in forming ozone in the upper troposphere, as a result of appropriate atmospheric and chemical conditions in this altitude region. The second reason is the higher influence ozone has on the greenhouse effect at higher altitudes.
The dependence on location and season is also true for the *GWP* value of H_2O emissions. Just like CO_2, water does not absorb the incoming radiation from the Sun, but absorbs the infrared radiation reflected from the Earth's surface. As

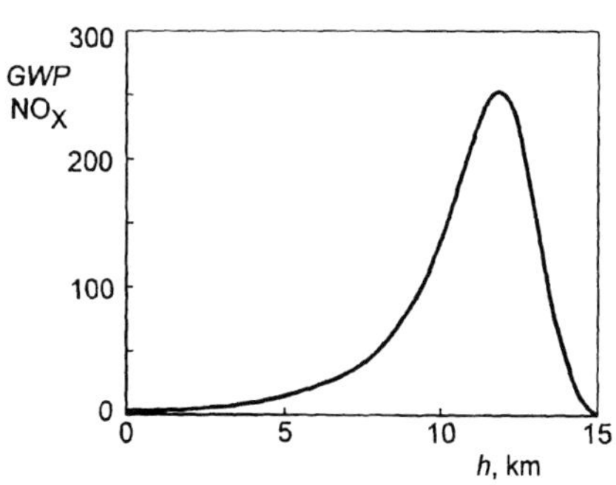

Figure 12.5-2. GWP-NO_x versus altitude.

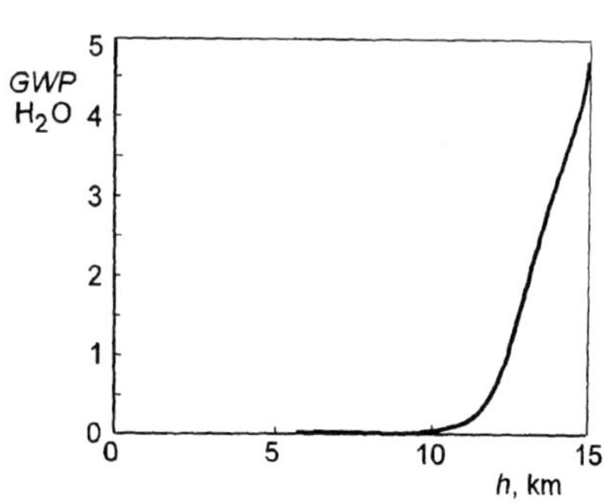

Figure 12.5-3. GWP-H_2O versus altitude.

shown in Figure 12.5-3, in summer atmosphere and in mid-latitudes, *GWP*-H_2O values also strongly depend on the altitude. At altitudes up to 10 km, H_2O emissions have no effect at all, whereas above 10 km the contribution of H_2O becomes rather high. The reason for this behavior is based on the fact that H_2O shows a small residence time in the troposphere because it is admitted as a part of the hydrological cycle (see Chapter 1, Section 1.7). At higher altitudes the residence time becomes much larger, resulting in higher *GWP* values.

Multiplication of the total quantities of NO_x, H_2O, and CO_2 with their *GWP* value at the cruise altitude gives the equal amount of CO_2-harm caused by these gases.
Summing the values obtained for NO_x, H_2O, and CO_2, yields a parameter that represents the equivalent amount of CO_2-harm caused by the greenhouse gases during a particular mission. This parameter, expressed in grams per flight, has been referred to as *total GWP*, and follows from the summation:

$$GWP_{tot} = EI_{CO_2} \times M_f + EI_{NO_x} \times M_f \times GWP_{NO_x} + EI_{H_2O} \times M_f \times GWP_{H_2O} ,$$

where M_f is the mass of the cruise fuel.
Of course, GWP_{tot} only may provide a first indication of the effect of the emissions on global warming.

Within the first type of optimization in Ref. 71 the aspect ratio has been varied between 7 and 16 for a fixed wing area (845 m^2), payload-range specifications (8000 nm or 14816 km at 84000 kg), flight Mach number (0.85), and initial cruise altitude (10668 m).

The impact on GWP_{tot} *and DOC* is shown in Figure 12.5-4. It turns out that choosing the *DOC*-optimum as a design point results in an aspect ratio of A = 9.9 (wingspan b = 91.5 m), providing a reduction of 2.4% in direct operating cost and 11% in total Global Warming Potential relative to the baseline airplane (A = 7.53). Clearly, the optimum aspect ratio for GWP_{tot} is relatively high (A = 12.2), which makes this solution hardly practical.

Within the second type of optimization, starting from a specified set of thrust loading - wing loading combinations, the aspect ratio and cruise altitude have been varied simultaneously. In this case, the study predicts that the *DOC*-optimum aspect ratio has to be A = 8.9 (b = 86.7 m) and H = 9022 m. However, the gain in *DOC* with respect to A = 9.9 and H = 10668 m will be virtually negligible.

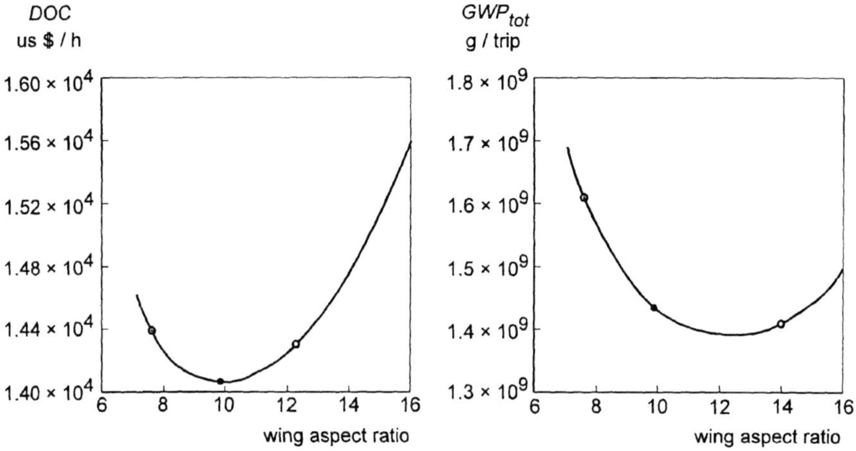

Figure 12.5-4. Direct operating cost and total Global Warming Potential versus wing aspect ratio.

As discussed in Ref. 71, to comply with the actual 80 m wingspan limit on the ground and take advantage of the higher aspect ratio in flight, consideration should be given to the introduction of a folding wing tip system.

Before closing this section we may repeat that the global warming values of NO_x and H_2O are highly dependent on altitude, location, and season (Ref. 1). In this respect, one of the difficulties in predicting the environmental impact is that our knowledge and understanding of the influences of gas emissions from aircraft on the atmospheric processes is still very limited so that many data are inaccurate and need a continuous updating. As an example, Figure 12.5-5 gives modifications of the relations in Figures 12.5-2 and 12.5-3, showing that according to the latest insights the variations have been decreased by a factor 5.5 (Ref.78).

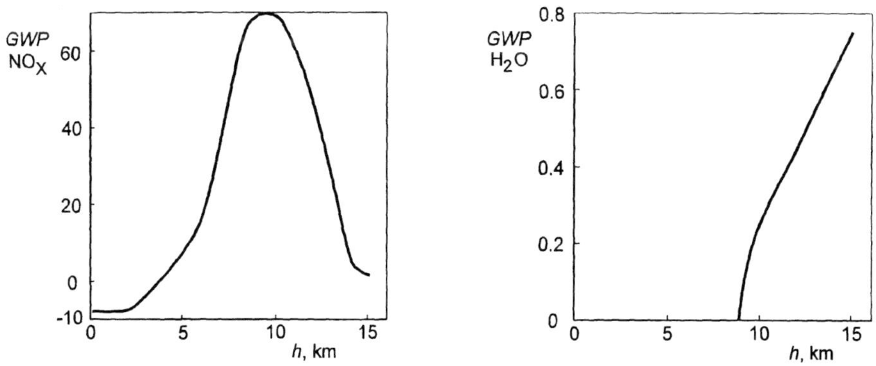

Figure 12.5-5. GWP versus altitude.

12.6 WEIGHT REDUCTION

The overall configurations of the modern subsonic long-haul airplanes manifest only little alterations to the designs of the first generation jet transports, the Boeing B 707 and its rival, the Douglas DC 8 (Figure 12.6-1). Almost all intercontinental airliners introduced since the beginning of the jet age around 1960 are characterized by swept wings and wing-mounted gas turbine engines.
They fly at altitudes near the tropopause, and in the narrow Mach-number range between 0.82 and 0.85. Complex flaps and slat systems are used to slow down the airplane to the low airspeeds required for takeoff and landing.
Clearly, the most noticeable innovations have been in the size (number of passenger seats) and the number of engines. However, this must be seen in connection with vital improvements in fuel consumption, significant advances in drag reduction, changes to the structural materials, and the adoption of modern avionics and computer control technologies. Especially, the development of the original pure turbojet engine into the more fuel efficient and lighter turbofan engines has resulted in very impressive improvements in performance and economics.

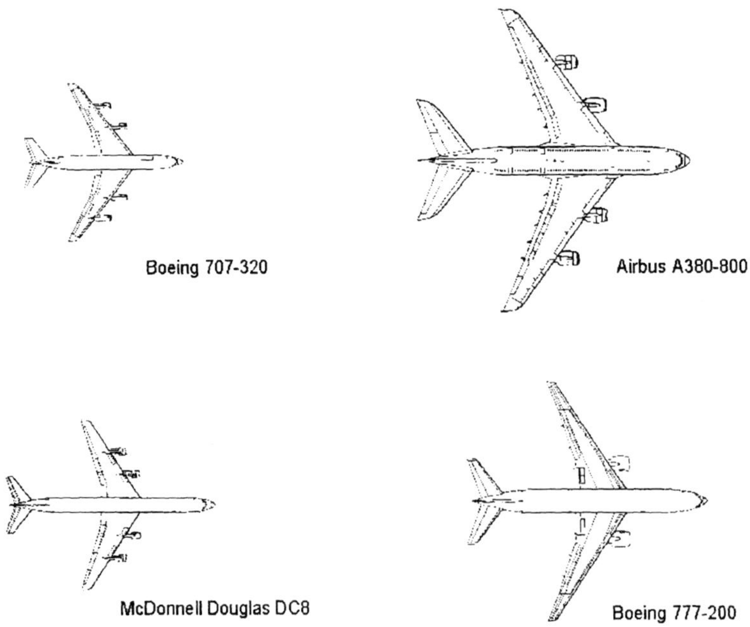

Figure 12.6-1. Typical airliner layouts (drawn to scale).

As shown in Figure 12.6-1, the B 707 and DC 8 needed four engines to cover transoceanic distances. Conform to the airworthiness requirements, even in the case of two inoperative engines airplanes must be able to fly. Today's turbofan engines are so powerful and reliable that most of the current long-haul airplanes, such as the Airbus A 330 and the Boeing B 777, have only two engines. Also the B 787 "dreamliner", announced by Boeing, and its competitor, the Airbus A 350 to be developed by Airbus, will be powered by two turbofan engines.

A special example of a modern four-engine transport airplane is the just unveiled Airbus A 380-800 (see also Figure 12.5-1). Although, this is newest and the largest airplane ever built, at first sight it seems to have the same classic configuration as the earliest jetliners.

The philosophy behind its design is that the availability of substantial larger aircraft leads to a more efficient use of airspace with less congestion. In this way the environmental impact in terms of noise and fuel burn may be limited, in spite of the anticipated increase in air traffic (Ref. 76).

Looking to the future, a reduction of the operational empty weight (OEW) may be considered as another possibility for fuel conservation. From the relationship for level, steady flight, we have:

$$T = D = \frac{C_D}{C_L} W \, . \tag{12.6-1}$$

This equation shows that a reduction of the weight of the airplane directly leads to a lower required thrust, and so to a lower fuel consumption for a given trip length. At a given payload, this results in a lower takeoff weight (see also Equation (11.13-2)).

When investigating weight-reduction programs, in the first place, application of new materials in airframe structures is the obvious target as most of the weight is in the airframe.

Today, alloys of aluminum are the most widely used materials in airplane structures. Of interest for the future, are newly developed aluminum alloys, titanium, and composite materials. Especially, the composite materials, made up of different parts, may bring about a decrease in the empty weight of airplanes.

A promising composite for primary and secondary airplane structures is the glass reinforced laminate, developed at Delft University of Technology. This material, called "glare," is a member of the fiber metal laminates (FML) family. Glare consists of alternating layers of thin aluminum alloy sheets and strong glass fiber prepreg layers embedded in an epoxy resin (Ref. 77).

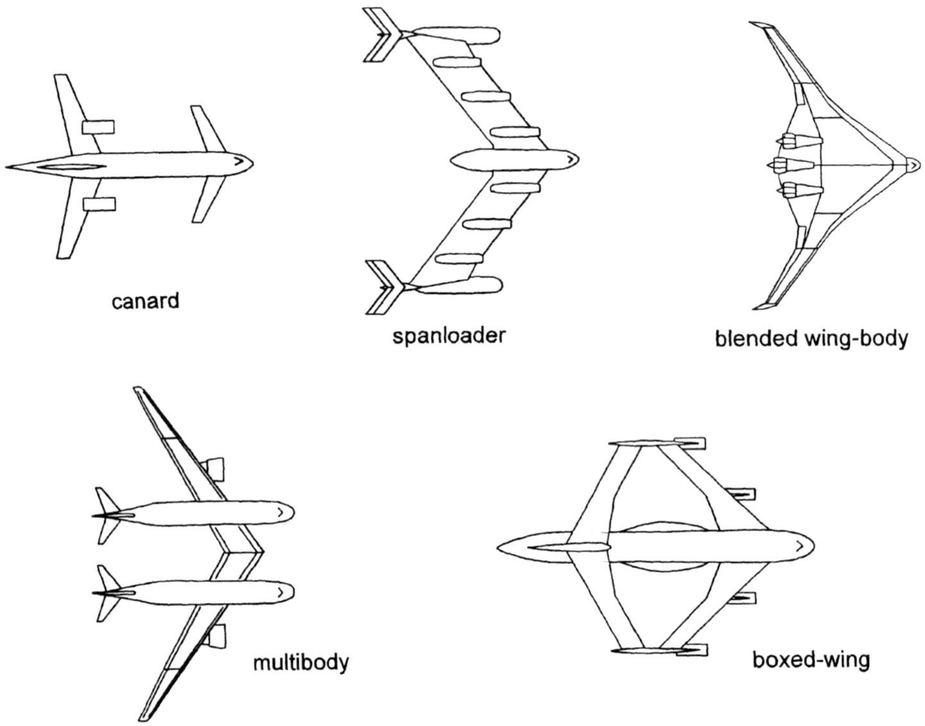

Figure 12.6-2. Unconventional airplane concepts.

In addition to the application of laminar flow control and new materials, aerodynamic improvements and weight reductions may be possible through the development of radical new transport airplane concepts.

Figure 12.6-2 gives an outline of some unconventional design concepts for long-haul transport airplanes.

The canard layout avoids tail-down forces, which are the causes of trim drag, i.e., the drag due to adjusting the trim tabs placed at the trailing edges of the primary control surfaces of an airplane, so that it flies in the correct attitude without manual effort. The canards are small wings situated in front of the main wing. They produce an upload, which stabilizes the airplane in its pitch angle and supports the wing lift. A design problem related to this configuration is the occurrence of the trailing vortices from the forward planes, which may affect the flow over the wing and fin. Nowadays, it's common knowledge that the canard concept is unprofitable.

In the spanloader concept the payload is carried in the wing. In this configuration the wing structure experiences decreased bending moments, resulting in a weight saving. In addition, the large wing aspect ratio leads to reduced induced drag and so to an increased lift-drag ratio.

The blended-wing body concept is characterized by an integration of the fuselage with the wing. In this way a smaller wetted area is created, and thereby a lower drag. The concept will produce lift over the entire body, which may lead to higher aerodynamic efficiency. This configuration is likely to carry portions of the payload within the wing, and therefore also may significantly reduce the bending moments and so the structural weight. Estimates on this layout predict a substantial reduction in fuel consumption.

The twin-fuselage configuration consists of two separate bodies located on the wing. This layout offers improvements due to the higher effective wing aspect ratio, providing a reduced parasite and induced drag with the result that an increased lift-drag ratio is obtained. The setup also reduces the wing root bending moments, which may save airplane weight. Therefore, multi body layouts can attain improved performance compared to that which can be attained by the individual airplanes.

Finally, the boxed-wing configuration has wings forming a closed rectangular lifting system. This configuration also offers a higher wing aspect ratio and wing efficiency factor, thus decreasing the induced drag. Because of increased stiffness, also, a reduced structural weight may be obtained.

From this concise sketch, it will be clear that extensive research and considerable financial investments are required before solid conclusions can be drawn about the relevance of these novel concepts to future airplane development. In other words, these new configurations can only be seen as long-term solutions.

12.7 ALTERNATIVE FUELS FOR FUTURE TRANSPORT AIRPLANES

The fuels used for civil and military jet-driven airplanes are all kerosene-type fuels. The normal fuel burnt in the engines of commercial airplanes is designated Jet A-1, involving well-defined properties (see Ref. 1). The standard properties of civil aviation kerosene as specified by ICAO for aircraft turbine engine emission testing, were given earlier in Table 10.4-1. Within the detailed specifications there is hardly a chance for decreasing emissions from current aviation kerosene. Therefore, besides the measures for fuel conservation mentioned in Section 12.3, one may think on the addition of chemical additives to jet fuel or replacing the existing jet fuels by new, so-called *green fuels*. Certainly, fuels other than the usual fossil fuels should be given consideration

fuel	lower heating value, J/kg	mass density, kg/m^3	energy density, J/m^3	boiling point, °C
kerosine	4.3×10^7	800	34.4×10^9	260
hydrogen LH_2	11.9×10^7	68	8.1×10^9	-253
methane LCH_4	5.0×10^7	420	21.0×10^9	-164

Table 12.7-1 Fuel properties.

for future air and ground transportation, not only because of offering emission benefits, but also because of the fact that the world's crude oil resources are limited.

At this point we mention that the addition of ethanol, also known as ethyl or grain alcohol, C_2H_5OH, to petrol yields a gas that when used as a fuel in internal-combustion engines shows less CO_2 emissions.

Focusing on aero-engines, it appears that for the replacement of conventional aviation kerosene, especially, liquid hydrogen, LH_2, liquid methane, LCH_4, and liquid natural gas (LNG) are potential candidates. Of these, liquid hydrogen is considered to be the most promising fuel in future commercial aviation.

The use of LNG, a mixture of hydrocarbon gases consisting largely of methane (CH_4) and ethane (C_2H_6) and having a LHV of 3.5×10^7 to 5.5×10^7 J/kg, may be seen as a meaningful interim solution for domestic air services, in particular, in regions where vast reserves of natural gas are available.

Some of the more important fuel properties of liquid hydrogen and methane are compared to kerosene in Table 12.7-1, showing that the lower heating value of hydrogen is almost three times the specific energy content of kerosene.

The only primary combustion product of liquid hydrogen is water because due to the absence of carbon, we have the exothermic reaction:

$$2H_2 + O_2 \rightarrow 2H_2O \,. \qquad (12.7\text{-}1)$$

Therefore, the burning of LH_2 causes no emissions of carbon oxides and unburned hydrocarbons. That's why hydrogen also eliminates the problem of soot particle formation. However, independent of the type of fuel, nitrogen oxides always will be exhausted, because of the formation of thermal NO coming from the atmospheric air, which flows through the core of the engine. Since hydrogen can be combusted in a leaner mix than hydrocarbons, nitrogen oxides levels will be greatly reduced in comparison with kerosene. On the other hand, water vapor emissions which also are of environmental concern at cruise

altitudes will increase, but are expected to be not critical. Some aspects of contrail formation characteristics need confirmation by flight tests (Ref. 74).
To produce liquid hydrogen, we have to extract hydrogen from water by electrolysis, which requires a large amount of electricity. Another problem is to turn the gaseous hydrogen into a liquid, for which it has to be cooled to -253 ^{0}C (20 K).
Industrial hydrogen is obtained from the catalytic reaction of natural gas (CH_4) and steam (H_2O):

$$CH_4 + H_2O \rightarrow CO + 3H_2 , \qquad (12.7\text{-}2)$$

followed by the reaction:

$$CO + H_2O \rightarrow CO_2 + H_2 . \qquad (12.7\text{-}3)$$

Obviously, the use of liquid hydrogen only may offer environmental benefits if it can be obtained from water without application of fossil fuel to produce the required energy. Often, it is mentioned that hydrogen produced by using solar energy is a realistic possibility. From an environmental perspective, another alternative is perhaps the generation of electrical energy from the movement of the tides and ocean swell waves.
Due to the high heating value of liquid hydrogen, the weight of the fuel needed for traveling a given flight distance is much less. As a result, preliminary studies of hydrocarbon-fueled airplanes show increased range or payload as a direct result of the weight reduction obtained by using hydrogen. The lower weight also allows for a shorter takeoff distance and faster climbing.
Certainly, a major problem imposed by the use of liquid hydrogen instead of fossil jet fuels is the consequent substantial impact upon airplane design, resulting from the lower energy density of hydrogen compared with kerosene (see Table 12.7-1). For an equivalent energy content, the tank volume requirement is three to four times that of kerosene, resulting in high volume airframe designs. Other disadvantages are the required cooled, thermally insulated and pressurized tanks and a fuel system that must be able to withstand the cryogenic temperature at which liquid hydrogen exists.

Hydrogen can also be combined with oxygen in an electrochemical reaction to produce electricity by means of a *fuel cell* (Ref. 78). However, this device cannot be considered an alternative for the propulsion of aircraft. Currently, a lot of research is executed on the development of small, compact fuel cells for road vehicles. A practical application is the European “fuel cell bus” project.

Table 12.7-1 indicates that the heating value of liquid methane is only 40% that of hydrogen. On the other hand, its mass density is a factor six greater, thereby making the storage problem easier. Although in terms of energy, liquid methane is less favorable than liquid hydrogen, it remains an alternative fuel to kerosene. For the same energy production, burning methane yields less CO_2 and more H_2O than burning kerosene.

In conclusion, we may state that liquid hydrogen exhibits the technical promise to become an alternative aviation fuel. A main obstacle facing hydrogen-fueled aircraft, is the storage problem and the general requirement for aviation fuels that they must be widely available and must have the same quality everywhere.

12.8 SUSTAINABLE DEVELOPMENT

Sustainable development may be described as the strategy to seek for product development approaches, by which we obtain a steadily improving quality of life for everyone, now and in the future. This aim should include all natural, life-giving elements around us, in particular the air we breathe, as well as the concern for our well-being.

In the early days of aviation, environmental considerations used to come a long way after other topics such as safety, performance, and economics. Nowadays, environmental considerations have become important design criteria in order to prevent environmental damage, which is the precondition for maintaining sustainable mobility.

The aviation industry has a history of innovation and technological improvement that has had a powerful effect on improving efficiency and reducing nuisance, such as noise and emissions impact, over the last four decades of jet travel.

Returning to Figure 12.3-2, we see that improvements in engine design have been more than halved the SFC during the cruise. In addition, also the emitted acoustic energy level between the 1960s and the present day has been reduced in the same order of magnitude (Figure 12.8-1).

Doubtless, the most important reason for the progress in specific fuel consumption and noise level is the development of the bypass and turbofan engine.

More reductions in environmental impact are certainly needed in the future, in particular, to offset the expected aviation traffic growth. Therefore, the aviation industry will have to continue developing sustainable technology to minimize noise, reduce air pollution, and safe fuel energy.

This may be possible in the following essential ways:

- By improving the efficiency in the way airplanes are operated.
- By improving the air traffic management (ATM) systems.
- By improving the designs of airplanes and airplane components.
- By developing new combustor technology.
- By introducing new fuels.
- By introducing new airplane concepts.

It is evident that the existing legislation on engine pollutant emissions during takeoff and landing has improved the air quality around airports.
Notwithstanding, future modified ground operations, and alterations in departure and approach flight procedures may further reduce noise and gas emissions near ground level.
However, contrary to the control over the emissions in the vicinity of airports and the increasing knowledge of stratospheric chemistry, we still lack detailed scientific understanding about the true environmental effects of airplane exhaust gases emitted directly into the atmospheric regions just below and above the tropopause. In particular, great uncertainties exist in the effects of nitrogen oxides and water vapor emissions.
Therefore, one of the main problems facing us, is the ability to determine the impact of the emissions produced by the subsonic airliners flying at their optimal cruise altitudes near the tropopause.
In this light, much detailed research work must be executed, aimed at improving predictions of aviation induced changes in climate and stratospheric ozone depletion in order to advise airplane designers and manufacturers how to reduce future environmental impact.

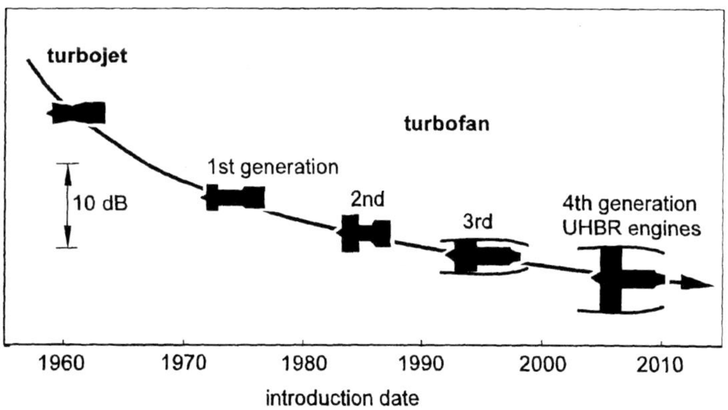

Figure 12.8-1. Noise reduction through engine design.

Especially, reduction of NO_x emission is a great technical challenge to engine manufacturers, since NO_x is formed in the high temperature zones of the combustion chamber, mainly through the reaction of atmospheric nitrogen with atmospheric oxygen (thermal-NO formation). Accordingly, the sustainable development of the aviation industry will require that the highest priority is given to achieving an improved combustion process. In Ref.1 it is stated that internationally research programs are in progress with goals to reduce LTO-cycle NO_x emissions by up to 70% from today's regulatory standards. Also, a 20% improvement in fuel efficiency is projected by the year 2015 to 40 to 50% improvement by 2050 relative to the airplanes manufactured nowadays.
Likewise, substantial research programs on aviation-generated emissions in relation with air traffic control are required before any attempt to optimize operational procedures, without destroying economic benefits of air travel, can be made. In this respect, it is urgently needed that our knowledge of emission indices and *GWP* values at the cruising altitudes is completed.
With regard to possible solutions by means of the introduction of alternative fuels and radical new airplane concepts, as much research is needed. It is clear that any result is likely to cover a long development time. It is to be hoped that in a much shorter time period, workable solutions can be realized by full development and airline introduction of new concepts for low-emission combustors.

In closing this book, it is worth stating again that for any answer to environmental problems, it is essential to have knowledge of the many branches of science involved, such as combustion, chemistry, the atmosphere and its processes, and how these are changing under the influence of aviation. Also an understanding of the performance of airplanes, the power plants, and the air traffic management systems, is of vital importance. In conclusion, knowledge gained through education must be considered to be of paramount significance in dealing with sustainable development.
Therefore, we may quote the father of all sciences, the famous Greek philosopher Socrates (469-399 BC):

"GAINING KNOWLEDGE IS ESSENTIAL IN SURVIVING"

Apparently, there is nothing new under the Sun. In other words, there has always been a strong demand for knowledge as a prerequisite in solving any kind of problem. Hopefully, this book is a reasonable attempt to relieve that need, at least insofar as it concerns elementary knowledge on the subject of aircraft pollution.

REFERENCES

1. D.H. Lister and J.E. Penner (Report Coordinators), *Aviation and the Global Atmosphere*, Report by the Intergovernmental Panel on Climate Change (IPCC), Cambridge University Press, UK, 1999.
2. E. Roberts Alley & Associates, Inc., *Air Quality Control Handbook*, McGraw-Hill, New York, 1998.
3. Anon., *Saving Our Planet; Challenges and Hopes*, The state of the environment (1972-1992), United Nations Environment Programme, Nairobi, 1992.
4. Anon., *U.S. Standard Atmosphere, 1962*, U.S. Government Printing Office, Washington, D.C., 1962.
5. G. Baumbach, *Air quality control*, Springer-Verlag, Berlin Heidelberg, 1996.
6. Anon., ICAO/Annex 16 to the Convention on International Civil Aviation, Environmental Protection, Volume II, *Aircraft Engine Emissions*, First edition, 1981.
7. A.E. Gill, *Atmosphere - Ocean Dynamics*, Academic Press, Inc., New York, 1982.
8. G.J.J. Ruijgrok, *Elements of Airplane Performance*, Delft University Press, Delft, The Netherlands, 1996.
9. A. Upgren and J. Stock, *Weather. How It Works and Why It Matters*, Perseus Publishing, Cambridge, Massachusetts, 2000.
10. J.V. Iriborne and H.-R. Cho, *Atmospheric Physics*, D. Reidel Publishing Company, London, 1980.
11. J.F.R. McILveen, *Basic Meteorology*, Van Nostrand Reinhold (UK) Co. Ltd, Berkshire, 1986.
12. C. Dykes, *Operating factors affecting the design of future civil transport aircraft*, Second Int. Aeron. Conf., New York, 1949.
13. O.G. Sutton, *Micro Meteorology*, McGraw-Hill Book Company, Inc., New York, 1953.
14. D.M. Driscoll, *Windchill*, Weatherwise, December 1987.
15. Anon., *International Standard Atmosphere (ISA)*, International Organization for Standardization, ISO 2533, 1975.
16. Anon., *U.S. Standard Atmosphere Supplements, 1966*, Environmental Science Services Administration, National Aeronautics and Space Administration, United States Air Force, U.S. Government Printing Office, Washington, D.C., 1966.

17. Anon., *U.S. Standard Atmosphere, 1976*, U.S. Government Printing Office, Washington, D.C., 1976.
18. U. Schumann (Editor), *Air Traffic and the Environment - Background, Tendencies and Potential Global Atmospheric Effects*, Proceedings of a DLR International Colloquium, Koln, Germany, Lecture Notes in Engineering, DLR 60, Springer-Verlag, Berlin Heidelberg, 1990.
19. D.C. Giancoli, *Physics for Scientist and Engineers*, Prentice-Hall Englewood Cliffs, New Jersey, U.S.A., Second Edition, 1988.
20. Anon., *Chemistry in the Atmosphere, A strategy for European research into global environmental issues*, Commission of the European Communities, Directorate- General XII for Science, Research and Development, EUR 15351 EN, Brussels, September 1993.
21. Anon., *Protecting the Earth's Atmosphere - An International Challenge*, Interim Report of the Study Commission of the 11th German Bundestag (1989) *"Preventive Measures to Protect the Earth's Atmosphere"*, Edited by: Deutscher Bundestag, Referat Offentlichkeitsarbeit, Bonn, Germany, 1989.
22. G.J.J. Ruijgrok, *Elements of Aviation Acoustics*, Delft University Press, Delft, The Netherlands, 2004.
23. M. Allaby, *The Nature of Atmosphere and the Climate*, Facts On File, Inc., New York, U.S.A.,1992.
24. J.T. Houghton, G.J. Jenkins, and J.J. Ephraums, (eds.), IPCC 1990, Climate Change: *The IPCC Scientific Assessment*, Cambridge University Press, Cambridge, UK.
25. J.T. Houghton, L.G. Meira Filho, J. Bruce, Hoesung Lee, B.A. Callander, E. Haites, N. Harris, and K. Maskell, (eds.), IPCC 1995, Climate change 1994: *Radiative Forcing of Climate,* and An Evaluation of the IPCC IS92 Emission Scenarios, Cambridge University Press, Cambridge, UK.
26. J.T. Houghton, L.G. Meira Filho, B.A. Callander, N. Harris, A. Kattenberg, and K. Maskell, (eds.), IPCC 1996, Climate Change 1995: *The Science of Climate Change*, Cambridge University Press, Cambridge, UK.
27. J.T. Houghton, Y. Ding, D.J. Griggs, M. Noguer, P.J. Van der Linden, X. Dai, K. Maskell, and C.A. Johnson, (eds), IPCC 2001, Climate Change 2001: *The Scientific Basis*, Cambridge University Press, Cambridge, UK.
28. E. Boletis, *Aviation in the Transport System*, Fokker Contribution to the AECMA Study on the Environment, 1993.
29. E. Boletis, *Air Traffic Operations and the Environment: Facts and Concerns,* Year book of the Netherlands Association of Aeronautical Engineers (NVvL), 1994.

30. K.D. van den Hout, M.G.M. Roemer, and P.J.H. Builtjes, *NO_x in the Global Troposphere: Emissions, Concentrations and Model Calculations*, TNO Report P90/223, TNO, Delft, The Netherlands, 1990.
31. S.R. Turns, *An introduction to Combustion, Concepts and Applications*, McGraw-Hill, Inc., New York, 1996.
32. U. Schumann and D. Wurzel (Eds.), *Impact of Emissions from Aircraft and Spacecraft upon the Atmosphere*, Proceedings of an International Scientific Colloquium, Koln (Cologne), Germany, DLR Mitteilung 94-06, 1994.
33. H.S. Johnston and G. Whitten, *Reaction of Ozone with Nitrogen Oxides at High Altitudes*, Paper 2 in AGARD-CP-125, April 1973.
34. Anon., *Aeroengines and Propulsion*, Imech E Seminar Publication 1996-12, Mechanical Engineering Publications Limited, London, 1996.
35. E. Meijer, *Modelling the impact of subsonic aviation on the composition of the atmosphere*, Koninklijk Nederlands Meteorologisch Instituut (KNMI),-Technische Universiteit Eindhoven, 2001.
36. K.S. Law and R.G. Penkett, *The upper troposphere / lower stratosphere ozone programme: contributing to research on aviation impact*, The Aeronautical Journal, Volume 104, Number 1037, 325-329, July 2000.
37. P.J. Crutzen, *The influence of nitrogen oxides on atmospheric ozone content*, Q. J. R. Meteorol. Soc., 97, 320-325, 1970.
38. H.I.H. Saravanamuttoo, G.F.C. Rogers, and H. Cohen, *Gas Turbine Theory*, Pearson Education Limited, England, Fifth edition, 2001.
39. N. Cumpsty, *Jet Propulsion*, Cambridge University Press, United Kingdom, 1997.
40. D. Fielding and J.E.C. Topps, *Thermodynamic Data for the Calculation of Gas Turbine Performance*, Aeronautical Research Council, R. & M. No. 3099, HMSO, London, 1959.
41. P.J. McMahon, *Aircraft Propulsion*, Pitman Publishing, London, 1971.
42. J.C. Samuels and B.M. Gale, *Effect of Humidity on Performance of Turbojet Engines*, National Advisory Committee for Aeronautics, Technical Note 2119, Washington, June 1950.
43. K. Eftekhari Shahroudi, *Development and Validation of a Computer Assisted Design Methodology for Gas-Turbine-Based Aircraft Engines*, Ph.D thesis, Delft University of Technology, Delft University Press, Delft, The Netherlands, 1994.
44. H. Wittenberg, *Prediction of Off-Design Performance of Turbojet and Turbofan Engines,* Paper 4 in AGARD-CP-242, October 1977.
45. P.G. Hill and C.R. Peterson, *Mechanics and Thermodynamics of Propulsion,* Addison - Wesley Publishing Company, Inc., New York, 1965.
46. S. Pal Arya, *Air Pollution Meteorology and Dispersion*, Oxford University Press Inc., 1999.
47. E.H.J. Pallet, *Aircraft Instruments,* Pitman, London, 1981.

48. A.H. Lefebvre, *Gas Turbine Combustion,*Taylor and Francis, Phiadelphia, 1998.
49. S.C.A. Kluiters, *A New Combustor and Emission Model for the Gas Turbine Simulation Program GSP,* Memorandum VH-98-010, National Aerospace Laboratory NLR, Amsterdam, The Netherlands, 1998.
50. J.L. Kerrebrock, *Aircraft Engines and Gas Turbines*, The MIT Press, second Edition. Cambridge Massachusetts, 1992.
51. W.S. Blazowski, D.E. Walsh, K.D. Mach, *Operating and Ambient Condition Influences on Aircraft Gas Turbine* NO_x *Emissions*, Journal of Aircraft, Vol. 12, No. 2, February, 1975.
52. Anon., ICAO / Annex 16 to the Convention on International Civil Aviation, Environmental Protection, Volume II, *Aircraft Engine Emissions*, Second edition, International Civil Aviation Organization (ICAO), Montreal,1993.
53. Anon., *ICAO Engine Exhaust Emissions Data Bank*, DOC 9646 - AN / 943, First edition, International Civil Aviation Organization (ICAO), Montreal, 1995.
54. F.W. Lipfert, *Correlation of Gas Turbine Emissions Data*, ASME Paper 72 - GT - 60, March 1972.
55. A.J.M. van der Velden, *Aerodynamic Design and Synthesis of the Oblique Flying Wing Supersonic Transport*, Ph.D thesis, Stanford University, SUDAAR 621, June 1992.
56. A.H. Levebvre, *Fuel Effects on Gas Turbine Combustion-liner Temperature, Pattern Factor and Pollutant Emissions*, Journal of Aircraft, Vol. 21, 1984.
57. Anon., *Air Quality Control Handbook*, E. Roberts Alley & Associates, Inc., MacGraw - Hill, New York, 1998.
58. C.E. Padilla, *Optimizing Jet Transport Efficiency, Performance,Operations & Economics*, McGraw-Hill, New York,1996.
59. P. Clark, *Buying the Big Jets: Fleet planning for airlines*, Ashgate Publishing Ltd, Aldershot, England, 2001.
60. M.E. Eshelby, *Aircraft Performance - Theory and Practice*, Arnold, London, 2000.
61. B.W. McCormick, *Aerodynamics, Aeronautics, and Flight Mechanics*, Second edition, John Wiley, New York, 1995.
62. Anon., ICAO PANS-OPS - *Procedures for Air Navigation Services - Aircraft Operations*, Doc 8161-OPS/611, Volume I, *Flight Procedures*, Fourth Edition, Part V, *Noise Abatement Procedures*, International Civil Aviation Organization, Montreal, 1993.
63. L.R. Jenkinson, P. Simpkin, and D. Rhodes, *Civil Jet Aircraft Design*, Edward Arnold (Publishers) Limited, London, 1999.

64. Anon., *Long range aircraft requirements*, Association of European Airlines (AEA), AEA Publication, Brussels, 1989.
65. J. Grey (editor), *Aircraft Fuel Conservation: An AIAA View*, American Institute of Aeronautics and Astronautics, New York, 1974.
66. Anon., *Lost range, fuel and time due to climb and descent: aircraft with turbo-jet and turbo-fan engines,* Engineering Sciences Data Unit (ESDU), Item Number 74018, London, 1974.
67. S. Stewert, *Flying the Big Jets, 4th edition - Flying the Boeing 777*, Airlife Publishing Ltd, Shrewsbury, England, 2001.
68. D.A.C.M. Bremmers, *The Low NO_x Flight*, Master thesis, Delft University of Technology, Faculty of Aerospace Engineering, August 1999.
69. R.D. Hager and D. Vrabel, *Advanced Turboprop Project*, National Aeronautics and Space Administration, SP-495, NASA Scientific and Technical Information Division, Washington, DC, 1988.
70. M. Volders and R. Slingerland, *Environmental Harm Minimization during cruise-flight for long-range Preliminary Aircraft Design*, American Institute of Aeronautics and Astronautics (AIAA), Paper 2002-6803.
71. J.L. Dalhuijsen and R. Slingerland, *Preliminary Wing Optimization for very large Transport Aircraft with Wingspan Constraints*, American Institute of Aeronautics and Astronautics (AIAA), Paper 2004-0699.
72. E. Torenbeek, *Synthesis of Subsonic Airplane Design*, Delft University Press, 1982.
73. S.L. Baughcum, *Scheduled civil aircraft emission inventories for 1992: database development and analysis*, NASA CR-4700, Hampton: NASA, Langley Research Center, 1996.
74. H.G. Klug, Liquid Hydrogen Fueled Aircraft - System analysis, CRYOPLANE, *Global Warming Potential Data for Simplified Sensitivity Analysis and Optimization*, Task technical report no. 1.1-5, 2001.
75. H.G. Klug, *CRYOPLANE - Liquid Hydrogen Fueled Aircraft*, Year book of the Netherlands Association of Aeronautical Engineers (NvvL), 1997.
76. A. Garcia, *Aviation and the environment - The balanced approach for sustainable development*, The twelfth Dr. Albert Plesman Memorial Lecture, Delft University of Technology, December 3rd, 1999.
77. C.A.J.R. Vermeeren, Th. Beumler, J.L.C.G. de Kanter, O.C. van der Jagt, and B.C.L. Out, *Glare Design Aspects and Philosophies*, Applied Composite Materials **10**: 257-276, 2003, Kluwer Academic Publishers, The Netherlands, 2003.
78. F. Svensson, *Potential of reducing the environmental impact of civil aviation by using liquid hydrogen*, Ph.D. thesis, Cranfield University, UK, 2005.

INDEX

F

G

H

Q

R

S